434462

WITHDRAWN

Statistical and Mathematical Aspects of Pollution Problems

STATISTICS

Textbooks and Monographs

A SERIES EDITED BY

D. B. OWEN, *Coordinating Editor*
Department of Statistics
Southern Methodist University
Dallas, Texas

PETER LEWIS
Naval Postgraduate School
Monterey, California

PAUL D. MINTON
Virginia Commonwealth University
Richmond, Virginia

JOHN W. PRATT
Harvard University
Boston, Massachusetts

VOLUME 1: The Generalized Jackknife Statistic
H. L. Gray and W. R. Schucany

VOLUME 2: Multivariate Analysis
Anant M. Kshirsagar

VOLUME 3: Statistics and Society
Walter T. Federer

VOLUME 4: Multivariate Analysis: A Selected and Abstracted Bibliography, 1957 - 1972 (out of print)
Kocherlakota Subrahmaniam and Kathleen Subrahmaniam

VOLUME 5: Design of Experiments: A Realistic Approach
Virgil L. Anderson and Robert A. McLean

VOLUME 6: Statistical and Mathematical Aspects of Pollution Problems
edited by John W. Pratt

OTHER VOLUMES IN PREPARATION

Satellite Symposium on Statistical Aspects of Pollution Problems, Harvard Business School, 1971.

Statistical and Mathematical Aspects of Pollution Problems

edited by JOHN W. PRATT

Graduate School of Business Administration
Harvard University
Boston, Massachusetts

MARCEL DEKKER, INC.   New York

COPYRIGHT © 1974 by MARCEL DEKKER, INC.

ALL RIGHTS RESERVED

Neither this book nor any part may be reproduced or transmitted in any form or by any means, electronic or mechanical, including photocopying, microfilming, and recording, or by any information storage and retrieval system, without permission in writing from the publisher.

MARCEL DEKKER, INC.

270 Madison Avenue, New York, New York 10016

LIBRARY OF CONGRESS CATALOG CARD NUMBER: 73-90771

ISBN: 0-8247-6132-4

Current printing (last digit):
10   9   8   7   6   5   4   3   2

PRINTED IN THE UNITED STATES OF AMERICA

CONTENTS

List of Contributors xiii

Preface xv

Part I: COLLECTION AND DISSEMINATION OF BASIC DATA

1. GLOBAL ENVIRONMENTAL MONITORING 3

    Sören Svensson

    1. Introduction 3
    2. Monitoring - Basic Considerations 5
    3. Environmental Problems 5
    4. Environmental Variables Appropriate for Monitoring . . 7
    5. The Design of a Global Environmental Monitoring System 10
    6. Operational and Institutional Arrangements 12

    DISCUSSION - L. A. Clarenburg 15

2. A SIMPLIFIED STATISTICAL PROCEDURE TO OPTIMIZE
   THE SITING OF AIR POLLUTION MEASURING STATIONS . . 21

    André Junod

3. STATISTICAL STUDIES OF URBAN AIR POLLUTION —
   SULFUR DIOXIDE AND SMOKE 25

    Michael Legrand

    1. Introduction 25
    2. Collection and Publication of Pollution Data 27
    3. Statistical Studies of Pollution Data 32
    4. Frequency Distribution Curve 32
    5. Conclusion 36

4. THE DESIGN AND OPERATION OF THE NATIONAL
   AEROMETRIC DATA BANK 39

   J. W. Overbey II and R. P. Ouellette

   1. Introduction 39
   2. System Operation 40
   3. Data Editing and Validation 40
   4. Analyses 43
      References 59

DISCUSSION - D. W. Moody 61

Part II: INDICES, ECONOMIC AND POLICY PERSPECTIVES

5. TREATING EXTERNAL DISECONOMIES —
   MARKETS OR TAXES? 65

   David Starrett and Richard Zeckhauser

   1. Tractable Externality Problems 65
   2. Resistant Externality Problems 70
   3. Conclusion 83
      References 83

6. ENFORCEMENT MEASURES IN WATER QUALITY
   CONTROL 85

   Murray Stein

   1. Introduction 85
   2. Present Enforcement Authorities ... 86
   3. Administration Legislative Proposals ... 87
   4. Statistics in Water Quality Control ... 88
   5. EPA Budget 89
   6. The Economy and the Environment ... 90
   7. Conclusion 90

CONTENTS

7. PROGRESS IN ENVIRONMENTAL QUALITY 93

   Thomas C. Winter, Jr.

   0. Summary 93
   1. Introduction: Federal Institutional Arrangements 94
   2. Other Federal Actions 95
   3. Needed Legislation 96
   4. Trends 97
   5. Costs 98
   6. Information Requirements 99
      Reference 101

8. DEVELOPMENT OF ENVIRONMENTAL INDICES 103

   Robert Pikul

   0. Summary 103
   1. Indices and Indicators of Environmental Quality 103
   2. Composite Grouping of Indices 106
   3. Example: Air Quality Index 115
      References 121

9. A STATISTICAL SYSTEM FOR ESTIMATING THE
   DETERIORATION OF THE HUMAN ENVIRONMENT 123

   R. Hueting

   1. Environmental Goods: Functions by Component 123
   2. Quantitative, Spatial, and Qualitative Competition of
      Functions 124
   3. Indicators 127
   4. Shadow Prices of Environmental Functions 128
      References 132

Part III: MATHEMATICAL MODELS

10. PENALIZATION OF THE ENVIRONMENT DUE TO STENCH: A Study of the Subjective Experience of the Population 135

    L. A. Clarenburg

0.	Summary	135
1.	Introduction	136
2.	Penalization	136
3.	Mathematical Model	137
4.	Testing of the Model	141
5.	Conclusions	156
6.	Applications	156
	References	160

DISCUSSION - Gerald van Belle 161

11. ON THE ESTIMATION OF POLLUTION-CAUSED GROWTH REDUCTION IN FOREST TREES 167

    Rolf Sundberg

0.	Summary	167
1.	General Introduction	167
2.	The Local Problem	168
3.	The Regional Problem	173
	References	175

12. A MATHEMATICAL MODEL FOR PARTICULATE DEPOSITION IN THE RESPIRATORY SYSTEM 177

    H. D. Landahl

0.	Summary	177
1.	Introduction	177
2.	Factors Determining Deposition	178
3.	Calculation of Deposition in the Respiratory Tract	179
4.	Redistribution of the Deposited Particles	183
	References	186

DISCUSSION - Stanley V. Dawson 189

CONTENTS   vii

AUTHOR'S REPLY TO COMMENTS BY STANLEY V. DAWSON -
H. D. Landahl   191

13. SOME BASIC PROBLEMS OF ESTUARINE WATER QUALITY
    CONTROL   193

    J. R. West and D. J. A. Williams

    1. Introduction   193
    2. The Nature of the Physical Problems   194
    3. One-Dimensional Representation of Estuarine Mixing   196
    4. Data Acquisition and Analysis   200
    5. Conclusions   209
       References   211

14. MODELLING CURRENT MOVEMENTS WITHIN
    BODIES OF WATER   213

    D. Harverson

    1. Introduction   213
    2. Outline of Techniques   214
    3. Correlation of Wind with Induced Current   214
    4. Variations in the Basic Drift   215
    5. Sinusoidal Representation of Current   216
    6. Spatial Extension of the Data   218
    7. Calibration and Use of the Model   219
       References   220

Part IV: STATISTICAL STUDIES USING REGRESSION AND
              RELATED METHODS

15. DOES AIR POLLUTION SHORTEN LIVES?   223

    Lester B. Lave and Eugene P. Seskin

    1. Introduction   223
    2. Method   224
    3. Results   225
    4. Discussion   234
    5. Conclusion   241
       References   242

DISCUSSION - William Fairley   245

CONTENTS

16. THE USE OF MULTIPLE REGRESSION EQUATIONS FOR THE ASSESSMENT AND SHORT-TERM FORECASTING OF URBAN AREA BACKGROUND POLLUTION 249

   M. Benarie, D. Badellon, T. Menard, and A. Nonat

   0. Summary 249
   1. Introduction 250
   2. The Observed Data 250
   3. Variables Used in the Multiple Regression Equations . 251
   4. Multiple Linear Regression Equations 256
   5. Uses of Representation by Multiple Regression Equations 264
      References 273

17. ANALYSIS OF HEALTH EFFECTS DATA: SOME RESULTS AND PROBLEMS 275

   V. Hasselblad, W. C. Nelson, and G. R. Lowrimore

   1. Introduction 275
   2. Data Processing 276
   3. Analysis of Repeated Measurements 276
   4. Categorical Dependent Variables 277
   5. Threshold Estimation 277
   6. Other Problems 280
      References 282

18. A STATISTICAL MODEL OF A POLLUTED ESTUARY . . . 283

   D. W. Mackay and J. Gilligan

   1. Introduction 283
   2. General Nature of the Situation 284
   3. The Effect of Tidal Range 288
   4. Multiple Regression Analyses 288
   5. Importance of Variables 291
   6. Conclusions 294
      References 295

CONTENTS

19. SOME STATISTICAL ANALYSES OF WATER QUALITY
    IN THE DELAWARE RIVER 297

    Robert V. Thomann

    1. Introduction 297
    2. Available Data 297
    3. Statistical Analyses 298
    4. Results of Analysis 301
    5. Summary 310
       References 310

    Part V: OTHER STATISTICAL METHODS

20. ASPECTS OF SAMPLING ATTITUDES TOWARDS SOLID
    WASTE PROBLEMS 315

    Albert J. Klee

    1. Goals and Objectives 315
    2. Information Sought 315
    3. Some Concepts Used in the Study 316
    4. The Survey of Householders 317
    5. The Survey of City Officials 320
    6. Other Interviews 321
    7. Study of Organizations 321
    8. Data Processing 322
    9. The Case Studies 324
    10. Analysis of Organizations 324
    11. Results 325
        References 326

21. STATISTICAL PROBLEMS IN AEROSOL STUDIES 329

    Gerald van Belle

    0. Summary 329
    1. Introduction 329
    2. Statistical Distributions 330
    3. Comments 332
       References 335

DISCUSSION - S. Dawson 337

22. EVALUATION OF ACTIVATED SLUDGE TREATMENT
    PLANT PERFORMANCE 339

    Jerzy Ganczarczyk

    0. Summary 339
    1. Introduction 339
    2. Activated Sludge Treatment Plant Control Data 340
    3. General Appraisal of Control Data 341
    4. Automation of Activated Sludge Treatment Plants . . . 343
    5. Analysis of Plant Performance Data 344
    6. Conclusions 348
       References 350

23. COMBINING MULTIPLE ATTRIBUTE OUTCOMES
    INTO AN OVERALL INDEX 353

    Clifford J. Maloney

    1. Introduction 353
    2. Example 354
    3. Scales 355
    4. Probability 356
    5. Quantal Regression 357
    6. Individual Sickness Score 361
    7. Construction of the Sickness Score 362
    8. Check of the Model 362
       References 363

24. A PROPOSAL FOR ECONOMICAL FIRST STAGE
    SCREENING FOR TUMORIGENS WITH A POSSIBLE
    "JOINT ACTION" BONUS 365

    Robert M. Elashoff, Milton Sobel, and M. A. Schneiderman

    1. Introduction 365
    2. A Probability Model for Joint Action of Compounds . . 367
    3. Decision Rules for Selecting Tumorigens 367
    4. Probability Inequalities to be Satisfied 369
    5. Comparisons Between the 2AAT and 1AAT Designs 369
    6. Inhibitors and Multiple Tumorigens 371
    7. Summary 373
       References 374

CONTENTS

25. SOME STATISTICAL ASPECTS OF EXPERIMENTS FOR
    DETERMINING THE TERATOGENIC EFFECTS OF
    CHEMICALS 375

    David G. Hoel

    1. Introduction 375
    2. The Teratological Experiment 376
    3. Some Probabilistic Considerations 377
    4. Some Statistical Considerations 379
        References 381

26. CLOSING COMMENTS 383

    Gerald van Belle

Index 387

LIST OF CONTRIBUTORS

D. BADELLON, Institut National de Recherche Chimique Appliquée, Vert-le Petit, France

M. BENARIE, Institut National de Recherche Chimique Appliquée, Vert-le Petit, France

L. A. CLARENBURG, Rijnmond Authority, Rotterdam, The Netherlands

STANLEY V. DAWSON, Harvard University, Boston, Massachusetts

ROBERT M. ELASHOFF, University of California at San Francisco, San Francisco, California

WILLIAM FAIRLEY, Harvard University, Cambridge, Massachusetts

JERZY GANCZARCZYK, University of Toronto, Toronto, Canada

J. GILLIGAN, Clyde River Purification Board, Glasgow, Scotland

D. HARVERSON, Edinburgh Corporation, Edinburgh, Scotland

V. HASSELBLAD, National Environmental Research Center, Environmental Protection Agency, Research Triangle Park, North Carolina

DAVID G. HOEL, National Institute of Environmental Health Sciences, National Institutes of Health, Research Triangle Park, North Carolina

R. HUETING, Netherlands Central Bureau of Statistics, The Hague, Netherlands

ANDRÉ JUNOD, Institut Suisse de Météorologie, Payerne, Switzerland

ALBERT J. KLEE, Solid and Hazardous Waste Research Laboratory, Environmental Protection Agency, National Environmental Research Center, Cincinnati, Ohio

H. D. LANDAHL, University of California, San Francisco, California

LESTER B. LAVE, Carnegie-Mellon University, Pittsburgh, Pennsylvania

MICHAEL LEGRAND,*Institut Royal Météorologique de Belgique, Brussels, Belgium

G. R. LOWRIMORE, National Environmental Research Center, Environmental Protection Agency, Research Triangle Park, North Carolina

---

*Present address: Institut d'Hygiene et d'Epidemiologie, Brussels, Belgium.

D. W. MACKAY, Clyde River Purification Board, Glasgow, Scotland

CLIFFORD J. MALONEY, Food and Drug Administration, Bethesda, Maryland

T. MENARD, Institut National de Recherche Chimique Appliquée, Vert-le Petit, France

D. W. MOODY, U. S. Geological Survey, Washington, D. C.

W. C. NELSON, National Environmental Research Center, Environmental Protection Agency, Research Triangle Park, North Carolina

A. NONAT, Institut National de Recherche Chimique Appliquée, Vert-le Petit, France

R. P. OUELLETTE, The MITRE Corporation, McLean, Virginia

J. W. OVERBEY II, The MITRE Corporation, McLean, Virginia

ROBERT PIKUL, The MITRE Corporation, McLean, Virginia

M. A. SCHNEIDERMAN, National Cancer Institute, Bethesda, Maryland

EUGENE P. SESKIN, The Urban Institute, Washington, D. C.

MILTON SOBEL, University of California at San Francisco, San Francisco, California

DAVID STARRETT,[*] Harvard University, Cambridge, Massachusetts

MURRAY STEIN, Office of Water Enforcement, Environmental Protection Agency, Washington, D. C.

ROLF SUNDBERG,[†] Stockholm University, Stockholm, Sweden

SÖREN SVENSSON,[‡] SCOPE Commission on Monitoring, Stockholm, Sweden

ROBERT V. THOMANN, Manhattan College, Bronx, New York

GERALD VAN BELLE, Florida State University, Tallahassee, Florida

J. R. WEST,[**] University of Dundee, Dundee, Scotland

D. J. A. WILLIAMS, University College, Swansea, Wales

THOMAS C. WINTER, JR., Council on Environmental Quality, Washington, D. C.

RICHARD ZECKHAUSER, Harvard University, Cambridge, Massachusetts

---

[*]Present address: Stanford University, Stanford, California.
[†]Present address: Royal Institute of Technology, Stockholm, Sweden.
[‡]Present address: Zoological Institute, Lund, Sweden.
[**]Present address: The University of Birmingham, Birmingham, England.

PREFACE

Pollution has become and will remain a problem of great concern to government, business, the public--and scientists. Statistical methods should play an important part in pollution studies and research, and conversely, playing its part should enrich the field of statistics in both methodology and applied experience. The papers in this book report on work being done of a statistical type, in the widest sense, on a number of kinds of pollution problems, and are representative of the best such work, insofar as the editor was able to identify it. They were written for a heterogeneous scientific audience and can be largely understood without specialized knowledge. As well as being of interest in themselves, then, they afford a broadly understandable survey of the state of the field, to the extent that this is possible in a book of such modest size. It has also proved possible, without impeding these uses of the book, to arrange the papers so as to provide an introduction to the application of statistical methodology in the context of pollution problems--not to each important statistical technique, but to each phase of the statistical method and some of the practical questions which arise therein, from the data collection program to the policy implications of the analysis. This will be explained more fully later.

The book originated in one of the Satellite Symposia sponsored by the International Association for Statistics in Physical Sciences (IASPS), an autonomous section of the International Statistical Institute (ISI), in connection with the 38th Session of the ISI which was held in Washington, D. C., August 10-20, 1971. The idea of the IASPS Satellite Symposia is well described in an article by P. A. P. Moran and J. Neyman,[*] who say that it was borrowed from the astronomers, although they attribute the name, oddly enough, to a statistical colleague. In brief, the purpose of the symposia was to foster cooperation between statisticians and substantive scientists on problems calling for skills of both types. A few sentences of their article are particularly relevant here:

> Of the interdisciplinary studies none seems as important as those relating to two international problems now widely discussed

---

[*] The IASPS and its Satellite Symposia, in The American Statistician, Feb. 1971, pp. 15-18, and the ISI Review, April, 1971.

in scientific literature and in the daily press, the problem of the
population explosion and the problem of environmental pollution.
Both are interdisciplinary to an unprecedented degree and both
relate to situations "at large" in which the establishment of
pertinent facts must depend on sampling surveys, on random-
ized experiments, etc., that is, on the use of statistical
methodology. The perusal of literature, often sensational and
nearly always controversial, suggests skepticism as to the uni-
versal reliability of the underlying statistical studies, often performed by nonspecialists. On the other hand, the mass of important physical as well as biological knowledge involved in the two
problems are overwhelming to statisticians.

Professor Neyman, as President of IASPS, asked me to organize a Satellite
Symposium on Statistical Aspects of Pollution Problems. With the primary
support of the National Science Foundation, it was held at the Harvard
Business School, August 31-September 3, 1971. I believe that it served its
purpose well. The participants were primarily responsible for this, of
course, but, as an organizational point, it seemed to be facilitated, especially the important aspect of informal contact and discussion, by the
location, including housing the participants together (in modest comfort but
not hotel style), at a slight remove from the physical and mental pollution
of Harvard Square.

The symposium was entirely invitational. I wanted papers by people
doing real work of quality and value, viewed by informed scientists as
leaders in such work, without regard to public notoriety or even to purely
disciplinary reputation. I was in no position to make a list of such people
myself, and I found that, in a field involving so many disciplines, it is
difficult to identify them even second-hand (which perhaps adds support to
the idea of a Satellite Symposium). I consulted too many people to name
here but their scope is indicated by two who particularly deserve mention
for help and helpfulness in getting me started: Myron B. Fiering, of the
Division of Engineering and Applied Physics at Harvard, and David G.
Kendall, a specialist in probability theory at Cambridge, England.

There was no advance commitment by anyone to any form of proceedings volume, but it was decided to undertake publication of all papers given
at the symposium which were available within a reasonable period and
were not being published elsewhere. A few of the written papers differed
from the spoken ones. None were "refereed," but two or three have been
completely or substantially rewritten by their authors since receipt, and
two or three have been considerably cut. The rest have been edited, mostly
slightly, a few more extensively, by the authors and/or editor, without any
intention to change the substance appreciably. In the rare cases where it

PREFACE

matters, the papers should be understood to apply as of the date of delivery. The discussion was excellent, but it seemed impractical to attempt to publish it, and only a small--but cogent--fraction appears here, essentially unedited. The following papers given at the symposium are not included in this book:

A mathematical model for pest control, Samprit Chatterjee, New York University; appearing in Biometrics, Dec. 1973.
Information-statistical evaluation of air pollution data samples, Camilo P. de La Riva, Landesanstalt für Immissions- und Bodennutzungsschutz des Landes Nordrhein-Westfalen.
A mathematical model for relating air quality measurements to air quality standards, Ralph I. Larsen, Environmental Protection Agency; published as Environmental Protection Agency Publication AP-89.
Invariant imbedding, nonlinear estimation, and the adaptive forecasting of stream pollution, E. Stanley Lee, Kansas State University.
Instabilities of regression estimates relating air pollution to mortality, Gary C. McDonald and Richard C. Schwing, General Motors Technical Center; appearing in Techonometrics, 15 (3), 463 (1973).
Advantages and limitations in storing, retrieval, and statistical evaluation of water quality data, Wolfgang Schmitz, Landesstelle für Gewässerkunde und Wasserwirtschaftliche Planung, Karlsruhe.
Evaluation of the APCO air quality display model calibration procedure, Max A. Woodbury, Duke University.

One area which would probably have been more heavily represented at the symposium but for conflicting meetings is biology and health. This gap was more than filled by a symposium in which Professor Neyman had a considerable direct part and which placed considerably more emphasis on airing controversies, with attendant advantages and disadvantages. Read all about it in the Proceedings of the Sixth Berkeley Symposium on Mathematical Statistics and Probability, Vol. VI, Effects of Pollution on Health (Lucien M. LeCam, Jerzy Neyman, and Elizabeth L. Scott, eds.), University of California Press, 1972. A critical survey of considerable literature by two participants in the IASPS symposium appears so useful that it deserves mention here, "Some Statistical Aspects of Environmental Pollution and Protection" by Gerald van Belle and Marvin Schneiderman, Florida State University Statistics Report M245, Nov. 1972, to appear in the ISI Review.

There are various ways to organize the field of pollution. A common one, which is useful for many purposes, is by medium: air, water, and soil, sometimes with additional categories such as biota, noise, and radiation. Another, obviously crosscutting, is by activity: power generation, transportation, etc. Stanley Dawson, one of the participants, separated the symposium papers neatly into four groups entitled Biological effects of air pollution and ingested substances, Measurement and analysis of air pollution, Waste disposal problems, and Strategies for control of environment. I chose a much more methodological organization for this book, however, for several reasons.

Much of the work described here, though without details if they are too complicated, is approaching the guts of a problem, and the real value of such work is often difficult to assess in isolation. The value of a data collection program depends on the inferences to be drawn from it and the effect of these inferences on policies and actions later implemented. The value of a mathematical model of the flow of a river depends on its use in this process. And so on. And the relative strength of much of the more specific work represented here is that it has a real use in this process. This is not to say that each paper considers everything from data collection to final decisions. That is ordinarily impossible to do in a modest space without becoming overly superficial. In fact, most of the papers rather clearly concentrate on a single aspect. But the ultimate goal is the improvement of the whole process of using data to reach decisions, rather than elaborating or refining one aspect for its own sake.

Organizing the book in accordance with this process therefore divides the papers naturally while emphasizing the importance of the whole process and clarifying the real purpose of the papers by relating them to it. More specifically, the process I mean is the now well-known statistical method (in whose elucidation, incidentally, Professor Neyman's role seems more and more to me to have been crucial). Each step will be discussed further later, but in brief: we have or obtain observations; we have or develop a model describing the probabilistic behavior of the observable data under various assumptions about the true state of the world; using the model and the observations, we make statistical inferences about the true state of the world; and, finally, decisions are made based on our inferences (we hope) and on other considerations of value, cost, feasibility, etc. The change to the passive voice accords with the fact that we (most of the authors, and I, if not the reader) make analyses as part of our profession, but will not make the final decisions. Despite this, however, the ultimate purpose of most of the work here is to aid decisions, not to advance pure science.

Reality may be much more complicated than this description, of course. In particular, we may recycle, from inference to further observation, for instance. And in designing an observational program or experiment, we should take at least informal account of all the subsequent phases

PREFACE   xix

of the process. But while considering them simultaneously we should not confuse them with one another.

The sections of the book relate to the process just described as follows. Parts I and II concern the collection and dissemination of basic data, and the general background and policy perspective to which decisions about the collection of data should be related. Part III concerns the development of models. Parts IV and V concern the drawing of statistical inferences from data using models.

To represent the remaining phase, I would have liked to have papers analyzing what decisions ought to be made, or what effects various decisions might have, on the basis of statistical inferences and other considerations including costs and the values that society might place on the possible outcomes. The absence of such papers at the symposium may represent bad luck or oversight. I suspect, however, that there is a dearth of such work, at least of a kind which takes explicit and analytical account of uncertainty. Likely reasons include the complexity of the decision problems and the diffused and political nature of decision making in this area. Also perhaps that decisions are not usually made until there is enough data to remove essentially all uncertainty (implying excessive delay or overspending on data). The mind's difficulty in encompassing uncertainty and the psychological tendency to block it may contribute. Unfortunately, the statistical concepts of "testing hypotheses" and "point estimation" may encourage this failing. Inferences incorporating a measure of uncertainty do not. Remember "confidence intervals"! Dare I suggest "posterior distributions"? I believe, and not just rhetorically, that both statistics and practice would gain if statisticians were more involved with this decision-making phase.

Comments follow on each Part in turn.

I. It seems natural to start with data--what data do you collect where, by what means, and how do you disseminate it or make it available? Part I consists of papers where these concerns dominate. They naturally emphasize large-scale, continuing programs and systems, but the possible efficiency of more specialized data-gathering efforts should not be forgotten. The most important question with large efforts is probably whether they should be undertaken at all. Admittedly my reading is far from comprehensive, but I cannot recall a single quantitative study of whether the gains from such an effort are likely to be worth the resources used, in pollution or any other field. A convincing job would be hard to do, but the only advantage of the rules of thumb one sometimes sees, stating an arbitrary percentage of the budget for data, is that they are cheap to compute. They certainly contradict what is known about optimal sample size in simple situations.

The first paper, by Svensson, sketches the thinking involved in the report of a high-level committee on environmental monitoring. This serves both to introduce the subject of pollution in its full international scope and to put forward some recommendations on data collection. These seem straightforward at first glance, but Clarenburg's bristling discussion promptly joins the issues, rapidly raising a host of difficulties.

The following two papers, by Junod and Legrand, illustrate local programs to collect air quality data, in areas of Switzerland and Belgium, Junod dealing particularly with the siting of the measuring stations and Legrand with the presentation of the data. (A more formal approach to the siting problem seems feasible, and likely to yield different rules from intuition.) Then Overbey and Ouellette discuss a data bank designed to collect centrally all such U. S. data, vast and disparate though it is, and to make it available in usable form. (They also illustrate its use, but if such illustrations appear to predominate, my editorial requests and inefficiency must be blamed for some distortion, as their real purpose was to describe the data bank.)

Other papers in the book also illustrate the nature and use of air quality data, especially the first three of Part IV. Some papers illustrating water quality data are the other two in Part IV and that by West and Williams (No. 13). All these later papers are, however, primarily concerned with analysis, rather than collection or dissemination per se. One other paper worth specific mention for relevance to Part I is that by Ganczarczyk (No. 22) who, in a very different situation (control of a sewage treatment plant), makes some pointed and possibly generalizable comments about data automation, quality, and usefulness.

II. A theoretically complete analysis of the design of an experiment or observational program involves considering, for each possible design, the model to be used in the analysis, the data which might be observed, and the resulting inferences, actions, costs, benefits, etc.--in short, the entire process through the final decision-making phase. And this must be done a priori, so that various possible outcomes must be considered with only subjective opinions available, not a posteriori when a particular outcome has occurred and evidence dominates opinion. More simply, how can you design an experiment without taking its analysis into account? Yet elementary courses in experimental design usually precede those in analysis, and one can go rather far in practical experimental design on the basis of informed common sense and rough notions of the decision problem. This may be true of purely observational programs too. Actually, what common sense handles fairly well is allocation of a given budget. It has little that is convincing to offer in setting the size of the budget, the total effort to expend in observation. Unfortunately, as mentioned earlier

PREFACE   xxi

in a more limited way, analysts have not yet tackled this question much in complex, practical situations. One hopes the day will come when they are ready to, and a later day when funders are ready to understand the results. For now, however, the decisions will be made on the basis of general background and policy.

For this and other reasons, the papers in this book with predominantly a policy perspective are related to Part I (data) rather than to Parts III-V (probabilistic modeling and inference). Part II contains these papers, and those concerned with indices, inasmuch as indices are not basic data, but are combinations whose definition may reflect policy, and are an important form in which policy makers absorb information.

Opening Part II, Starrett and Zeckhauser approach a policy question (pollution control by markets or taxes?) with economists' concepts and (deterministic) models. Though traditional assumptions and results fail, they explain and apply the concepts and way of thinking in a clear, elementary, and entertaining way in the context of pollution. Their consideration of economic efficiency may be juxtaposed with the consideration of enforceability in the following two papers by Stein and Winter, who represent the point of view of governmental policy (in the U. S.). Stein is concerned with water, while Winter discusses most other major topics. Next Pikul describes an effort under way to develop environmental indices. A frightening number may be required. Finally, Hueting presents a conceptual structure for the competition taking place in the use of the components of the environment for various purposes, giving particular attention to indices and to economic ideas.

III. The remaining papers in the book almost all concern the analysis of quite specific types of situation, and often of particular data. In Part III the dominant theme is the development of models, while in Parts IV and V it is the later stage of statistical analysis.

The first paper, by Clarenburg, illustrates excellently the complete process of data collection, probabilistic model building, statistical inference, and application to decisions. Particular steps can be questioned, but the process as a whole is very clear. Furthermore, of the individual aspects, models seem to me furthest developed. Thus the paper simultaneously serves well to introduce the whole body of analytical papers and fits into Part III. Van Belle's discussion, a fine example of reading both on and between the lines, provides much additional insight into what was actually done and raises many appropriate questions about it.

Clarenburg's specific subject is urban air pollution. The next two papers, also concerned with air pollution, touch two extremes of scale:

reduction in growth of forest trees (Sundberg) and deposition of particles in the human respiratory system (Landahl). The remaining two papers of Part III develop models for water movement. West and Williams treat an estuary one dimensionally (the Tay), while Harverson treats a bay two dimensionally (the Firth of Forth). For applications of standard statistical models in air pollution, see the first three papers of Part IV and those by van Belle and Maloney (Nos. 21 and 23); in water pollution, the final two papers of Part IV.

IV. As an alternative to building a special model for the problem at hand, it may be possible to use standard statistical models, some of which are very flexible. This offers obvious advantages, including well worked-out theory and, often, computer programs. The commonest statistical technique is simple cross-tabulation, where statistical inference beyond using observed proportions as estimates is often unnecessary and even more often omitted (except occasional ritual significance testing). Of the more advanced models and techniques, although some exist for contingency tables, regression is by far the most widely applied. Part IV contains five papers making heavy use of regression and related methods, other statistical methods being postponed to Part V. Both these parts naturally emphasize inference rather than model building, since prefabricated models are used.

Regression is such a powerful technique that it blinds us to its limitations. Deficient vocabulary compounds the problem. An increase of 1 unit in x is less easily said to "be accompanied by an increase in y of" beta units on the average than to "increase y by" this amount. Even such a correct phrase as "variance accounted for by x" is all too easily taken to mean "variance due to x," and how many can really remember in the heat of fray that the fraction of the variance of y accounted for by x coincides with the fraction of the variance of x accounted for by y? Or that the partial correlation of y with x_1 given x_2, x_3, ... equals that of x_1 with y? And when, in the physical nature of things, y cannot contribute causally to x but x can to y, even if the words speak only of association, it is hard to avoid attributing the association to the effect of x on y and to remember that the coefficient of x in a regression of y on x reflects not only any effect x has but also part of the effect of every omitted variable which is correlated with x.

To be explicit, with apologies for incomplete explanation, here is an approach I have found helpful. Suppose that the true causal model of y is

$$E(y|x,z) = \beta_0 + \beta_1 x + \beta_2 z,$$

so that β_1 measures the true effect of x on y, in which we are interested, and the only other variable having a true effect on y is z. Suppose that z is not observed, but its association with x, for whatever reason, leads to

$$E(z|x) = \alpha_0 + \alpha_1 x.$$

Then taking the conditional expectation of the first equation given x gives

$$E(y|x) = (\beta_0 + \beta_2 \alpha_0) + (\beta_1 + \beta_2 \alpha_1)x.$$

Thus in a regression model including x and excluding z, the coefficient of the included variable x differs from its true effect β_1 unless the excluded variable has either no effect or no association with x; the difference is the product of the effect of z on y and the coefficient of x in the regression of z on x. If x and z have more than one component, then β_2 becomes a row vector and α_1 a matrix, and the coefficient of an included variable represents its true effect if it has no association with any excluded variable having a true effect, given the other included variables. This can hold otherwise only if nonzero terms happen to cancel exactly.

The size of the effect of omitted variables can sometimes be bounded by plausible substantive arguments, with the help of the foregoing or some other technical approach. A model derived on the basis of theory has more causal plausibility than a purely observational relationship (which may push one toward custom-built rather than prefabricated models). And causation is supported if the coefficient of x is the same in different situations or populations (where the relation of z and x, measured by α_1 above, is presumably different) or when further variables are added (which has a similar effect). How much sameness is supported if it is accepted by a significance test is far from obvious, however, though it obviously depends on the power of the test to detect differences. High R^2 is not definitive: this can occur with cause and effect reversed or, more insidiously, through "spurious" correlations and "proxy" variables, especially for subsidiary causes.

Of course, the only sure way to measure causation is by experimentation, with controlled x's and appropriate randomization (making x ipso facto independent of all omitted variables). Unfortunately this road is often closed, and the alternative is arduous and hazardous. This is not to say regression should be avoided or ignored in nonexperimental situations. Other methods pose the same problems, perhaps hiding them even more, and serious analysis is always better than blind intuition or biased opinion. Nor is it to say, despite habits of thought derived from criminal law and statistical hypothesis testing, that it is always best to assume lack of causation until it is proved. Indeed, if this point is overlooked, it may sometimes lead to better ultimate decisions to overlook also the point that association does not prove causation. A little learning is a dangerous thing.

Lave and Seskin, in the first paper of Part IV, in connection with air pollution and mortality, provide a clear introduction to the regression model and the issues of association-causation-proof, though inevitably a misleading phrase or two must have escaped them (and me) which the sharp-eyed reader may find. Fairley's discussion carries the issues further.

In pure forecasting, the main purpose of the next paper by Benarie, Badellon, Menard, and Nonat, the usefulness of regression does not depend on causation, which therefore becomes peripheral or irrelevant. The inference to be drawn when large forecast errors occur raises similar issues, however, and Benarie et al. point out how, in their problem, the pattern of these errors affects the likelihood that the underlying situation has changed. Incidentally, it seems to me inappropriate to describe forecasting success by the significance level of a test of the null hypothesis that forecast and actual values are independent, because no one would be that pessimistic and because the significance level becomes extreme as more data is collected with no change in forecasting method or accuracy. Nevertheless, in correspondence Benarie informed me that in meteorology, although there is a lot of argument about how to do it, this is a common method.

In a third paper concerned with air pollution, Hasselblad, Nelson, and Lowrimore exemplify or touch on several extensions and relatives of regression. The remaining two papers of Part IV apply regression methods to water flow. Mackay and Gilligan study the Clyde estuary, while Thomann studies the Delaware--and demonstrates finally that the Scotch have no monopoly on water pollution analysis, at least. Thomann also demonstrates spectral analysis.

V. Papers predominantly concerned with statistical methods other than regression comprise Part V. In the first three, Klee describes a survey of attitudes on solid waste disposal, van Belle provides a one-man survey of statistical problems in aerosol studies, and Ganczarczyk discusses a number of aspects of activated sludge treatment plants.

The next three papers deal with three different kinds of problems involving dichotomous (binary, two-valued) data. Maloney introduces classical bioassay and indicates a special type of use it might have in pollution research. (Though he does not go on to such matters, analogues of the usual methods for continuous data have now been developed so far

PREFACE

both theoretically and computationally that the conceptual unification which has long been possible should no longer be blocked by practical difficulties. An excellent reference by a leader in this work is <u>Analysis of Binary Data</u> by D. R. Cox, Methuen, 1970.) Elashoff, Sobel, and Schneiderman discuss the advantages and problems of designing screening experiments in which chemicals are given two instead of one at a time. (I conjecture that a Bayesian approach to this problem would be illuminating and not equivalent.) And Hoel, also concerned with testing chemicals, demonstrates a simple and surprising way in which a harmful chemical can appear beneficial and conversely.

Finally, for the last word, I have chosen some bittersweet comments by van Belle on responsible reporting and the role of statisticians in pollution research. Some of my own thoughts on other aspects of the latter, and on other topics, are in "Statistical Aspects of Pollution Problems," <u>ISI Proceedings 44</u> (1971), Book 1, 181-6. Whatever I might have said here, I would certainly not have said as well as van Belle, so I will not try to top him.

Kyoto, Japan   John W. Pratt

ACKNOWLEDGMENTS

The National Science Foundation under Grant GP 30666 provided the main financial support for both the symposium and this book. Some additional and encouraging support was received from the National Bureau of Standards. The travel expenses of many participants, especially employees of U. S. government agencies, were paid for by their employers or other sources of funds, an important factor in paring the budget. My time in editing has been partly supported by sabbatical leave from Harvard and by the John Simon Guggenheim Memorial Foundation, with excellent working facilities provided by the Kyoto Institute of Economic Research of Kyoto University. All these sources of support are gratefully acknowledged.

David Lynn helped with many aspects of putting the program together. Sandra MacArthur, my secretary at the time, performed nobly a variety of tasks extending far beyond ordinary secretarial ones. Edward D. Rowley took unstinting care of local arrangements and preoccupied conferees. Among them, they took everything off my hands I could hope for, and I am most grateful to them.

Finally, although I have not sought the authority of the authors to dedicate this book formally, I would like to dedicate at least my part in it to Jerzy Neyman, in recognition of his leading role in bringing statistics to its present stage of maturity and in conceiving the IASPS and its Satellite Symposia and bringing them into being by force of intellect, diplomacy, persistence, and sheer will.

John W. Pratt

Statistical and Mathematical Aspects of Pollution Problems

Part I

COLLECTION AND DISSEMINATION OF BASIC DATA

1

GLOBAL ENVIRONMENTAL MONITORING*

Sören Svensson†

SCOPE Commission on Monitoring
Stockholm, Sweden

1. INTRODUCTION

Concern with the environment stems from man's preoccupation with his health and well-being, traditionally centered around his attempts to obtain shelter and food and to avoid disease and natural disasters. Man's efforts have always been directed toward increasing his control of the environment, most often for the purpose of increased food production. However, such environmental modifications as land drainage and deforestation for agriculture have often caused harmful side effects, e.g., loss of soil fertility, loss of topsoil, or extension of arid zones, processes that sometimes are irreversible or costly to correct. Over the last one hundred and fifty years, and particularly over the last few decades, the traditional agents of deliberate or unintentional modification, such as fire, grazing, and direct human labor, have been reinforced by modern technology and the waste products of that technology.

We can recognize four general categories of environmental phenomena of major concern to man: natural disasters, epidemics, agricultural and other biological productivity and water supply, and undesirable side effects resulting from modern urban/industrial technology.

---

*This paper is based on SCOPE Report No. 1, Stockholm, 1971, prepared by the Commission on Monitoring of the Scientific Committee on Problems of the Environment of the International Council of Scientific Unions, and submitted to the United Nations Conference on the Human Environment, Stockholm, 1972.

†Present address: Zoological Institute, Lund, Sweden.

All these phenomena have been subject to intensive attention. We already have warning systems for natural disasters such as cyclones and tsunamis, and we are on our way to a better understanding of those geophysical phenomena that generate earthquakes and volcanic activity. Public health authorities have been set up all over the world to investigate, evaluate, and control disease and other human health problems. In most countries considerable efforts have been made to improve and control biological productivity and water supply, although not always very efficiently. The problems of hazardous side effects of modern urban/industrial technology have only recently been subject to any considerable attention, and that only in a small number of advanced nations.

Most of the current activity in the field of environmental control is national in character. Efforts to deal with man's impact on the environment and the repercussions on himself on a world-wide scale are uneven and uncoordinated. This is not only because virtually all governments regard their exploitation of natural resources as their own exclusive business, but in part also because most aspects of this field are relatively new or have become serious only during the last thirty years or so. Concern with these problems seems to depend on population density, degree of industrialization, consumption of products by urban societies, intensity of agriculture, and standards of living, public education, and environmental awareness. However, as contaminants from industrialized countries spread around the globe, and as more and more areas are developed, more and more nations must face these problems. The prevailing tendency to internationalization of social and economic development also implies that the problems must be dealt with on an international scale through joint efforts by many nations.

The current awareness of the truly international character of many environmental problems has lead to a number of actions in the international field. These actions have been developed both on intergovernmental and nongovernmental levels. The most prominent so far is the decision by the United Nations to hold a Conference on the Human Environment in Stockholm in 1972. Within the international scientific sector the most significant event was the establishment of the Scientific Committee on Problems of the Environment (SCOPE) by the International Council of Scientific Unions (ICSU).

SCOPE was asked by Mr. Maurice Strong, Secretary-General of the UN Conference, to "prepare a report recommending the design, the parameters and technical organization needed for a coherent global environmental monitoring system making maximum use of available capabilities of existing and planned national, regional and international networks, together with such data collection and processing centres as may be required." This request was accepted by SCOPE early in 1971; and the responsibility for

1. GLOBAL ENVIRONMENTAL MONITORING

the preparation of the report was given to the Commission on Monitoring of SCOPE. A special secretariat was set up in Stockholm under the auspices of the chairman of the Commission, Dr. Bengt Lundholm.

A brief review of the content of the report follows. Although the report was compiled by scientific experts, it does not depart far from the prevailing views on political levels in most countries. I am therefore satisfied that the proposal for a global environmental monitoring system as outlined in the SCOPE report is fairly realistic, relevant and feasible from both scientific and political points of view.

2. MONITORING - BASIC CONSIDERATIONS

A total system for environmental control includes three basic types of activity. The first involves measurements and observations directed toward a description of the state of the environment and its changes. The second is the evaluation and analysis of environmental data to determine possible trends and to develop a warning system related to preset criteria. This specifically includes such functions as predictions of the environmental consequences of planned actions, descriptions of the global budget of contaminants, the analysis of ecosystems, determination of environmental criteria for specific pollutants, and the formulation of recommendations for action. The third activity is that of action designed to avoid environmental deterioration but in the overall context of achieving environmental management in the most beneficial way.

It can be discussed whether only the first, both the first and the second, or all three types of activity should be included in a global monitoring system. Final action is, however, something that must be decided on and implemented by each government. Therefore we understand by "monitoring" in the SCOPE report a scientifically designed system of continuing measurements and observations. However, when we talk about a "monitoring system" we also include the evaluation and interpretation phase. There are as yet no proposals to include the third or action phase.

The remaining sections of this paper correspond to the four main chapters of the report.

3. ENVIRONMENTAL PROBLEMS

In a report of this kind it is important to identify those particular environmental problems that are relevant to monitoring in a more or less global or international context. Furthermore, there must be a consensus among scientists that the problems recommended for inclusion in the study are of critical importance, either in the sense that the present state of the environment creates immediate and widespread concern or in the sense that present developments may lead to deep concern in the future.

The report states that an environmental problem arises whenever there is a change in the quality or quantity of any environmental factor that directly or indirectly effects the health and well-being of man in an adverse manner. In an effort to isolate and identify a set of central critical and relevant problems a number of criteria were used, namely: number of people involved, geographical distribution of the problem, duration of the problem (short- or long-term effects), degree of irreversibility of the effects, degree of impact on health, standard of living, social structure and economy, and degree of international significance of the problem.

The consensus of the Commission's survey of the wide range of environmental problems recognized in the world today in relation to the aforementioned criteria was that a fairly limited number of problem areas were of paramount importance. These problem areas were defined as follows:

1. potentially adverse climatic change resulting from human activities;
2. potentially adverse changes in biota and man from contamination by toxic substances including radionuclides;
3. potentially adverse changes in biological productivity caused by improper land use (reduced soil fertility, soil erosion, extension of arid zones, etc.);
4. potentially adverse changes in the growth, structure, and distribution of the human population;
5. changes in the subjective human perception of the environment;
6. eutrophication of waters;
7. decreasing freshwater resources;
8. natural disasters.

A number of human activities have the potential to induce climatic changes. Among the most important of those we are aware of today are the increase of atmospheric carbon dioxide due to the burning of fossil fuels; the increase of atmospheric turbidity caused by introduction of dust and aerosols into the atmosphere; changes in the Earth's reflectivity (albedo) due to changes in the cloudiness caused by, e.g., the operation of aircraft in the upper atmosphere or attempts at deliberate weather modification; and direct heat output from energy production in cities and industrial areas. It is not envisaged that any of these factors will induce measurable climatic changes of global significance in the near future. In spite of this the very important consequences of climatic changes that we can foresee, if and when they occur, make it necessary to maintain constant and careful surveillance of the problems.

1. GLOBAL ENVIRONMENTAL MONITORING

The problem that is considered to be most urgent, particularly in industrialized nations but also in an increasing number of developing nations, is that of the adverse effects of chemical contamination of the environment. The problems created by a number of toxic substances, such as heavy metals (mercury, lead, cadmium, etc.), chlorinated hydrocarbons (e.g., DDT and PCB), and petroleum products are well known and recognized all over the world. This problem is considered urgent because profound effects on biota all over the world have already been observed many times, and the properties of many of the toxic chemicals (their persistence, their global distribution, and their accumulation in organisms and food chains) make them particularly hazardous in the environment.

Improper land use and the consequences thereof, loss of soil fertility, erosion, extension of arid zones, and major changes in the vegetation, etc., constitute a problem of paramount importance because detrimental changes in this category may have considerable impact on the economy and the availability of food and other resources in many nations or whole regions of the world, with direct repercussions on the nutritional and health status of man. In spite of the fact that most problems in this category may be considered local for the time being, the similarity of the problems in different parts of the world, in combination with the fact that changes in the economy in one part of the world will very probably affect the economy in other parts of the world, make a monitoring program for land use relevant.

The problem of uncontrolled growth of the human population and of other changes in the population, particularly the strong and increasing tendency toward urbanization, constitute one, if not the, major factor responsible for the creation of environmental problems. It might be said that the size of the human population is the volume control to all other environmental problems created by man. However, the Commission does not recommend that any new major monitoring program be created for the registration of the size and structure of the human population. It is satisfied that the United Nations will continue and intensify its current activities in this field.

The other four problem areas are for different reasons considered less important or less relevant to global monitoring than the first three. It was accordingly concluded that the report should concentrate on the first three major problems: climatic change, chemical contamination, and land use.

4. ENVIRONMENTAL VARIABLES APPROPRIATE FOR MONITORING

The variables that need to be measured or observed in relation to these three major problem areas may be grouped under four headings:

1a. physical and chemical data from the atmosphere pertinent to climatic-change potential;

1b. physical and chemical data from air, water, soils, and biota pertinent to human health and welfare;

2a. physical, chemical, and biological data reflecting the state of human health;

2b. biological data reflecting the performance of biological systems.

We are of course most interested in those changes in the environment that are caused by human activities. It is nonetheless essential to monitor a number of natural processes in order to evaluate accurately and correctly the relative contribution by man to the observed changes. For example, it is necessary to monitor the output of particles and gases to the atmosphere from volcanic activity in order to evaluate the relative importance of air pollution caused by man.

A basic concept in the monitoring process is the cause-and-effect relationships between different variables. In order to be able to take the appropriate measures to solve a problem it is often essential to know the cause or causes creating the problem. Observations of, for example, decreases in a number of animal populations in an area serve as a valuable and indispensable signal that a detrimental process of some kind is working in the community; but as such they do not normally reveal anything about the causes behind it.

The four headings above can be related to causes and effects. Variables under 1a and 1b can be characterized as cause variables, in the sense that changes in them may indirectly (the climatic variables) or directly (the toxic substances) cause detrimental effects in plant, animal, and human populations. The variables under 2a and 2b constitute effect variables, which characterize states and changes in the structure and function of plant and animal populations, ecosystems, and human health and well-being.

In order to assess possible causes of climatic changes, the report recommends that first priority be given to measurements of atmospheric turbidity (aerosol content), carbon dioxide, and solar radiation. For the evaluation of the effects on climate it recommends that collection of standard meteorological data be extended. At a later stage it will be necessary to include more detailed measurements of the vertical distribution and size distribution of aerosols, to collect rawindsonde data, to study the vertical flow of carbon dioxide, and to measure ozone, water vapor, and trace gases in the stratosphere.

1. GLOBAL ENVIRONMENTAL MONITORING

It is a very complicated matter to monitor physical and chemical data from air, water, soils, and biota, particularly the occurrence of toxic chemicals. The number of substances released to the biosphere by man is enormous and increases every year. However, it seems possible to identify a small number of substances with very high priority on the basis of their toxicity, their high levels of incidence, and their persistence in the biosphere. Accordingly it is recommended that highest priority be given to measurements of mercury, lead, cadmium, DDT and its degradation products, and PCB. At a later stage a number of other variables should be considered for inclusion in the program: petroleum products; persistent organochlorine compounds other than DDT and PCB; chlorinated aliphatic hydrocarbons; chlorinated phenoxyacetic acid derivatives; relevant compounds in the cycles of sulfur, nitrogen, phosphorus, and carbon; certain metals (As, V, Zn, Se, Cr, Cu, Be, Ni, Mn); organophosphorus compounds; and oxygen in water. All measurements should be carried out in all relevant media.

It is a very difficult and extensive task to determine the state of and possible changes in human health. The Commission was convinced that a very broad and comprehensive health monitoring program should not be recommended, since it could not see any ways or means by which such a program could be successfully implemented in a majority of nations today. Instead, it recommended that a very restricted number of informative health indicators be monitored, with the knowledge that even this restricted program cannot be fully implemented worldwide. It recommended the collection of data on life expectancy, age structure of populations, and excess crude mortality, and also of data on the frequency of certain age-linked and other relevant diseases and disorders known or suspected to be induced by long-term ingestion of trace amounts of environmental contaminants, special attention to be paid to diseases of the blood and of the cardiovascular system and to certain forms of cancer. It was particularly stressed that human health data should be collected in populations carefully selected to represent various age groups and degrees of exposure to urban, industrial, and intensively agricultural conditions, more specifically to represent populations under very high exposure (occupational), high exposure (urban-industrial or intensively agricultural), medium exposure (rural in densely populated regions), and low exposure (remote regions) to contaminants.

The determination of the performance of biological systems involves numerous possible variables because of the great complexity of biological systems. It is therefore essential to find those biological variables that most efficiently provide reliable information about effects on biota. However, our existing knowledge is too limited to do this at once. There is a great need for pilot studies and basic research into the structure and function of biological systems before a detailed monitoring program can be designed.

In the process of isolating suitable variables that would be both relevant and amenable to measurement, the Commission considered biome studies; the distribution of vegetation types; species diversity; primary productivity, biomass, and growth rate; the size and distribution of species populations; specific population characteristics (reproductive success, mortality, age structure, and migration); physiology, ontogeny, and pathology; genetics; behavioral responses and mental performance; phenology; and registration of short-lived biological phenomena.

After analyzing the present possibilities for each item it was concluded that with the existing resources and knowledge only the following biological actions should be recommended: registration of vanishing or endangered ecosystems, registration of vanishing or endangered vertebrates, studies of population size and distribution of certain groups of birds, and registration of short-lived biological phenomena. It was, however, also recommended that research programs directed toward the development of practical programs be started immediately for registration of global distribution of major vegetation types (preferably by satellite sensing); for monitoring of species diversity of soil organisms, marine algae, and air plankton; for measurements of specific population characteristics (reproductive success, life expectancy, mortality, and age structure); for the study of the pathology of selected standard plants; and for phenology.

As a consequence of the urgent need for better understanding of the structure and function of ecosystems, high priority was given to the implementation of a number of so-called biome studies. Such studies should concentrate on both basic and applied research into the energy dynamics and integrated function of ecosystems in representative areas in all major biome types of the world. The purposes of these studies would accordingly be to develop mathematical and statistical models with predictive capacities in the management of the ecosystems and to identify those variables it would be most useful and efficient to include in the monitoring system.

5. THE DESIGN OF A GLOBAL ENVIRONMENTAL MONITORING SYSTEM

From the viewpoint of economy, a variable should not be measured at more sites, transects, or areas than necessary. On the other hand, as many variables as possible should be measured at the same location in

1. GLOBAL ENVIRONMENTAL MONITORING

order to support an integrated evaluation and the correlation of different variables.

The concept of a network of monitoring sampling areas or stations spread over the world has been widely discussed. It must, however, be understood that many variables cannot be measured at only a limited number of stations. This is the case with large-scale vegetation surveys and with studies on the distribution of animal populations. Many physical and chemical measurements, including meteorological measurements, should, however, be made at specific stations or in comparatively small sampling areas. Since one of the major aims of the monitoring process is to establish long-term trends, the arguments favor the station concept, i.e., the location of the measurements at the same point or points for long periods of time.

Another important concept in the design of a monitoring system is the baseline concept. It is of extreme importance to establish, as soon as possible, the current baseline for different environmental properties, that is, the current background conditions in remote and comparatively unaffected regions of the world. But it is also important to carry out measurements in those areas where people normally live and under extreme exposure conditions where effects can be expected to appear first. Therefore the Commission recommended the establishment of

Ai. networks of reference stations or sampling areas, both baseline areas (low exposure stations) and regional areas (medium exposure stations);

Aii. networks of high-exposure areas;

B. other monitoring systems (regional and global surveys).

The network of baseline stations should initially comprise about ten very remote stations for the establishment of globally significant climate conditions and changes, and at least the same number of baseline stations for the establishment of chemical background conditions and changes. The biome studies should represent all the major biome types, namely, tundra, coniferous forest, deciduous forest, tropical forest, savannah, thorn scrub, grassland, desert, open ocean, upwelling area, coastal shelf, estuary, and epicontinental semienclosed sea. It is recommended that at least ten terrestrial stations be established initially for this purpose. Baseline stations for climatic change, for chemical contamination, and for biome studies should be combined wherever possible in order to reduce the total number of baseline stations.

The so-called regional stations have two purposes. The first is that the measurements should describe the general background conditions in areas with average population density and average exposure conditions (generally normal rural areas). The measurements should be representative for a region, normally a natural geographical or topographical region (a drainage basin, a special vegetation zone, etc.), but not for a considerable part of the globe. The second is that the data obtained should assist in the establishment of dynamic budgets for different substances in the biosphere.

The high-exposure areas have a more specific purpose, directly connected with the early detection of detrimental effects of exposure. Such stations should therefore be located in large cities, heavily industrialized areas, intensive mining regions, or areas with intense farming. A special type of high-exposure station is the rivermouth station on a highly polluted river.

A number of special monitoring systems must be created independently of the networks of areas or stations for the monitoring of such variables as global albedo, global vegetation patterns, land use, erosion, distribution of plants and animals, etc. Some of these variables are best monitored by remote sensing from satellites or aircraft. A special kind of monitoring is the registration of so-called short-lived phenomena. Many such phenomena, e.g., sudden and extensive bird deaths, algal blooms, oil spills, etc., either are supposed to be indicative of possible detrimental developments that might deserve greater attention in the future or are directly hazardous.

6. OPERATIONAL AND INSTITUTIONAL ARRANGEMENTS

A correct and complete evaluation of environmental data is only possible when the environment is treated as a unity. The interactions of the processes in space and time make it necessary to consider information from all media of air, water, soil, and biota including man. Traditionally, however, the approach to environmental problems has been medium-oriented: air pollution, water pollution, etc. Different organizations and agencies still often have responsibility for each separate medium. This is the case in the United Nations system of specialized agencies, for example, and within many national governments. Sometimes the situation is very complicated, with a great number of governmental agencies involved and very little coordination between them.

In order to create an efficient organizational design of a global environmental monitoring system it is therefore essential to achieve both

1. GLOBAL ENVIRONMENTAL MONITORING

an appropriate international form of cooperation and an integration of different medium-oriented activities, whether on a national or international level. International cooperation and coordination are basic requirements for, among other things, international standardization of measurements, methods, and data distribution and presentation. International agreements and recommendations for action also call for strong coordination of national efforts and programs.

Whatever kind of international cooperative organizational framework can be established, the most important unit in the monitoring system is the single and sovereign nation. Clearly it will not be possible to force "cooperation" onto any nation, and the interest within a nation in taking part in an international program is strongly influenced by the importance for the nation itself of acting on the problems and activities in which the international organization finds it suitable to engage. The local, national problems will always be given the highest priority.

The Commission recommends the establishment in each country of a monitoring office, which would have the double function of coordinating the different, often medium-oriented activities within the nation and of serving as the link between the nation and the international monitoring organization. This would also apply to intergovernmental and other international organizations. By previous agreement this function as correspondent might involve the handling of selected material emerging from a nationally controlled reference station, of health monitoring data, or of relevant information obtained in some other agreed manner. By international agreement, this office would not only transmit certain data to any intergovernmental coordinating machinery, but also receive as required all data from the global system.

The establishment of national monitoring offices and intergovernmental agencies is, however, not enough to achieve efficient international cooperation. There is a strong and recognized need for a central body at a high level, capable of coordinating different agreed-upon international activities. It seems that such a body should be set up under the United Nations at a level above that of the specialized agencies. The basis for the work of this body should be the wide range of competent current and planned environmental monitoring programs. These include territorial monitoring activities, which are the sole concern of each national government; regional programs, where a shared resource or region is monitored collaboratively by those governments directly affected; and the UN agencies and other intergovernmental or nongovernmental programs concerned with climatic change, human health and toxicology, marine conditions, radioactivity, education, and training.

This central coordinating monitoring unit should have the following functions:

1. to delineate programs and to continuously supervise global environmental monitoring activities in order to make certain that the system operates with maximum efficiency and relevance and that optimum output is achieved;

2. to coordinate current monitoring programs and to recommend new activities in order to ensure that the requirements of the system are satisfied;

3. to standardize the methods of observation in order to ensure comparability of data;

4. to take the necessary steps to provide the monitoring system with appropriate means for data handling and dissemination;

5. to evaluate advice from its independent scientific advisory bodies, the UN agencies, other international organizations, and nations; and

6. to report to UN on the state, needs, and results of the global environmental monitoring system.

It must be stressed that an international monitoring system must eventually be an integrated part of a much more comprehensive framework for policy, research, and action in the field of environmental affairs in general, and that monitoring cannot operate with maximum efficiency without close connection with these other environmental activities. Today there is no organization with the competence of coordinating all global environmental activities. The report therefore also recommends that the United Nations seriously consider using an existing body or establishing a separate body directly responsible to the General Assembly for the purposes mentioned above.

DISCUSSION *

L. A. Clarenburg

Rijnmond Authority
Rotterdam, The Netherlands

1. On page 5 of the report I read: "The term 'monitoring' is used throughout the Report to mean a 'scientifically designed system of continuing measurements and observations'." I am afraid that the report is not a pure scientific document; it seems as if politics are intermixed with science. As I will show, some parts of the report give the impression that since there is a political demand for action, there should be a global monitoring system - later we will see how to use it. Let me make it clear that I am certainly in favor of global monitoring, but not as unconditioned as is proposed in the report.

2. I would like to start with the last section, Section 8. The first sentence reads: "A correct and complete evaluation of environmental data is only possible when the environment is treated as a unity." I could not agree more, if we have the same understanding of the notion "environment." To my knowledge, our living environment consists of the working, social, and spatial cultural or physical environment; together they make up our living environment. Now let us look more closely at the spatial cultural or physical environment: there we can distinguish the microsphere (the house), the mesosphere (the direct neighborhood of the house, local facilities), the macrosphere (the region, higher-order facilities, traffic and transportation, recreation), and the biosphere (air, water, and soil pollution, noise, and calamities). We do know that about 80% of all illness is mental or stress-induced dysfunctioning. The densely packed environment of urban areas imposes

---

* These comments refer to the full text of SCOPE Report No. 1, Stockholm, 1971. All quotations and page and section numbers are from the report, not from Paper 1 of this book.

an enormous stress on humans. We have been able to measure
uniquely the alienation caused by environmental deficiencies, the
deficient environment frustrating the basic human needs (motives),
which are the same in all of us. As health is defined by WHO as
a situation of complete physical, mental, and social well-being,
it is my conviction that the environment of today attacks in the
first instance the mental and social dimensions of health. Environmental pollution is just one of the many problems of the physical
environment, and presumably not even the most important. In my
opinion, it would have been desirable to recognize the relative
importance of global monitoring. This is the more so, because
the propositions for the global network of monitoring stations,
when carried out, would cost a considerable sum of money. The
report states (page 69): "There is little doubt that the least
wasteful final design will be arrived at by evolving from a modest
beginning."

3. There is every reason for a "modest beginning." Here begins my
second comment. Let me quote the section on "Evaluation,"
(page 71, Section 8.3): "The most critical point in relation to the
monitoring activities is the evaluation of the results." That is all.

To my mind any study on the design of a global monitoring system
should start with a large chapter on Evaluation of Data. (a) What
do we want to know? (b) How can we extract this information
from data? (c) What kind of data do we need for that purpose?
(d) What reliability should our data have, then? (e) What techniques should we use, then? (f) How many stations will finally
yield the information at the required reliability level? In the
present report none of these considerations (which are basically
statistical in nature) are found. (I hope I am rousing your interest
in these important and difficult problems.) Instead I read (page
55): "The stations selected should be representative of fairly
large areas of the globe." This is easily stated. But how do we
know whether a station is really representative of an area? What
do we know about natural variabilities of the background (baseline) values of the parameters to be measured? "To obtain
representative measurements of bio-environmental conditions, at
least two stations should eventually be located in each of these
biome types." Why two? "A total of not less than ten terrestrial
baseline areas should be established." Why ten? These figures
are stated; they do not follow from any analysis and thus are
subject to perpetual haggling. Moreover, space correlation between pollution levels in various biome types could eventually
reduce the number of stations required. I would plead therefore

DISCUSSION

    to add to recommendation 2, page 12, a point 7 at least, reading: "Develop techniques to extract relevant trends from the data."

4. My third remark is related to the aim of the global monitoring system. Page 5 has: "Initially the monitoring process will help to detect significant, potentially harmful changes early enough to take avoiding action and eventually predict harmful environmental trends before they occur." Quite a mouthful: "significant," "potentially harmful," "early enough." A "significant" change is a change beyond the background noise. And here is another problem for you, not touched in the report: how do we detect small and slow changes at global scale - our environment only allows for small changes - as distinguished from background noise? This detection should on top of that be "early enough." If you think you have had your problem for today, let us read on pages 7 and 15: "... relevant diseases and disorders, known or suspected to be induced by the long term ingestion of trace amounts of environmental contaminants." It would not cost me much trouble to name a hundred environmental factors, all inducing disease or disorder in humans. My question to you, unanswered in this report, is: how do we extract information on one single potentially harmful factor - out of a hundred, say, all affecting human health - significantly and early enough? I would have wished the present report more analytical and less descriptive.

    I therefore do not concur with this sentence on page 11: "We have determined that a global environmental monitoring system is desirable, timely and feasible." My doubts are especially related to the word "feasible," and I believe this is the appropriate place, in front of professional statisticians, to express my doubts.

5. This brings me to my fourth remark. The network is intended to detect trends, as I just quoted. The report distinguishes three areas: base-line areas, regional areas, and high-exposure areas. Let a contaminant have mean value 1 in the first area, 25 in the second, and 200 in the third. If appropriate analytical tools are available it seems easier to me to detect significantly and early enough a change from 1 to 2 than from 200 to, say, 210. I would have wished that the report would have stressed the greater importance of the base-line stations over the other stations. One could get a wrong impression; I quote from page 57: "the minimum number of high exposure stations should be approximately three times the number of base-line stations." I can say from my own experience that the pollution within a city can be adequately estimated using mathematical models, so why three times as

many fairly predictable stations? The trend is of importance, not the actual level; adverse trends will induce countermeasures, adverse levels have to be prevented.

6. With my emphasis on base-line stations, not all problems have been solved because (page 58): "Information on substances emitted to the environment is important to any global monitoring system and is especially needed for global budgeting." Unfortunately the report is only dealing with man-induced emissions. My questions are: What is the relative contribution of pine woods to the hydrocarbon content of the atmosphere, and how do pine woods contribute to the ozone formation? What are the effects of sandstorms, volcanic eruptions, pollen, material blown from the ocean? And another question, not touched by the report: how do these major causes of pollution contribute to the variability at a base-line station?

7. I may end up by quoting pages 25 and 26: "Four considerations relevant to the basic system must be stressed initially:

   "a. The system which is finally evolved will be costly and time-consuming and occupy the full attention of a large amount of scientific and technical manpower. For this reason the functioning of the monitoring system must be kept under careful and continuous review. This will also assist it to develop a rapid and flexible response to any new environmental situation which appears to merit investigation.

   "b. The most optimistic results that can be obtained from the monitoring process is that nothing significantly harmful is happening to the human environment, although beneficial trends may be recognized. Thus the justification for the system, as in law enforcement or any other 'watchdog' activity, is a negative one."

As the report states, the system proposed will be costly and will require great effort, while the results are doubtful, at least at our present state of knowledge. Now I disagree with the report as to the functioning of the law as a "watchdog" over global pollution. I believe that international law at the United Nations level on the protection of the environment is a highly effective means of preserving our globe. Therefore I would not call enforcement of a law a negative factor; establishment of an international law to preserve the environment, and enforcement at an international level will undoubtedly serve the aim of the monitoring network: to prevent the occurrence of significant potentially harmful changes of the environment.

8. In conclusion, neither the need, nor the feasibility of a global monitoring system have been proved in the report. The significance of the present report is in the first instance political, in the last instance scientific.

I recommend that the feasibility of a global monitoring system be proved by a thorough mathematical analysis. A strong international law is a highly effective means of preservation of the global environment.

2

A SIMPLIFIED STATISTICAL PROCEDURE TO OPTIMIZE THE SITING OF AIR POLLUTION MEASURING STATIONS

André Junod

Institut Suisse de Météorologie
Payerne, Switzerland

Air pollution measuring stations can be sited in three fundamentally different ways.

1. The stations may be placed at those sites where sensitive receptors (e.g., crops) are present and/or where the background level of the relevant pollutant is already high.

2. The stations may be located according to an estimated concentration distribution, relying on the knowledge of the emission distribution as well as that of the dispersion climate.

3. The stations may be distributed regularly, at the nodes of a square grid, for instance.

Each of these methods has some specific application. The first one has, however, the least explanatory power for understanding the emission-concentration relationship. While the third method is in principle the most powerful because it provides the most systematic and complete data, it often incurs high expenses because of the dense station coverage of the area studied and the huge amount of data to be handled. It appears that the second siting method has in fact many advantages, including adaptability to the topographical and meteorological features of the region, as well as economy.

The statistical procedure described in this paper belongs to the second class of siting method. It applies essentially to those cases where

the pollution sources to be monitored are scarce and not too near to each
other. Moreover this procedure is easier to operate when the pollutant
under consideration is specific.

   As a typical example, let us consider the situation that arose in a large
Alpine valley - the Rhône Valley upstream of Lake Geneva - when several
aluminum-producing plants and phosphate fertilizer factories, all active
emitters of gaseous fluorine compounds, settled there. It is well known
that fluorine represents a serious hazard for vegetation, especially for
some sensitive plants, such as apricot, that are widespread in that valley.
The local authorities set up a network of fluorine-measuring stations,
locating them first in those orchards where fluorine caused the most
severe necrosis (siting method of the first kind). Fairly good correlation
was found between measured fluorine concentrations (on a monthly basis)
and amount of damage. The local authorities did not succeed, however, in
ascribing a determined part of the damage to a given factory, due to the
lack of knowledge of the emission-concentration relationship with respect
to each source.

   In an effort to improve the effectiveness of the concentration measuring network, it was decided to redesign the network by taking into account
the dispersion climate of the region, as determined by its main topographical and meteorological features. In a first phase, a limited but critical
section of the Rhône Valley was investigated. This particular area is
characterized by an about 90° elbow of the main valley, by the outlet of an
important side valley, and also by the presence of two factories close to
each other, an aluminum plant and a phosphate fertilizer plant. It should
be noted that the emission conditions of two such plants are quite different,
the aluminum plant acting like a low line source, the phosphate plant like
an elevated point source. This means that separate evaluation of the
contribution of each plant to concentrations at any point is worth a try,
although the distance between plants amounts to some hundred meters
only. On the other hand, a fairly complicated wind regime should be
expected in this region, although some previous measurements revealed
a predominance of a few topographically induced wind sectors. Generally
speaking, the dispersion of pollutants in the atmosphere depends essentially on the wind field and the temperature field, the vertical gradient of
the latter ("lapse rate") being a measure of the thermal stability of the
layer considered. A "dispersion situation" at a given time (or during
a given period) could be defined as the set of relevant wind and lapse rate
data at this time (or during this period). With the aid of these primary
data a corresponding set of directly usable parameters can be derived,
i.e., $\bar{\theta}$, $\bar{u}$ (mean wind direction and speed) as well as C_y, C_z, n (Sutton's
diffusion parameters) or σ_y, σ_z (standard deviations), as they are needed
in the bivariate normal distribution formula giving the pollutant concentration in the lee of a source.

2. OPTIMIZING SITING OF MEASURING STATIONS

In the present case the problem was not to determine the distribution of instantaneous fluorine concentrations but to evaluate the mean (or integrated) concentration distribution at the ground over periods of one month, because the concentrations are actually measured with static absorbers (Harding devices), the filters being changed once a month. The meteorological data at hand come essentially from a wind recorder located in the immediate vicinity of the fluorine sources. In a first phase, lapse rate measurements are not performed at the site, but the diffusion parameters are estimated by using the trace width of the wind direction record ($\sim \sigma_\theta$), as in a well known Brookhaven study. Then, for every month, statistics of hourly dispersion situations are worked out, based on 12 wind direction sectors (width $30°$), 10 wind speed classes (width 1 m/sec), and 4 stability classes (labile, neutral, stable, very stable) with corresponding values of C_y, C_z, n (or σ_y, σ_z), a total of 480 theoretically, if not meteorologically, possible cases.

It can be seen straightaway that such monthly statistics display a remarkable accumulation in the most frequent situations, so that more than 80% of the actual dispersion situations belong to a mere 50 of the 480 possible sets. In the simplified procedure reported here the computation of the monthly mean concentrations takes into account only these 80%, the most frequent situations. The error thus introduced remains relatively small, since only three of the twelve wind sectors are retained, while the discarded hourly situations (20%) are distributed rather evenly in all sectors. In short, this simplified procedure consists of cutting off the statistical "noise" of the pertinent distribution in order to reduce the volume of calculations to a still representative minimum. In the present case the concentration computations are restricted to one downstream $30°$ sector and to two adjacent upstream sectors. In each of these sectors the axis concentration is calculated as well as the lateral contribution of the adjacent sector, if any. The spacing of evaluated points along any axis amounts actually to 200 m, but could be given another value if needed. All these calculations are easily carried out with a programmable desk computer.

The resulting (computed) concentration distribution displays some interesting features, such as two concentration maxima along the axis, well marked upstream of the sources, less apparent or even blended downstream. Furthermore, if some monthly variation of the concentration distribution occurs in every sector, this does not profoundly affect the general course of concentration against distance from the sources. At the present stage it is not possible to carry out a systematic comparison of computed and measured concentrations, owing to the wide scattering of the actual measuring stations. However, the few relevant data at hand show a reasonably good agreement with computations up to about 5 km from the sources. The contemplated adjustment of the primary measuring network to the calculated concentration distribution requires the shifting - or, when this is not allowed, the addition - of several stations in order to ensure a ground coverage proportional to the estimated hazard.

A final remark: the purpose of this preliminary report is not to discuss a large data set, but primarily to bring out some basic ideas about a simple way to optimize the siting of air pollution measuring stations by taking into account the emission-concentration relationship.

3

STATISTICAL STUDIES OF URBAN AIR POLLUTION - SULFUR
DIOXIDE AND SMOKE

Michael Legrand*

Institut Royal Météorologique de Belgique
Brussels, Belgium

1. INTRODUCTION

Some years ago, the Public Health Ministry of Belgium decided to organize, with the collaboration of several institutions, principally the Institut Royal Météorologique de Belgique, a semiautomatic network composed of 180 stations measuring air pollution by sulfur dioxide and smoke on an urban scale.

From the technical point of view, the recommendations of the OECD have been followed. The apparatus (Fig. 1) is very simple: a little electric pump aspirates about 2 m^3 of air daily through a Whatman filter No. 1 to retain the smoke, a bottle containing 50 ml of water with 0.3% H_2O_2 at pH 4.5 to retain the sulfur oxides, and a gas meter to measure the volume of air treated. Such an apparatus is, in fact, composed of eight filter-bottle circuits and one automatic commutator that switches to the next circuit each day at 9 a.m.

The actual network is presented in Fig. 2. There are 25 stations in Brussels, 26 in Antwerp, 20 in Liege, 15 in Ghent and 15 in Charleroi. The other stations have been placed in towns of less importance or at meteorological stations equipped with anemometric instrumentation.

In the large urban areas, we tried to place the stations at a mean distance of one kilometer from one another, but this was very difficult for topographic and administrative reasons, and because the stations must

―――――――――
*Present address: Institut d'Hygiene et d'Epidemiologie, Brussels, Belgium.

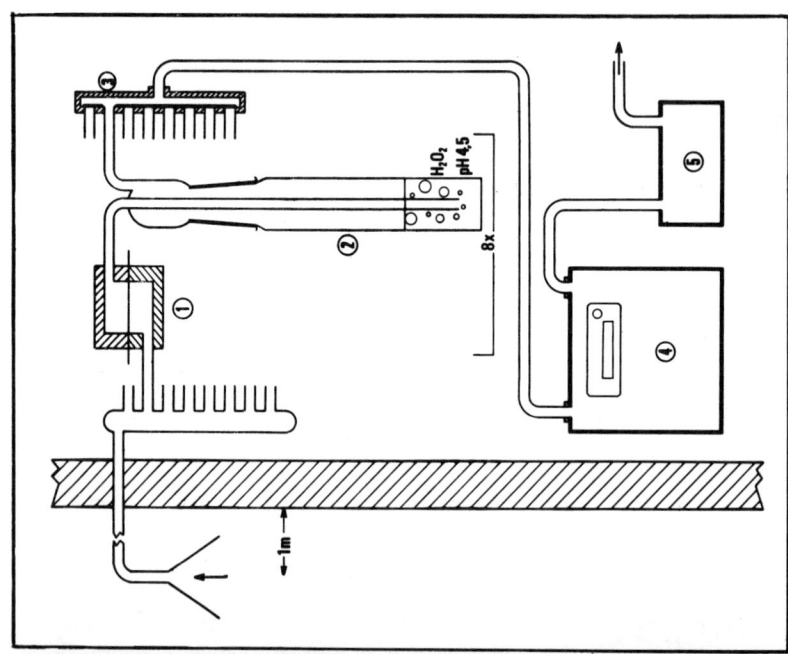
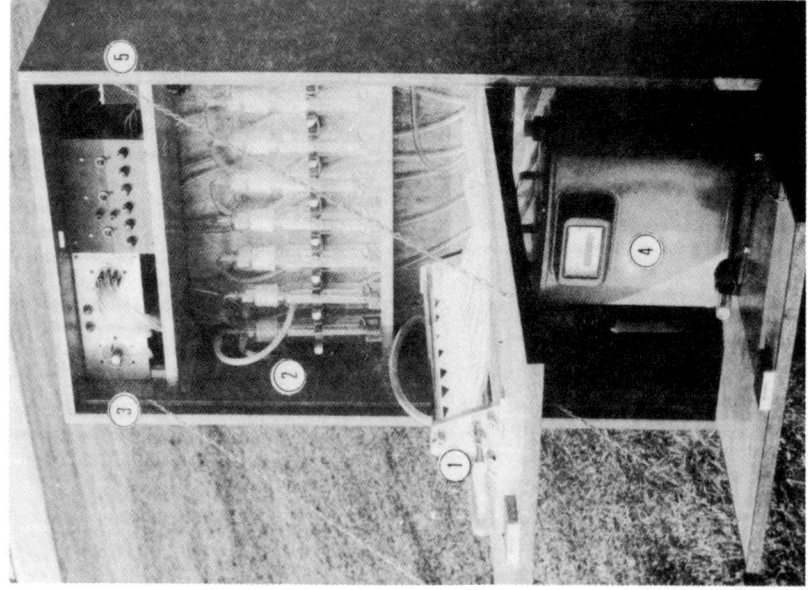

Fig. 1. Photograph and diagram of the sampling apparatus; 1 - filter; 2 - bottle; 3 - automatic commutator; 4 - gas meter; 5 - pump.

3. URBAN AIR POLLUTION   27

be located at places that are not subject to the direct influence of heavy pollution sources, or local sources not representative of the general urban pollution.

2. COLLECTION AND PUBLICATION OF POLLUTION DATA

Each month the data collected from the pollution network are centralized, verified, and processed by an IBM 360/44 so as to give daily smoke and SO_2 concentrations in micrograms per cubic meter of air. These are published in a monthly review compiled by the Institut Royal Météorologique de Belgique, wherein the extremes, the monthly averages, and the daily mean concentrations in the five great urban areas of Belgium and in some other locations are tabulated (Fig. 3). Besides these tables, the review also provides graphical representations of the daily pollution in each of the five great cities.

Fig. 2. Actual pollution measuring network.

PROVINCE D'ANVERS   NOVEMBRE 1969

CU AU	801 POLITIEBUREEL 1 KIOSKPLAATS HEBOKEN			802 RIJKSWACHTKAZERNE 180 BOOMSESTEENWEG WILRIJK			803 OOSTERVELD KLINIEK ST.AUGUSTINUS WILRIJK			804 SCHOOL 13 EGGESTRAAT MORTSEL		
	FUMEE ROOK	SO2 SO2	F/SO2 R/SO2	FUMEE ROOK	SO2 SO2	F/SO2 R/SO2	FUMEE ROOK	SO2 SO2	F/SO2 R/SO2	FUMEE ROOK	SO2 SO2	F/SO2 R/SO2
1- 2	37	166	0.22	27	116	0.23	20	120	0.17	31	165	0.19
2- 3	25	133	0.19	17	311	0.05	11	96	0.12	16	93	0.17
3- 4	30	253	0.12	13	236	0.06	15	177	0.08	18	88	0.20
4- 5	23	85	0.26	15	173	0.09	20	62	0.32	31	125	0.25
5- 6	61	216	0.28	70	243	0.29	61	183	0.33	75	305	0.24
6- 7	86E	229	0.37	93E	353	0.26	81E	167	0.48	86E	241	0.36
7- 8	52	85E	0.61	58	135	0.43	45	103	0.44	47	168	0.28
8- 9	26	146	0.18	14	158	0.09	13	108	0.12	23	128	0.18
9-10	17	122	0.14	9E	112	0.08	8	69	0.12	16	79E	0.20
10-11	28	174	0.16	14	218	0.06	10	81	0.12	23	104	0.22
11-12	17	187	0.09	12	68	0.18	11	41E	0.28	16	91	0.18
12-13	22	128	0.17	19	175	0.11	13	78	0.17	19	110	0.18
13-14	59	249	0.24	48	161	0.30	45	155	0.29	60	189	0.32
14-15	24	168	0.14	27	127	0.21	15	71	0.21	27	165	0.17
15-16	15	142	0.11	12	117	0.10	7E	56	0.12	13E	126	0.10
16-17	13E	113	0.12	9E	58E	0.15	7E	54	0.13	14	128	0.11
17-18	31	229	0.13	32	164	0.19	26	126	0.20	47	201	0.23
18-19	46	259	0.18	25	191	0.13	26	124	0.21	47	174	0.27
19-20	24	179	0.13	14	248	0.06	22	90	0.24	31	168	0.19
20-21	57	244	0.23	25	234	0.11	37	151	0.24	57	267	0.21
21-22	39	156	0.25	48	147	0.33	30	91	0.33	50	183	0.27
22-23	50	222	0.23	42	226	0.19	39	147	0.27	50	226	0.22
23-24	39	169	0.23	39	183	0.22	34	102	0.34	34	200	0.17
24-25	34	202	0.17	42	203	0.21	28	159	0.17	63	195	0.32
25-26	32	313	0.10	45	203	0.22	30	159	0.19	60	250	0.24
26-27	44	205	0.22	48	242	0.20	13	137	0.09	47	243	0.19
27-28	33	197	0.17	33	310	0.11	33	148	0.22	52	221	0.23
28-29	18	129	0.14	12	126	0.10	15	83	0.18	33	122	0.27
29-30	71	305	0.23	52	221	0.24	49	168	0.29	62	297	0.21
30- 1	56	426E	0.13	46	364E	0.13	41	324E	0.13	49	455E	0.11
MCY.	36	194	0.15	32	194	0.16	26	121	0.21	39	183	0.21

CU AU	809 SCHOOL 38 QUELLINSTRAAT ANTWERPEN			810 GEMEENTEHUIS MOORKENSPLEIN BORGERHOUT			811 BRANDWEERKAZERNE 54 WATERBAAN DEURNE			812 LINKER OEVER 86 HALEWIJNLAAN ANTWERPEN		
	FUMEE ROOK	SO2 SO2	F/SO2 R/SO2	FUMEE ROOK	SO2 SO2	F/SO2 R/SO2	FUMEE ROOK	SO2 SO2	F/SO2 R/SO2	FUMEE ROOK	SO2 SO2	F/SO2 R/SO2
1- 2	64	145	0.44	37	182	0.21	32	201	0.16	31	59	0.53
2- 3	47	151	0.31	24	166	0.14	19	142	0.14	16	43	0.38
3- 4	75	150	0.50	24	185	0.13	21	145	0.15	13	36	0.35
4- 5	79	139	0.57	32	160	0.20	35	114	0.30	14	34E	0.42
5- 6	151	115	1.31	73	329	0.22	82	247	0.33	57	148	0.38
6- 7	172	495	0.34	93E	433	0.21	117	394	0.30	93E	272	0.34
7- 8	103	209	0.49	66	214	0.31	70	199	0.35	60	141	0.42
8- 9	62	179	0.35	49	212	0.23	21	172	0.12	12	58	0.22
9-10	46	119	0.39	26	136	0.19	17	105	0.16	11E	58	0.18
10-11	81	182	0.44	40	218	0.19	21	160	0.13	14	106	0.14
11-12	43	122	0.35	20	98E	0.20	21	98E	0.21	12	143	0.09
12-13	77	173	0.45	43	212	0.20	19	142	0.13	14	92	0.16
13-14	140	254	0.55	66	280	0.23	67	206	0.32	60	172	0.35
14-15	115	259	0.44	26	190	0.14	27	136	0.20	23	155	0.15
15-16	65	157	0.42	15E	176	0.09	19	117	0.16	11E	84	0.12
16-17	65	149	0.44	15E	141	0.11	15E	129	0.12	11E	125	0.08
17-18	170	298	0.57	49	245	0.20	63	248	0.25	35	110	0.32
18-19	162	349	0.46	49	288	0.17	52	217	0.24	37	101	0.37
19-20	120	285	0.42	33	193	0.17	43	230	0.19	20	86	0.23
20-21	170	391	0.44	52	323	0.16	45	354	0.13	46	179	0.26
21-22				88	229	0.38	54	208	0.26	36	202	0.18
22-23				50	324	0.15	58	266	0.22	44	249	0.18
23-24				36	195	0.18	48	178	0.27	47	117	0.41
24-25				41	237	0.18	45	211	0.21	39	83	0.46
25-26				57	284	0.20	134E	264	0.51	47	186	0.25
26-27				57	378	0.15	48	310	0.15	50	169	0.30
27-28				53	308	0.17	51	354	0.14	68	104	0.65
28-29	95	214	0.45	47	190	0.25	44	233	0.19	15	40	0.36
29-30	117	272	0.43	56	301	0.19	73	302	0.24	48	97	0.50
30- 1	106	513	0.21	56	566E	0.10	69	510E	0.14	58	450E	0.13
MCY.				45	246	0.18	47	219	0.21	34	129	0.26

1969 NOVEMBER **PROVINCIE ANTWERPEN**

+ 805 ANTWERPEN POLITIEBUREEL ST.BERNARDSESTEENWEG			+ 806 BERCHEM GEMEENTEHUIS 152 GROTE STEENWEG			+ 807 DEURNE LUCHTHAVEN			+ 808 ANTWERPEN STADSLABORATORIUM 41 SCHILDERSSTRAAT			
FUMEE ROOK	SO2 SO2	F/SO2 R/SO2	FUMEE ROOK	SO2 SO2	F/SO2 R/SO2	FUMEE ROOK	SO2 SO2	F/SO2 R/SO2	FUMEE ROOK	SO2 SO2	F/SO2 R/SO2	VAN TOT
35	155	0.23	35	133	0.26	36	128	0.28	32	136	0.23	1- 2
22	139	0.16	26	252	0.10	19	79	0.24	25	161	0.16	2- 3
20	187	0.11	28	202	0.14	20	185	0.11	21	134	0.16	3- 4
22	130	0.17	37	144	0.26	24	154	0.16	27	118	0.23	4- 5
66	241	0.27	105	288	0.36	67	215	0.31	70	284	0.25	5- 6
74E	131	0.57	127E	408	0.31	84E	174	0.48	95E	361	0.26	6- 7
52	179	0.29	86	183	0.47	47	77E	0.60	70	244	0.29	7- 8
24	300	0.08	38	164	0.23	22	116	0.19	21	150	0.14	8- 9
15	54E	0.28	33	146	0.23	15	77E	0.20	20	103E	0.19	9-10
24	116	0.21	38	253	0.15	22	81	0.28	28	203	0.14	10-11
15	164	0.09	30	96E	0.32	24	84	0.29	21	124	0.17	11-12
22	85	0.26	29	158	0.19	19	169	0.11	21	145	0.15	12-13
59	225	0.26	80	212	0.38	48	135	0.36	55	255	0.22	13-14
24	147	0.16	39	144	0.27	28	127	0.22	28	177	0.16	14-15
15	128	0.12	16E	115	0.14	13	111	0.12	14E	133	0.11	15-16
12E	107	0.11	17	112	0.16	10E	133	0.07	14E	128	0.11	16-17
43	164	0.26	80	243	0.33	40	184	0.22	43	175	0.25	17-18
52	211	0.24	64	233	0.28	43	160	0.27	43	219	0.20	18-19
35	129	0.27	50	244	0.21	20	198	0.10	32	175	0.18	19-20
49	249	0.19	72	258	0.28	33	295	0.11	55	273	0.20	20-21
38	158	0.24	96	188	0.46	33	209	0.16	51	235	0.22	21-22
52	225	0.23	86	279	0.31	41	168	0.25	62	323	0.19	22-23
30	132	0.23	72	220	0.33	31	137	0.23	51	205	0.25	23-24
46	169	0.27	77	249	0.31	33	161	0.21	43	205	0.21	24-25
49	207	0.24	86	312	0.28	44	185	0.24	62	286	0.22	25-26
52	230	0.22	88	312	0.28	31	187	0.17	55	362	0.15	26-27
50	282	0.18	60	287	0.21	41	194	0.21	52	269	0.19	27-28
23	104	0.22	44	135	0.32	31	132	0.24	30	134	0.23	28-29
60	231	0.26	67	275	0.24	68	208	0.33	52	228	0.23	29-30
50	431E	0.12	63	430E	0.15	53	403E	0.13	55	528E	0.11	30- 1
37	180	0.21	58	222	0.26	34	162	0.21	41	215	0.19	GEM.

+ 813 ANTWERPEN STADHUIS GROTE MARKT			+ 814 ANTWERPEN ZWEMBAD 1 ALFONS ENGELSSTRAAT			+ 815 DEURNE SCHOOL 173 BOSHOVESTRAAT			+ 816 ANTWERPEN LUCHTBAL SCHOOL 6 COLUMBIASTRAAT			
FUMEE ROOK	SO2 SO2	F/SO2 R/SO2	FUMEE ROOK	SO2 SO2	F/SO2 R/SO2	FUMEE ROOK	SO2 SO2	F/SO2 R/SO2	FUMEE ROOK	SO2 SO2	F/SO2 R/SO2	VAN TOT
41	125	0.33	41	194	0.21	40	191	0.21	28	72	0.39	1- 2
29	93	0.31	31	204	0.15	25	244	0.10	18	42E	0.43	2- 3
26	101	0.26	24	178	0.13	25	200	0.12	20	44	0.45	3- 4
22	109	0.20	28	150	0.19	43	214	0.20	30	205	0.15	4- 5
67	254	0.26	62	293	0.21	93	342	0.27	67	191	0.35	5- 6
95E	337	0.28	92E	371	0.25	128E	446	0.29	91E	263E	0.35	6- 7
67	203	0.33	64	182	0.35	73	202	0.36	67	145	0.46	7- 8
27	152	0.18	27	208	0.13	26	198	0.13	22	95	0.23	8- 9
19	93	0.20	19E	137	0.14	22	144	0.15	17	63	0.27	9-10
33	181	0.18	27	231	0.12	28	207	0.13	24	112	0.22	10-11
20	148	0.14	21	153	0.14	24	120E	0.20	20	92	0.22	11-12
33	147	0.23	23	195	0.12	24	154	0.15		83		12-13
67	242	0.28	64	250	0.26	83	279	0.30		162		13-14
41	197	0.21	36	175	0.21	34	179	0.19	31	100	0.31	14-15
24	147	0.17	25	159	0.15	17E	127	0.13	15E	83	0.18	15-16
17E	126	0.14	23	159	0.14	21	141	0.15	15E	81	0.19	16-17
53	153	0.35	50	206	0.24	59	297	0.20	41	149	0.27	17-18
63	207	0.31	44	258	0.17	52	284	0.18	43	118	0.37	18-19
31	162	0.19	23	201	0.11	49	250	0.19	24	75	0.33	19-20
60	283	0.21	57	273	0.21	97	426	0.23		204		20-21
64	209	0.31	50	208	0.24	97	286	0.34		123		21-22
76	318	0.24	64	299	0.21	115	404	0.28	60	188	0.32	22-23
54	197	0.27	42	198	0.21	63	217	0.29	33	75	0.45	23-24
47	111	0.43	44	193	0.23	67	325	0.21	31	91	0.34	24-25
61	78	0.77	47	265	0.18	102	399	0.26	50	156	0.32	25-26
54	62E	0.86	39	325	0.12	102	515E	0.20	40			26-27
64	147	0.44	40	301	0.13				51			27-28
25	190	0.13	37	135E	0.28				23			28-29
64	212	0.30	66	307	0.22				48			29-30
72	518E	0.14	70	540E	0.13				58			30- 1
47	183	0.26	42	231	0.18	58	261	0.22	37	120	0.31	GEM.

Unité / Eenheid) μg.m^{-3}

Fig. 3. A page from the monthly pollution review.

TABLE 1

Results Obtained from a One-Year Period of Observation in Antwerp

Time period	Station 808			Station 812			Station 826		
	Smoke	SO_2	$\dfrac{Smoke}{SO_2}$	Smoke	SO_2	$\dfrac{Smoke}{SO_2}$	Smoke	SO_2	$\dfrac{Smoke}{SO_2}$
Average pollution levels									
April	32	186	0.17				36	133	0.27
May	28	156	0.18	29	101	0.29	28	259	0.11
June	20	126	0.16	21	87	0.25	18	121	0.15
July	15	109	0.14	17	67	0.25	22	156	0.14
August	18	139	0.13	25	70	0.35	36	250	0.14
September	31	135	0.23	33	95	0.35	44	264	0.17
October	48	210	0.23	48	118	0.41	59	207	0.29
November	71	299	0.24	68	186	0.37	81	239	0.34
December	93	345	0.27	96	207	0.46	60	352	0.17
January	67	290	0.23	57	180	0.32	65	214	0.30
February	94	335	0.28	67	187	0.36	78	182	0.43
March	88	296	0.30	79	166	0.48			
Summer	24	142	0.17						
Winter	77	296	0.26	69	174	0.40	64	243	0.27
Year	51	219	0.23						

3. URBAN AIR POLLUTION

Maximum daily concentration

Time period						
April	46	285	65	167	74	512
May	66	255	80	262	51	379
June	45	220	62	200	63	564
July	32	258	48	106	57	408
August	37	370	52	164	46	560
September	62	209	72	223	106	1084
October	101	336	114	220	103	593
November	182	637	188	387	163	550
December	286	726	342	558	261	767
January	113	487	103	333	137	813
February	206	628	154	498	146	528
March	168	490	154	308	131	830

Number of days over a given level

Level						
100	42	322	33	191	30	249
150	12	233	10	114	5	179
300		79	1	11		86
500		9		1		27
750						4
1000						1
1500						

3. STATISTICAL STUDIES OF POLLUTION DATA

The daily results from each station are registered on magnetic tape so that they can be used at the end of the year. We consider a yearly period beginning on April 1 and ending on March 31 of the following year, so that the first six months of the period roughly correspond to the warm season, the last six months to the cold season.

If we look at Fig. 4, which shows the evolution through the year 1968 of the mean air pollution in Antwerp, we discover a fairly strong relation between SO_2 and smoke fluctuations, principally introduced by meteorological influences. In Belgium, it is likely that smoke and SO_2 are pollution indexes of similar validity.

For each yearly period, we have determined the monthly, seasonal, and yearly averages, monthly extremes, and pollution frequencies. Pollution maps have been drawn. Table 1 is an example of such a computation for three stations. In Fig. 5 we can see that the mean winter SO_2 concentrations in Antwerp are the highest in the center of the town and near the petrochemical installations located in the northwest.

In order to get a good idea of the danger of the persistence of given pollution levels for a number of consecutive days, we have determined a measure of the persistence of some pollution levels in the course of one year, represented by the number of distinct periods of exactly one, two, ..., up to ten consecutive days during which the pollution level exceeded a given concentration (see Table 2).

4. FREQUENCY DISTRIBUTION CURVE

In order to make a rough classification of the stations as a function of the type of pollution, we have computed statistical parameters by considering the frequency distribution curve at each station. This curve, when represented by a straight line on logarithmic probability paper, can be characterized by its slope and median value. A station having many high concentrations and many low concentrations will produce a high slope, indicative of great dispersion, while a station having almost constant concentration will give a distribution line with a low slope. On the other hand, the median value is not far from the mean level of pollution.

Maps giving the geophysical distribution of the slope and the median value across the five urban areas have been drawn (Fig. 6 and 7). We observe that the slope of the frequency distribution line tends to be lower

3. URBAN AIR POLLUTION

at locations where the source strength distribution is more or less isotropic. This is generally the case near the center of the urban areas. Where the source strength distribution becomes strongly anisotropic, the slope is much higher. Moreover, the slope also depends on the persistence and the frequency of certain wind directions. In Belgium, the dominant winds blow from the southwest. The median value, more directly

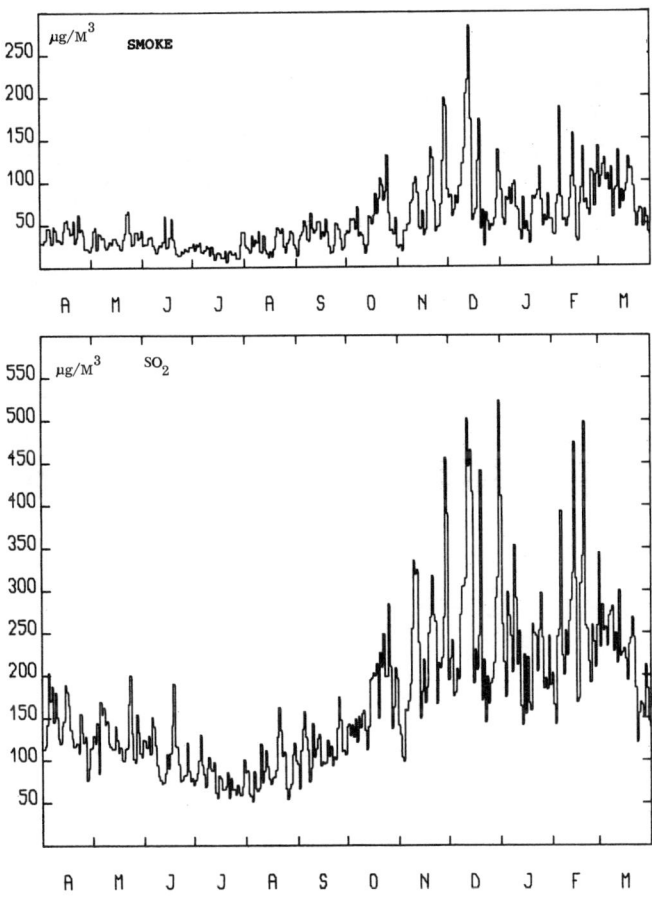

Fig. 4. Annual variation of air pollution in Antwerp.

associated with the mean pollution level, is highest near the center where most sources are to be found and decreases more or less rapidly toward the edges of the town.

Based on such a statistical analysis of pollution data, a distinction between industrial, urban, suburban, and rural stations could be developed. The pollution would then be characterized by two parameters only. Let us

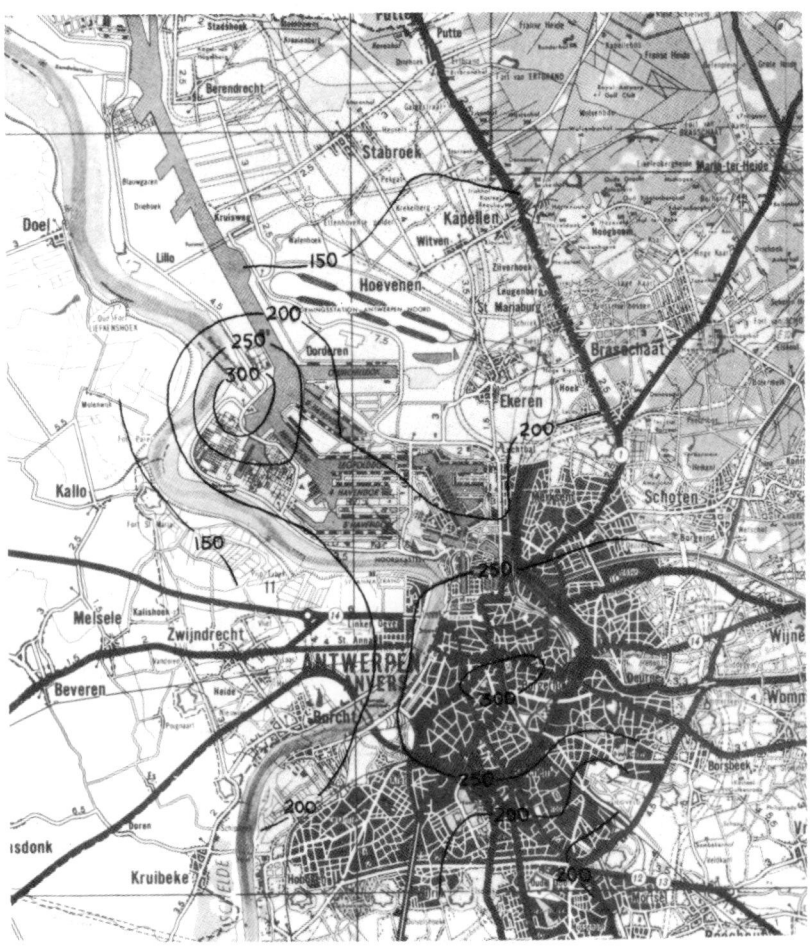

Fig. 5. SO_2 pollution map (Antwerp, Winter 1969).

3. URBAN AIR POLLUTION

consider, for instance, four stations located in Antwerp. The slope and median value for the winter period, 1969, are:

station number	median SO_2	slope SO_2	station type
810	288	0.18	urban
812	158	0.28	suburban
823	114	0.23	rural
826	252	0.29	industrial

Fig. 6. SO_2 median (Antwerp, Winter 1969).

TABLE 2

Persistence of Pollution Levels

| Pollutant | Level | \multicolumn{10}{c}{No. of periods of length} |
		1	2	3	4	5	6	7	8	9	10
Station 808, Municipal Laboratory, Antwerp											
Smoke	50	10	4		2	2	1	2		1	3
	100	7	6	2	1	1			1		
	200	1		1							
SO_2	150	14	9	3	2	1	1	1			5
	300	9	2	5	2	2	2		1		1
	500	5			1						
Station 809, School, Antwerp											
Smoke	50	4	1	5	2	1		1	2	1	4
	100	13	8	7		6	4	2			3
	200	7		1	1						
SO_2	150	9	4	1	2	1	1				4
	300	9	2	5	1	1	1	1	1		1
	500	5	2		1						

5. CONCLUSION

We have presented here a general survey of the statistical studies going on at the Institut Royal Météorologique de Belgique, using pollution data gathered in Belgium. For each station of the Belgian pollution network the monthly, seasonal, and yearly mean SO_2 and smoke concentrations,

3. URBAN AIR POLLUTION

the extremes and the frequencies, and the persistence of certain pollution levels are computed. From the frequency distribution line, we have determined the slope and median values and their geographical distribution across the great urban areas of Belgium, making possible a classification of the observing stations as a function of the pollution type.

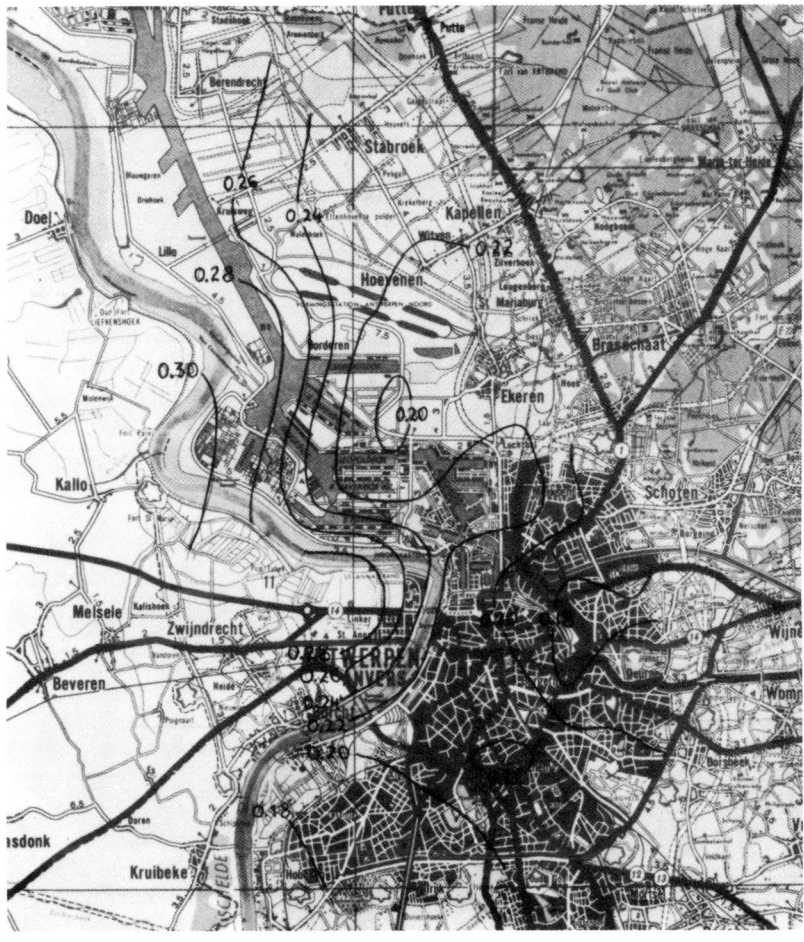

Fig. 7. SO_2 slope (Antwerp, Winter 1969).

4

THE DESIGN AND OPERATION OF THE NATIONAL AEROMETRIC DATA BANK

J. W. Overbey II and R. P. Ouellette

The MITRE Corporation
McLean, Virginia

1. INTRODUCTION

In September 1969, The MITRE Corporation, under contract to the Office of Air Programs of the Environmental Protection Agency, undertook the project of building a National Aerometric Data Bank as a central depository of all available aerometric data. The purpose was to assemble at a single location the aerometric data from such Federal programs as the Continuous Air Monitoring Program (CAMP) and the National Air Surveillance Networks (NASN) and from the air surveillance programs of state and local agencies.

The 250 individual nonfederal agencies contributing air quality data span a wide range as to the scope of their activity, the stage of development of their systems, and the completeness or availability of information. Historical data files are being kept in stages of development ranging from handwritten log sheets to well planned and efficiently organized computerized systems. The only consistency found among the individual contributors is the lack of consistency.

An integral part of the activity of collecting and analyzing air quality data is the development of a standardized, centralized data storage and information retrieval system designed to maintain efficiently the ever increasing amount of data available. Just as the scope of the data collection effort has expanded, the system for data storage and information retrieval has evolved from a simple system to the elaborate process necessary to maintain a large multifile data bank.

This paper describes briefly some aspects of the present system used by MITRE for creating, maintaining, inventorying, summarizing, and querying the data files containing pollutant concentrations measured by nonfederal agencies throughout the United States. It then gives some examples of statistical analyses the system makes possible.

2. SYSTEM OPERATION

The present system consists of six major activities: transformation of data from the form in which it is received into one compatible with the standardized form; maintenance of the master data sets; inventorying the data contained in the master data sets; summarizing those data into simple statistical descriptors; describing the individual sampling sites; and querying the summary data sets to respond to individual and unique requests for information regarding the air quality of specified locations.

Responses to requests for air quality information typically fall into two classes: fixed-interval, routine summaries and inventories of the data bank (performed quarterly), and nonroutine, individually tailored query-responses performed on an "as-received" basis.

Information contained in the quarterly reports is illustrated in Figs. 1-6. These are currently provided to federal, state and local agencies operating air quality monitoring programs.

Responses to nonroutine queries to the data bank generally involve supplying inventories of the contents of the bank, furnishing detailed descriptions of characteristics of sampling sites, producing reports containing precomputed statistical descriptors of ambient air quality, computing nonstored special-use statistical descriptors, or preparing basic observation dumps.

3. DATA EDITING AND VALIDATION

An essential part of maintaining the aerometric data bank is data validation and editing to compare observations submitted by individual agencies against logical values set with a prior knowledge of the air quality of the area. Values consistently falling below an anticipated low value or consistently falling above an anticipated high value should not be removed from the data bank, but rather should be flagged for a future

4. NATIONAL AEROMETRIC DATA BANK 41

```
05/14/71                    STANDARD REPORT NO. 1                                                           PAGE 1
                            AIR QUALITY DATA REPORT
                            STATE OF ILLINOIS

                        LOCATION: CHICAGO ILL        SITE: 03

    SAROAD NUMBER: 14122003                       STATE NAME: ILLINOIS
    SAMPLING ADDR: TAFT HS 5625 N NATOMA AVE      COUNTY NAME: COOK
    CITY NAME: CHICAGO ILL                        SMSA NAME: CHICAGO,ILL
    SMSA NUMBER: 0690                             AQCR NAME: CHICAGO
    AQCR NUMBER: 003                              LONGITUDE:    D.    M.    S.
    LATITUDE:    D.    M.    S.                   UTM EASTING:
    UTM NORTHING:                                 TIME ZONE: CENTRAL
    ECONOMIC ACTIVITY TYPE: UNIDENTIFIED          ELEVATION ABOVE SEA LEVEL:         FEET
    ELEVATION ABOVE GROUND: 49 FEET               TOPOGRAPHIC CODE: PLAIN
    REPORTING AGENCY TYPE: CITY                   CITY POPULATION(1000'S):
    SMSA POPULATION(1000'S): 6815                 AQCR POPULATION(1000'S):  7539
    COUNTY POPULATION(1000'S): 5399
    SUPPORTING AGENCY: CITY OF CHICAGO
```

	POLLUTANT METHOD INSTRUMENT INTERVAL & UNITS	TOTAL PERCENT OF YEAR COVERED	ARITHMETIC AVERAGE	STD DEV	EXTREMES MIN OBS	EXTREMES MAX OBS	FIRST QUARTER NUM OBS	FIRST QUARTER ARITH AVG	SECOND QUARTER NUM OBS	SECOND QUARTER ARITH AVG	THIRD QUARTER NUM OBS	THIRD QUARTER ARITH AVG	FOURTH QUARTER NUM OBS	FOURTH QUARTER ARITH AVG
YEAR														
66	SULFUR DIOXIDE CONDUCTIVITY BECKMAN 1-HR PPM	92	0.028	0.050	0.000	0.685	2033	0.049	1971	0.026	2013	0.014	2124	0.024
67	SULFUR DIOXIDE CONDUCTIVITY BECKMAN 1-HR PPM	75	0.037	0.064	0.000	0.955	2048	0.043	2053	0.047	1766	0.016	740	0.046
68	SULFUR DIOXIDE CONDUCTIVITY BECKMAN 1-HR PPM	80	0.040	0.074	0.000	1.305	2024	0.063	1633	0.040	1989	0.017	1367	0.036
69	SULFUR DIOXIDE CONDUCTIVITY BECKMAN 1-HR PPM	76	0.035	0.057	0.000	0.715	1050	0.049	1867	0.034	1870	0.021	1877	0.041

Fig. 1. Standard report 1.

Fig. 2. Standard report 2.

4. NATIONAL AEROMETRIC DATA BANK   43

in-depth examination. This type of editing can identify logical inconsistencies between the stated data characteristics and the reported values. For instance, order-of-magnitude discrepancies might be indicative of incorrectly specified units in which the observations are recorded. Automatic rejection or elimination of such data from the master data files is undesirable, since correction of data characteristics is more easily accomplished than re-entering of the data.

Observations of a given pollutant in a specified set of units and at an indicated interval between observations must fall within certain limits. In addition to high- and low-value checks, rate-of-change editing should detect values inconsistent with surrounding sets of time-ordered values. Such editing will identify both outliers of the group and lack of anticipated variation within the group. Abrupt changes in concentration values typically do not occur within short intervals of time. A single value which is much greater (or much less) than other values obtained at approximately the same time is highly suspect. Such outliers are likely to be the result of data transcription errors or incorrect specification of the data characteristics of that particular data. Similarly, long periods of no change in concentration values typically do not occur. Perturbations may be minor, but they do occur. A series of constant values is thus highly suspect, and observations should be flagged for future examination. The only exception to this is a long series of values below the sampling instrument's minimum detectable threshold. Automatic rejections of such values should be deferred until circumstances causing such a series can be determined.

Data should be examined to ensure that observations are reported sufficiently frequently during and uniformly distributed within the long-term interval for which summary statistics are to be calculated, in order that the statistics be based on observations "representative" of that time interval. Observations not meeting these criteria should be suppressed when summary calculations are being performed rather than culled from the master data files.

4. ANALYSES

A number of recent and on-going statistical research projects will be discussed. Only a rapid review will be presented, our aim being to document the variety and types of analysis feasible now that an operating data base is available.

As Fig. 7 indicates, two broad types of analyses are being directed to the National Aerometric Data Bank. The first is involved with defining a siting rationale for air pollution sensors and a statistical definition of the

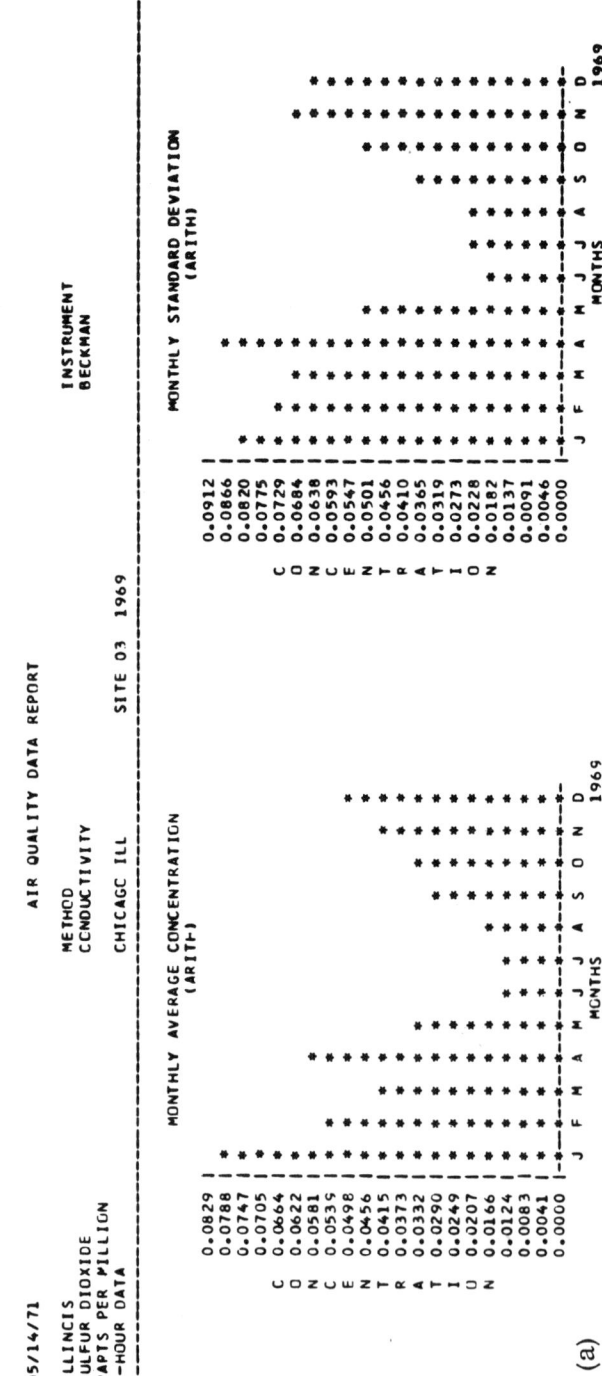

(a)

4. NATIONAL AEROMETRIC DATA BANK

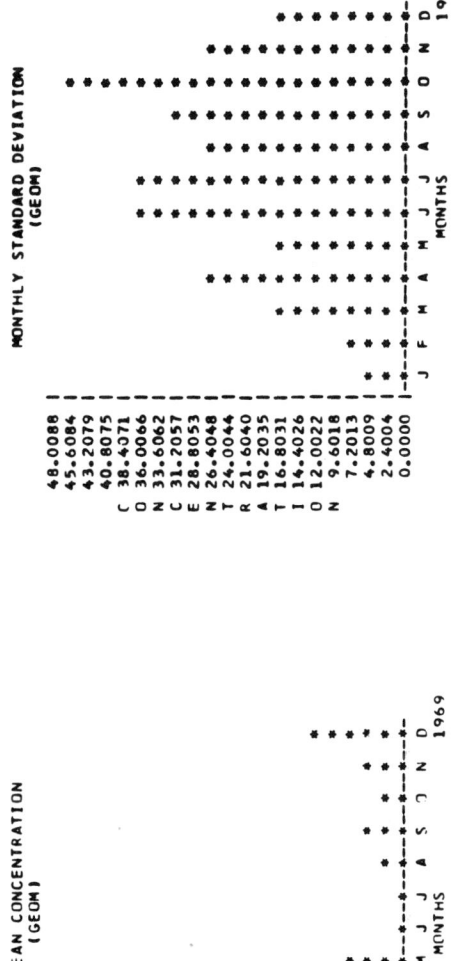

Fig. 3. Graphical display.

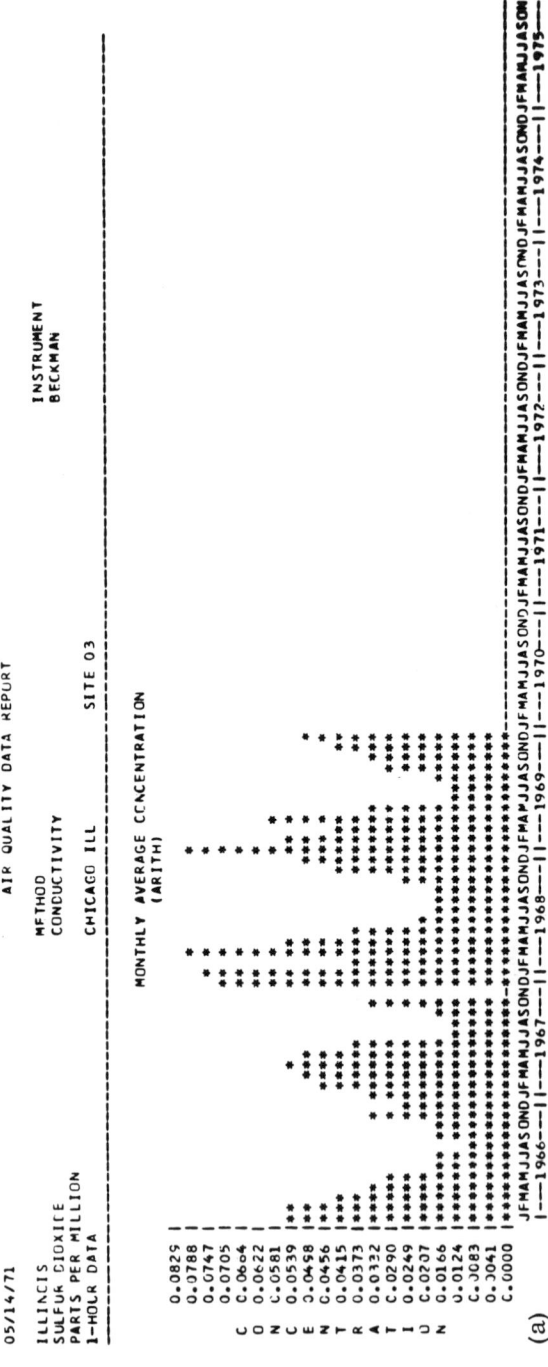

(a)

4. NATIONAL AEROMETRIC DATA BANK

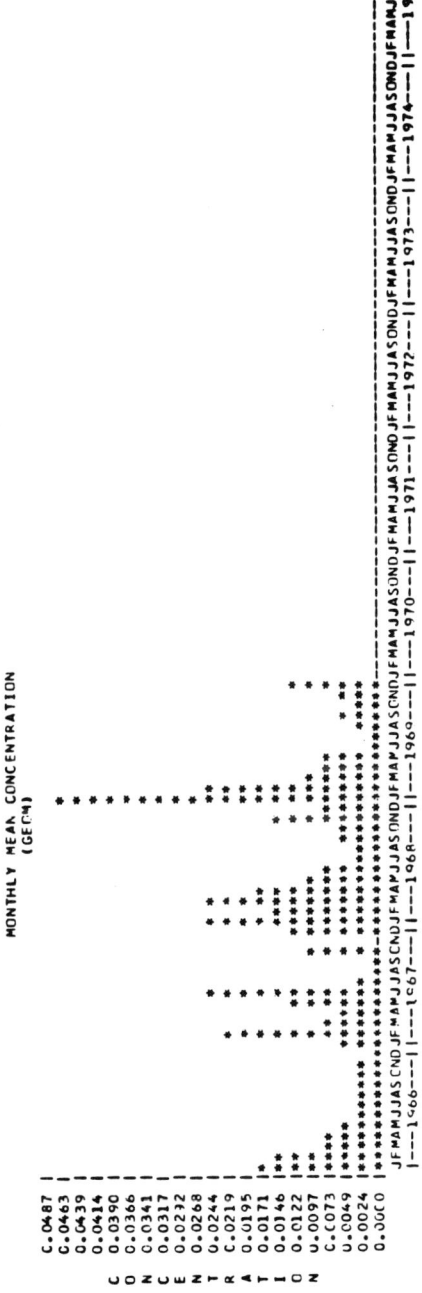

Fig. 4. Graphical display.

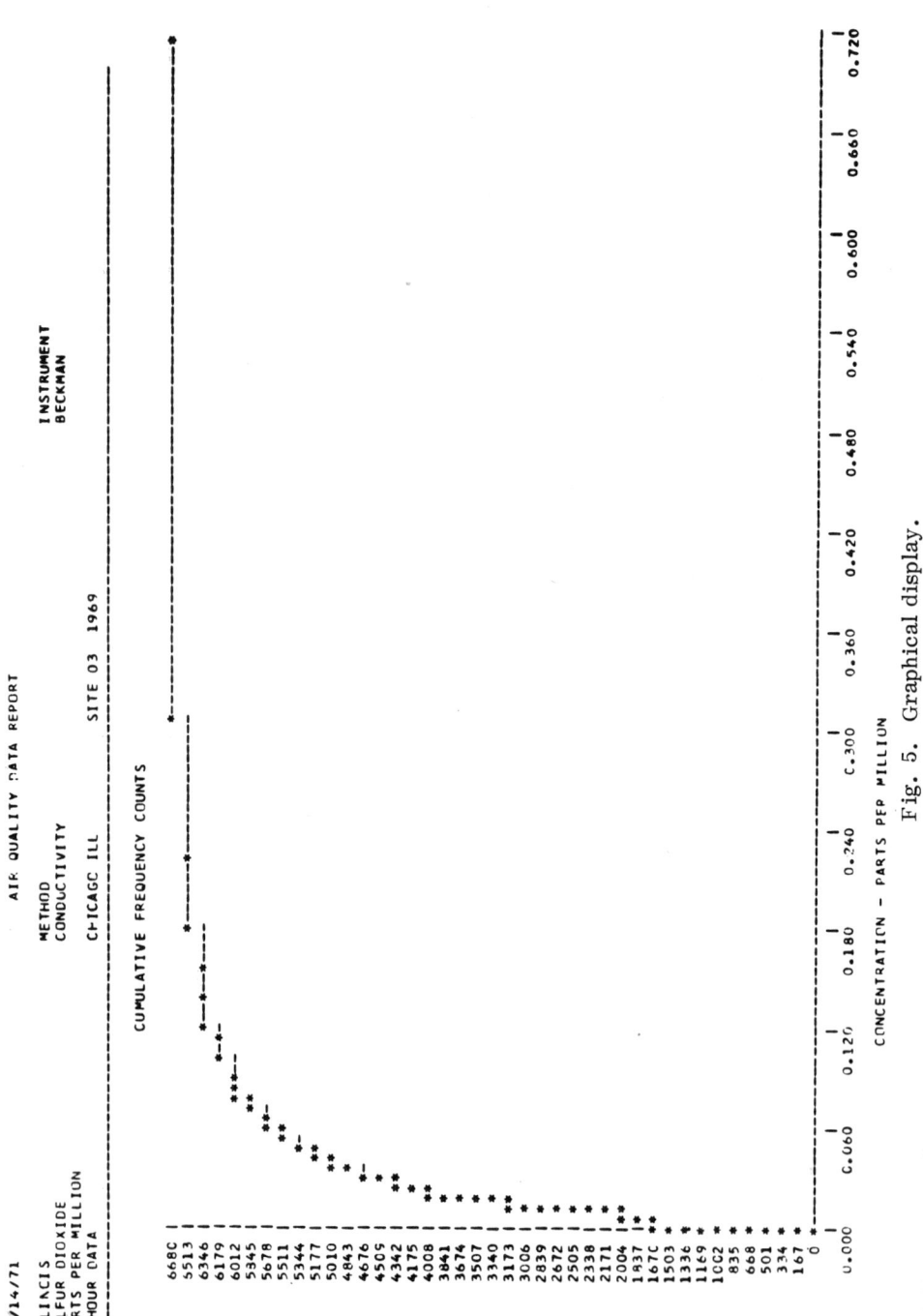

Fig. 5. Graphical display.

4. NATIONAL AEROMETRIC DATA BANK

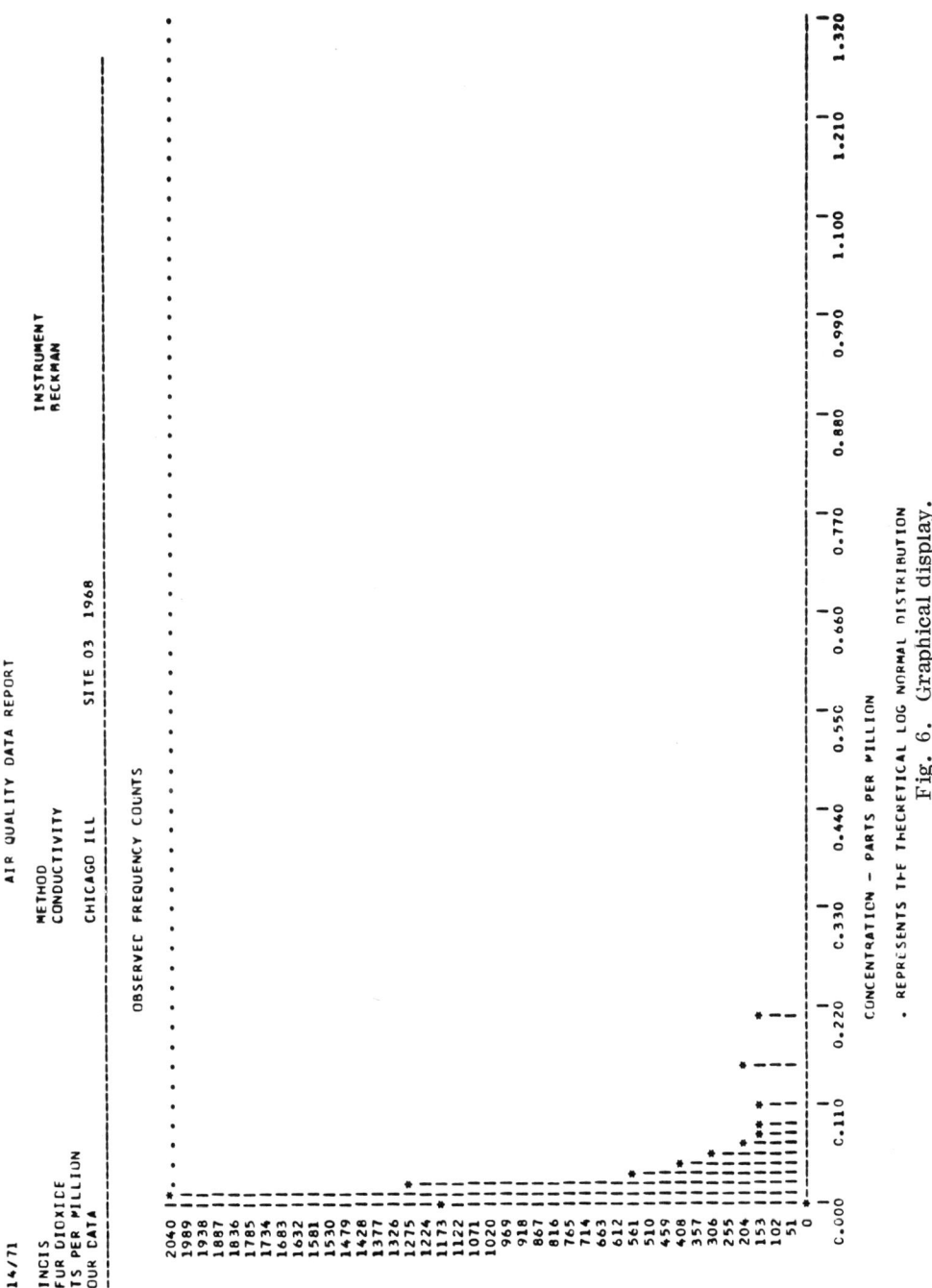

Fig. 6. Graphical display.

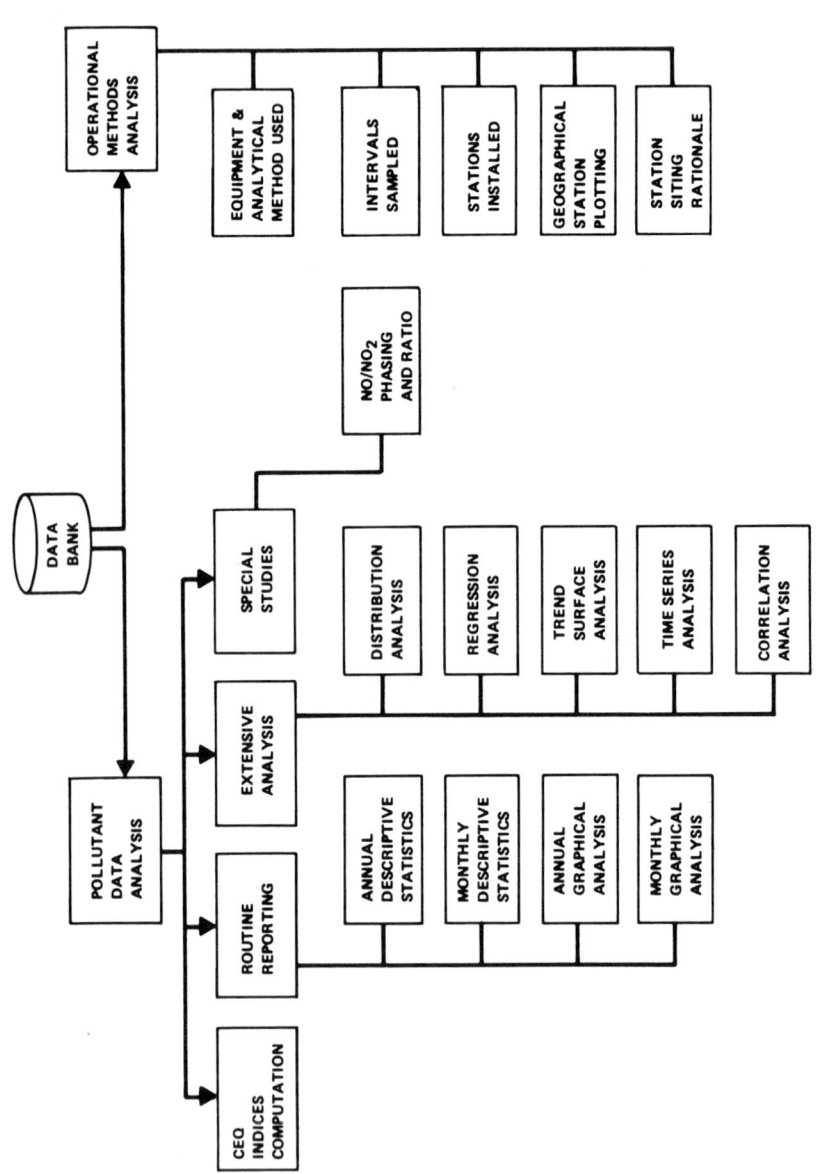

Fig. 7. Data analysis approach.

4. NATIONAL AEROMETRIC DATA BANK

methodology and instrument complement in use. The first findings from this study indicate that the instruments in use are unsophisticated and mostly of the integration type. It further indicates that only a handful of sites measure the six primary pollutants. The other side of the picture, and the most important aspect, is termed pollutant analysis, and this is broken down into computation of CEQ indices, routine reporting, extensive analysis, and special studies. Each will be described below.

The concept of developing indices to monitor trend and status has been developed for the President's Council on Environmental Quality. More than 110 indices have been conceptualized at this time. These are discussed by Pikul in Paper 8 of this volume. Routine reporting of the data bank is completed quarterly and elementary descriptor statistics are computed. These include arithmetic mean, arithmetic standard deviation, geometric mean, geometric standard deviation, minimum, maximum, percent available data, and percentiles. Extensive data analysis usually requires long uninterrupted series of observations or simultaneous observation on diverse pollutants. Some typical analyses in this category will be presented below. Special analysis usually involves the preparation of a special computer program to answer a technical question.

Some statistical applications of a large air quality data bank are listed in Table 1. We will briefly discuss three of them.

Air Quality Diffusion Model. This is essentially a bivariate Gaussian diffusion model. Given the emission patterns and local meteorological conditions, it is possible to compute long-term (1 year) ground-level concentration of pollutants at fixed receptor points. This model has been applied to the Cook County, Illinois, area and long-term averages for SO_2 and TSP for each of 1100 census tracts have been computed. These data are currently being correlated with biological data collected on draftees originating from the census tracts involved.

Air Pollution Abatement for Stationary Sources. The Air Quality Display Model (AQDM) developed by the Office of Air Programs was used by Golden and Mongan [1] to simulate the air quality of Cook County, Illinois. The AQDM is based on a Gaussian diffusion equation which describes the diffusion of a plume as it is transported from a continuously emitting source. The model computes the expected annual average concentrations at ground level at specified receptor points, which result from both point and area sources, as well as the percentage of the concentration contributed by each contaminant at the receptor points ascribable to each source. The AQDM represents a useful first approximation to the physical situation in areas with level topography.

TABLE 1

Some Statistical Techniques for Air Pollution Studies

Air quality diffusion model
Air pollution abatement for stationary sources
Computation of air quality indices
NO/NO_2 ratio
Trend surface analysis
Principal components analysis

Golden and Mongan concluded that emissions from power plants constituted 78% of the SO_2 emissions in the Cook County area in 1968. The elimination of SO_2 emissions from power plants at the beginning of 1968 would have produced an 11% reduction in the yearly average concentration of SO_2 at ground level in the most polluted area of Chicago during 1968. Although large reductions would have occurred in the outlying areas of Cook County, it was noted that the initial concentrations at these points were already well below those at which adverse effects from SO_2 have been observed (see Fig. 8).

Repeating the procedure with the concentrations expected in the absence of emissions from power plants, Golden and Mongan estimated that the 78% reduction of SO_2 emissions obtained by eliminating those from power plants would have produced a 14% reduction in the exposure of the population of Cook County to sulfur dioxide during 1968 (see Fig. 9).

Their analysis shows that even if SO_2 emissions from all power plants in the Cook County area had been completely eliminated in 1968, the area would still have been plagued by a sulfur dioxide pollution problem of considerable magnitude and extent. This can be understood by the following reasoning. To avoid high concentrations of sulfur dioxide at ground level in the vicinity of a power plant, the plants are usually equipped with tall stacks which help ensure that stack gases are significantly diluted before reaching ground level. On the other hand, space heating sources release their gases and particulate at the ground level where they are not readily diffused.

Trend Surface Analysis. The application of the AQDM for typical metropolitan areas requires extensive amounts of computer calculation time. A technique for obtaining the same estimates of ground-level

4. NATIONAL AEROMETRIC DATA BANK

21	23	24	26	28	30	33	36	40	44	46	44	40
21	23	26	28	30	33	36	39	43	49	52	49	44
21	24	27	29	32	35	38	43	48	56	59	56	48
22	24	27	30	30	38	42	47	54	64	69	64	54
22	25	28	31	34	39	44	51	63	76	83	78	60
22	25	28	31	34	40	45	53	67	86	109	96	75
21	25	29	32	34	39	44	64	70	90	110	119	86
20	24	29	32	34	37	43	54	72	91	110	110	87
20	23	29	33	34	36	41	56	64	79	89	93	81
20	24	30	36	34	36	41	48	56	62	76	80	78
19	24	29	39	37	37	38	42	47	53	60	62	64
19	25	30	46	37	36	36	39	43	49	67	58	63
17	25	31	37	39	37	35	36	37	43	56	50	50
17	24	33	43	37	33	32	33	34	37	40	40	41
17	23	34	35	32	30	30	30	31	33	34	34	34
18	21	32	31	29	29	29	29	29	30	30	31	29
16	20	25	28	27	27	27	27	27	27	26	26	26

Fig. 8. Result of air quality display model. Yearly average ground-level concentrations (micrograms per cubic meter) of sulfur dioxide resulting from emissions from all sources. Grid squares are 4.5 km on a side. Source: Golden and Mongan [1].

concentration but which requires less computation time is desired. Trend surface analysis is one such technique currently being investigated. In the limited examination we have undertaken, we have found that the use of trend surface analysis does provide estimates of comparable quality at a significant reduction in computation costs and in input preparation.

The objective of this technique is to fit a polynomial of degree n to a set of XYZ coordinates. A goodness-of-fit test is applied to the surface thus created, and an analysis of the variance explained by the linear, quadratic, and cubic components is presented. In our case, X and Y are geographical coordinates and the Z's are observed air quality measurements. In most cases, it is possible to explain most of the variance in the data by a polynomial of less than third degree. We have applied this technique to the Chicago-area air quality measurements for SO_2 and TSP since the data have been readily available. Using the 1965 SO_2 measurements, a cubic equation was fitted to the data. The equation is presented in Table 2. An analysis of variance (Table 3) indicates that some 68% of the variance is explained by the cubic surface. The contour lines of this cubic surface are shown in Fig. 10. An overlay of the Cook County outline provides some orientation. There are obvious similarities between a

15	16	17	19	21	23	26	29	32	36	37	35	32
15	13	18	20	22	26	28	31	35	40	43	40	35
14	16	18	20	24	27	30	35	39	47	48	46	39
15	16	19	21	24	30	34	39	44	54	58	53	44
15	16	18	20	24	30	36	42	52	66	71	67	52
15	15	17	19	23	30	36	44	56	76	99	84	63
13	15	17	19	22	30	34	44	58	79	101	107	74
12	13	13	18	21	26	32	44	60	81	102	99	74
11	11	14	17	20	24	30	46	55	68	79	83	69
11	12	14	17	19	23	29	37	45	52	66	71	62
10	11	13	27	20	23	25	31	36	48	50	52	55
10	11	14	22	19	21	23	27	32	39	57	49	54
9	11	14	23	18	19	20	23	25	33	46	40	40
10	11	13	26	20	18	18	21	23	26	30	30	31
10	12	14	17	16	16	18	19	20	22	24	25	25
9	12	14	15	15	15	16	17	18	20	20	22	20
9	11	14	13	13	14	15	16	16	17	17	17	17

Fig. 9. Result of air quality display model. Yearly average ground-level concentrations (micrograms per cubic meter) of sulfur dioxide resulting from emissions from all sources except electric power generating plants. Grid squares are 4.5 km on a side. Source: Golden and Mongan [1].

trend surface analysis and plot and the air quality diffusion model supported by an isocontour or SYMAP plot.

Principal Components Analysis. Principal components analysis is a technique for reducing the dimensionality of the observed space to a limited number of orthogonal components explaining a major part of the variance in the data.

Specifically, we solve the following determinantal equation:

$$(R - \lambda I) x = 0,$$

where R is traditionally a correlation matrix, I is the identity matrix, and the solutions λ_i and x_i are the eigenvalues and corresponding eigenvectors.

Traditionally, this technique has been applied to explaining the latent structure of a correlation matrix. There is no reason it cannot be applied to other matrices. Tables 4-6 present matrices of great interest. The

4. NATIONAL AEROMETRIC DATA BANK

TABLE 2

Trend Surface Analysis

Coefficients of linear equation:

$Z = -11.65632 + 0.96637X + 0.41745Y$

Coefficients of quadratic equation:

$Z = -34.67349 - 2.72514X + 4.88660Y + 0.02614X^2 + 0.01319XY + -0.04277Y^2$

Coefficients of cubic equation:

$Z = -242.04608 - 37.73093X + 47.82057Y + 1.14570X^2 - 1.26355XY$
$-0.02497Y^2 - 0.00702X^3 + 0.00362X^2Y + 0.00790XY^2 - 0.00335Y^3$

TABLE 3

Analysis of Variance

Surface	Linear	Quadratic	Cubic
Standard deviation	34.59	30.32	21.46
Variation explained by surface	0.50540977E 04	0.10880953E 05	0.20510852E 05
Variation not explained by surface	0.25130277E 05	0.19303422E 05	0.96735234E 04
Total variation	0.30184375E 05	0.30184375E 05	0.30184375E 05
Coefficient of determination	0.16744083	0.36048293	0.67951882
Coefficient of correlation	0.40919536	0.60040230	0.82432932

TABLE 4

Pollution Matrix

Pollutant	Pollutant			
	DUST	BR	MN	V
DUST	1.00	0.70	0.66	0.46
BR	0.70	1.00	0.59	0.27
MN	0.66	0.59	1.00	0.16
V	0.46	0.27	0.16	1.00

TABLE 5

Pollution Matrix

Pollutant	Station			
	Morton Grove	ANL	Taft	Lakeview
Dust	53	28	49	90
Br	0.086	0.024	0.15	0.15
Mn	0.22	0.34	0.28	
V	0.0022	0.0037		

first matrix (Table 4) is a product-moment correlation matrix, the second (Table 5) shows pollutant versus station; in this case, it is possible to plot the eigenvectors obtained over a map of the area. The third matrix (Table 6) adds the time dimension to the same data.

With 22 separate sets of observations at a single site in the Chicago area on twenty variables (Dust, Br, Mn, V, Al, Cl, Na, Hg, Cr, Zn, Fe, Sb, Sc, Co, La, Cs, Eu, Ce, Ag, Se), it is possible to explain 64% of the variance with the first eigenvalue and 90% of the variance with the first five roots (Table 7). The first eigenvector is unipolar and the elements of the vector are about equal. All other vectors are bipolar with high

4. NATIONAL AEROMETRIC DATA BANK

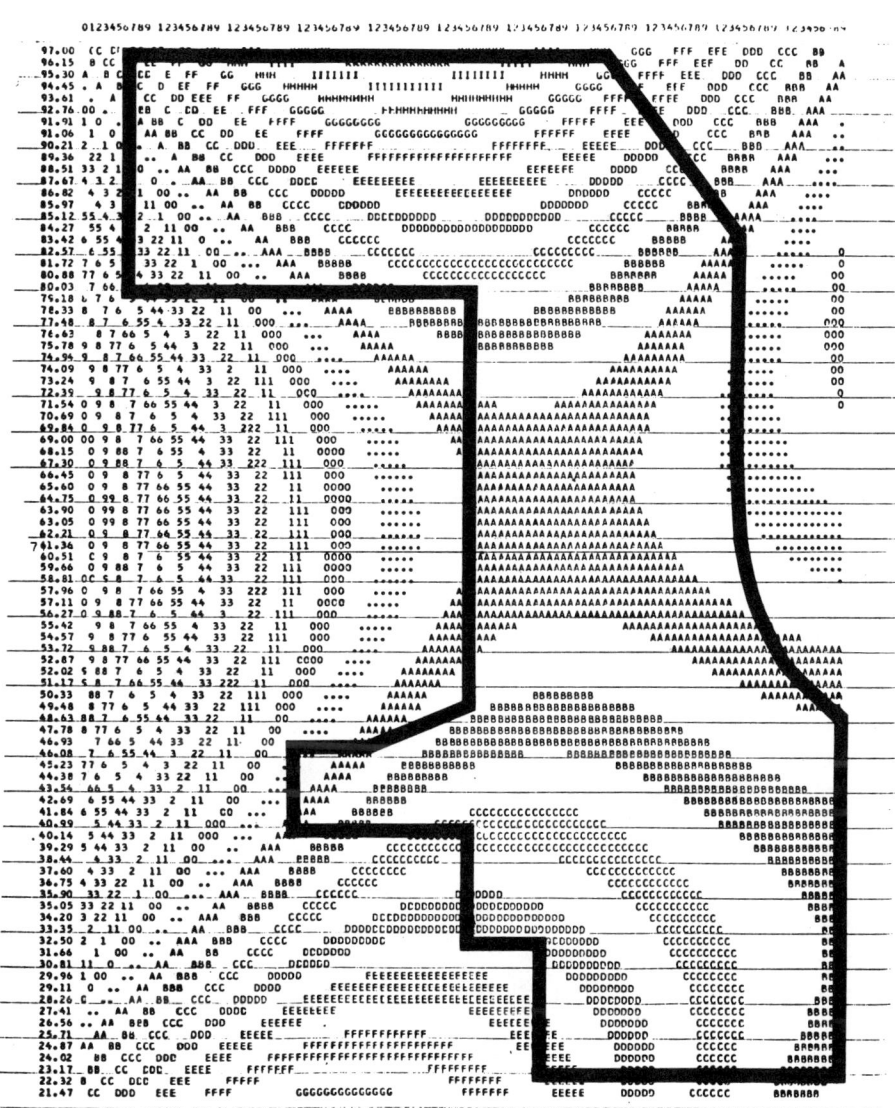

Fig. 10. Trend surface.

loadings for specific elements. The interpretation of these data is in progress.

These limited presentations of statistical analyses are given only to exemplify the kinds of studies that are possible with a large air quality data bank. We hope that more researchers will apply numerous techniques to the data to answer fundamental questions concerning air pollution phenomena.

TABLE 6

Pollution Matrix

Month	Station		
	Taft	Lakeview	Steinmetz
Jan	83	92	62
Feb	86	112	78
Mar	83	102	94
Apr	84	111	87
May	83	108	85

TABLE 7

Results of Principal Components Analysis

Eigenvalues	Cumulative % variance
12.75	64
1.88	73
1.55	81
0.97	86
0.63	87
0.58	92

4. NATIONAL AEROMETRIC DATA BANK

REFERENCES

1. J. Golden and T. R. Mongan, Sulfur dioxide emissions from power plants: their effect on air quality, Science, 171, 381-383 (1971).

2. L. J. Duncan, An inventory of air pollution control agencies and their data, MTR-4154, Rev. 1, The MITRE Corporation, December 1970.

3. J. Golden, Guidelines for the acquisition of validated air quality data, MTR-4143, The MITRE Corporation, June 1970.

4. R. P. Ouellette, D. M. Rosenbaum, and R. S. Greeley, Is there a system for pollution madness? Datamation, April 15, 1971.

5. R. P. Ouellette and R. S. Greeley, The design and use of an ambient air quality data bank, Second International Clean Air Congress, Washington, D. C., December 8, 1970.

DISCUSSION

D. W. Moody

U. S. Geological Survey
Washington, D. C.

Doctors Overbey and Ouellette describe the application of an existing generalized data management system (MARK IV) to the storage and retrieval of air quality data contributed to the National Aerometric Data Bank from over 250 individual nonfederal agencies. They encountered many difficulties in attempting to integrate data from so many divergent sources. Data on air pollution concentrations, for example, may be received on magnetic tape, decks of punched cards, and paper copies of site reporting forms or log records. Information coding schemes varied from agency to agency as did the frequency of observations; data formats differed; and the coding forms rarely provided sufficient information to identify the characteristics of all of the data elements. These problems had to be overcome by developing transformation algorithms for each data source to change the source data into the system standard coding scheme and formatting structure and to sort individual observations into separate files according to their frequency of observation.

They are to be commended for making use of an existing, commercially available data management system. Many such systems are available today which possess a wide range of capabilities. Generalized data management systems require very large expenditures of time and money to develop, and the designer of a new information system should carefully investigate existing systems before committing his resources to developing a special purpose system.

The flexibility of generalized data management systems and the availability of general purpose statistical programs and data display (graphs, charts, and maps) packages, which can process the retrieved data, have greatly reduced the effort and cost of establishing information storage and retrieval systems. As soon as the system designer attempts to incorporate the requirements of a number of user groups and to utilize data collected by a number of different agencies on a nationwide basis, however, problems of data consistency, format compatibility, and analytical methodology become acute, and the probability of introducing erroneous data into the system is greatly increased. The user of such data banks, particularly the model builder, should never be lulled into believing that the data are infallible.

On the other hand, the use of generalized storage and retrieval systems combined with a continuing dialog between the decision maker, the model builder, and the data collector permits new data elements to be added and old ones dropped as the objectives of the system are refined. Sensitivity analyses of management decision to errors in the input data can also provide much information about the type and accuracy of the data that should be collected. Thus, the information system can be designed to evolve in response to the changing needs of the decision maker. Unfortunately, this dialog rarely takes place.

Part II

INDICES, ECONOMIC AND POLICY PERSPECTIVES

5

TREATING EXTERNAL DISECONOMIES - MARKETS OR TAXES?

David Starrett*
and
Richard Zeckhauser

Harvard University
Cambridge, Massachusetts

Sometimes the parties to an economic transaction do not reap all of the benefits and costs involved: there are externalities. Handling an externalities problem is like fighting an undiagnosed illness. At some times it will yield to any of a number of familiar treatments. At others, it will be resistant to any that are known. In this paper, we first consider those pathologies that can be adequately treated by standard remedies. Then we turn to examine some more resistant cases, cases that may require new diagnoses and unconventional methods of treatment.

1. TRACTABLE EXTERNALITY PROBLEMS

Thrust a price theorist into a world with externalities and he will pray for second best - many firms producing and many firms and/or consumers consuming each externality, with full convexity everywhere. No problem for the price theorist. He will just establish a set of artificial markets for externalities, commodities for which property rights were not previously defined.[†] Decision units, being small relative to the market, will take price as given. The resulting allocation will be a competitive

---

*Present address: Stanford University, Stanford, California.
[†]Due to their elusive physical natures, most externalities have not developed property rights. Markets for externality rights are artificial in the sense that natural forces have been insufficient to foster their existence.

outcome of the classical type. If artificial markets do not appeal, an equally efficient taxing procedure is available.

Where the externalities are beneficial, the price theorist's life is particularly easy; the more intractable aspects of externality problems do not appear. We concentrate here on external diseconomies, the more interesting case, and the case of greater empirical import.

For either artificial-market or taxing procedures it is essential to determine who has initial rights. Our common-law upbringings would lead us to expect that rights should go to the sufferers. It turns out, however, that it is just as efficient to give the rights instead to those who dump the sewage and soil the air.

Upton Paper Mills and Downley Baths are our representative producer and recipient of an externality. Upton dumps organic residue into the very river that provides the water for Downley's swimmers. Downley of course screens and purifies, but not without cost.

1.1. Artificial Markets*

Consider first the working of an artificial-markets scheme. If Downley has rights, he sells Upton the right to pollute. Upton, being but one of many polluters, can have no effect on the total pollution level, and hence no effect on its price. He takes price as given and continues polluting up to the point where his marginal gain from polluting (what is in effect his marginal cleanup cost) just equals the going price. Downley will sell rights to pollute up to the point where his marginal loss in profits from the pollutant equals the going price. The equilibrium level of production for the externality, the pollutant, is established where the marginal loss to the recipient equals the marginal cost to the producer of not providing the externality, these equal marginal amounts being the equilibrium price of the pollutant. This price times the amount of pollutant Upton dumps is the competitive charge Upton must pay Downley.

Give Upton the right to dump, and Downley must pay him a competitive charge equal to marginal cleanup cost times the amount of cleanup he provides. At the equilibrium, the equation of the marginals is the same as before. This second scheme presents a difficulty, however; it requires a benchmark for original filth in order to determine just how much cleanup has been provided. One possible benchmark is the amount Upton would

---

*Allocation schemes of this type were first proposed by Lindahl [6]. They have since been discussed by Coase [4] and Arrow [2], among others.

5. TREATING EXTERNAL DISECONOMIES 67

dump if there were no Downleys, if there were no affected parties. For now we merely assume that the benchmark, N, can be established.*

A key question is whether shifting property rights here can drive either Upton or Downley out of business. If not, the marginal efficiency conditions will define a unique equilibrium, and the level of pollution will be the same whether Upton or Downley has rights. †

But the optimal allocations are quite different if one or the other of the firms ceases to function. If we give Upton rights, might Downley not decide to go out of business to avoid paying the charge for Upton's cleanup? Alternatively, might not a potential polluter who previously could not turn a profit because of high effluent charges now find it advantageous to go into business? When you consider that the size of N is arbitrary, it is evident that the answer to these important questions is yes. What is surprising, however, is that neither answer can be affirmative so long as the traditional convexity assumptions are satisfied.** For those interested we provide a mathematical demonstration of this assertion; others may wish to skip the next four paragraphs.

The assessment or payment of a competitive charge for an externality (as determined on an artificial market) to a firm with a convex production

---

*There is a third possibility that also leads to the same result. The government can have rights. It will sell these rights, say N of them, on the open market to both producers and recipients. Any right purchased by a recipient is not available to a producer, and thus reduces externality production by one unit.

† There is a more general way to look at this process of transferring rights and shuttling rents. An equivalent procedure is one where a standard level of externality production is set. Everything above that level requires a payment from producer to recipient and conversely. The case we call producers'-rights sets that standard level at N. The recipients'-rights case sets it at 0.

The argument given here that a change in property rights will not alter resource allocations only holds if <u>final demand prices do not change</u>. They might change if individuals with divergent tastes own the firms involved in the transfers, and if these firms are large relative to the rest of the economy. See Dolbear [5] for a discussion of the distributional effects of property rights transfers.

**This issue has a long but inconclusive literature. See, for example, Buchanan [3].

set can never determine whether it operates at a profit or at a loss. To see this, consider a simple case with a recipient firm (Downley in our model) that produces one output, b, using a single input, a, and that suffers from a single pollutant, z. The firm's production function may be written

$$b = f(a, -z),$$

with $f_1 > 0$, and $f_2 > 0$, where the subscripts refer to partial derivatives with respect to the first and second arguments. This function is assumed to satisfy the traditional convexity assumptions, i.e., it is concave throughout. We can measure the level of pollutant from the benchmark N so that the arguments of the production function are nonnegative. The translated production function,

$$b = g(a, N-z),$$

is clearly concave if f is.

The price of the output is normalized to unity. If p_a and p_z are the prices to the firm of the input and pollutant, respectively, then (regardless of whether it is an Upton- or a Downley-rights scheme) the marginal conditions

$$p_a = f_1(a, -z) = g_1(a, N-z), \text{ and}$$

$$p_z = f_2(a, -z) = g_2(a, N-z)$$

must be satisfied. In what follows, we suppress the arguments of f_1, g_1, f_2, and g_2.

The properties of concave functions imply that

$$g(a, N-z) - g(0, 0) \geq g_1 a + g_2 (N-z) = p_a a + p_z (N-z).$$

If recipients have rights, Downley's profits, π, will be

$$\pi = g(a, N-z) - p_a a \geq p_z (N-z) + g(0, 0).$$

To achieve the outcome of the producers'-rights case, we need merely transfer $p_z(N-z)$ from recipient to producer. After the transfer the recipient firm, Downley, makes $g(a, N-z) - p_a a - p_z(N-z)$, a still greater profit than he would make if he went out of business and got $g(0,0)$. Indeed, Downley has no incentive to change his behavior in any way.

The amount $p_z(N-z)$ of the transfer that shuttles back and forth depending on who has rights is a pure rent that returns to the fortunate

5. TREATING EXTERNAL DISECONOMIES

party. As such it cannot reverse either a profitable or unprofitable status for the externality recipient. A completely parallel argument establishes this same result for a firm, such as Upton, which is an externality producer.

Note that as N becomes larger, this result becomes increasingly counterintuitive. Surely the amount of rent that can be passed back and forth without forcing a shutdown decision of either the externality recipient or producer is finite. It would hardly seem possible that it could depend on an arbitrarily chosen benchmark N. The apparent paradox is explained in the second half of the paper, where we show that for sufficiently large N, full convexity is logically impossible; without convexity the preceding analysis is not relevant.

1.2. Taxes*

It is sometimes proposed that a taxing scheme be employed to overcome externality difficulties. With taxing schemes, the party with rights quotes its marginal loss to the government, which then makes the appropriate collection, the level of the externality times the marginal loss inflicted.† Given the conditions of (1) many-participant markets (assuring "honest" preference revelation), and (2) full convexity, the only difference between the artificial-markets scheme just described and a taxing scheme is that the latter does not provide positive payment for the party that has rights; the government makes a profit. A simple transfer from the government to the party with rights is all that is necessary to turn the taxing-scheme outcome to the one that would have resulted had an artificial market been established. The allocation of resources, following the argument of the previous subsection, is clearly the same. Furthermore, shutdown or operate decisions are not affected. Whether a firm is or is not paid the appropriate tax (the value on a competitive market of the loss it suffers) cannot reverse the profitability or unprofitability of its operation.

The most familiar proposal in the pollution world for this sort of tax scheme involves the imposition of effluent charges, the charge being a specific example of a unit tax on externality producers. It is sometimes proposed that instead of setting a per unit charge on effluent, a certain number of pollution rights should be auctioned among the potential polluters. Except for the mechanical way the equilibrium is reached, the two schemes are identical. The results they produce will be indistinguishable if the

---

*The first general discussion of tax schemes in this context was given by Pigou [8]. Further discussion can be found in Meade [7] and Coase [4].

† For a discussion of the mechanics of such a scheme, see Aoki [1].

quantity of rights auctioned under the latter scheme is just equal to the amount of pollutant dumped given the particular effluent charge established for the former scheme. They are both, in effect, unit-tax schemes.* We conclude, then, that the four schemes, artificial market or tax, recipient or producer having rights, all produce identical allocations of resources, assuming many participants and full convexity.†

In the discussion that follows, we detail some frequently occurring situations in which the assumption system that supported the foregoing analysis is not applicable. The equivalences we just developed will no longer hold, and, for the most part, the schemes we have presented will not lead to efficient outcomes.

2. RESISTANT EXTERNALITY PROBLEMS

Sometimes, alas, we find ourselves in third- and fourth-best worlds, where the externalities problems are not of the convenient type treated thus far, and new, more resistant externality problems arise. First we discuss briefly the problem of noncompetitive behavior, a problem which has been noted elsewhere by Arrow [2], due to thinness of markets and the public bad nature of the externality. Then we analyze difficulties involving nonconvexities. Finally we discuss parallel aspects of these difficulties when consumers are externality recipients.

2.1. Thinness of Markets

We have treated Upton and Downley, our representatives of the many producers and recipients, as buyers and sellers of the same externality. For this treatment to be meaningful on the selling side, it must be possible to have specific amounts of the externality conveyed privately to individual buyers. This would be the case, for example, with solid waste which could be disposed at any of a number of dumping grounds.

---

*There are (unfortunately) many frequently proposed schemes that do not achieve equivalent or efficient results even under our ideal conditions. For example, the oft-heard proposal that total costs of damage should be divided in proportion among the externality producers responsible for it, a scheme which is essentially average-damage taxation, is inefficient for the same reason that average-cost pricing is inefficient.

† We have demonstrated these results only for a particularly simple case. Starrett [11] has shown elsewhere that these results hold in a very general, abstract framework.

5. TREATING EXTERNAL DISECONOMIES 71

More commonly, unfortunately, any quantity of an externality received by one recipient is received by all. Thus, in the Upton-Downley example, Upton's effluent serves as a public bad for all downstream firms. The functioning of a market requires that the sellers compete with each other on the supply side. When the externality is a public bad, what is supplied by one (dumping rights in this instance) must be supplied by all. A traditional competitive market cannot exist.*

To achieve the equivalent of a competitive outcome, we would have to set up a separate competitive market for pollution rights between Upton and each one of the downstream firms. If each seller-recipient would take price as given in his particular market, and if Upton would act competitively and take as his given price the sum of these individual market prices, Upton would wind up paying a price per unit of externality equal to the sum of marginal downstream losses. To have such a price at the margin is the appropriate efficiency condition for the production of this public good (reduction in the level of externality). Given the public nature of the externality, the amount contracted on each market must be the same (the prices will differ in general). The simplest way to reach the efficient point may be to vary quantity on every market simultaneously and to ask each downstream firm to announce its going price, its marginal losses. Obviously, any scheme of this sort would require a substantial amount of centralized administration.

This scheme, as well as the tax proposals, will run into further difficulties given that there are not many producers and recipients dealing in the same commodity. Participants will find it in their interests to behave strategically. If recipients have rights, they will each be empowered to block any producer's production of the externality. A producer will be permitted to produce the minimum amount of the externality for which he buys rights from all sellers. Individual buyer-seller negotiations will probably result. Each seller will be responsible for but a small portion of the total price a producer (or all producers should they band together) must pay for increasing externality production. He will thus have little effect on the total amount the producer demands, and will have an incentive to sell at an extortionary price, a price well above his marginal losses. The likely outcome is that much less than an optimal

---

*There can still be competition on the demand side, all producers competing with each other to dump, the total amount dumped being the sum of what is dumped by each.

amount of externality rights will be sold, less than an optimal amount of externality production will take place.*

If producers have rights, the strategic shoe is on the other foot. A reduction in the amount of externality produced will benefit all recipients. The usual public goods, free-rider problem arises. Unless the recipients are capable of coordinating their activities, they will purchase less than an optimal amount of the public good (reduction of the externality).

Participants also have incentives to act noncompetitively in the tax schemes. If there are many downstream firms, they have an incentive to overstate their marginal losses to the government, since this may lead to a lower pollution level and there is nothing to lose. (Recall that recipients are not compensated in the tax scheme.)

2.2. Nonconvexities

Let us return to the river and assume that thinness-of-markets difficulties do not arise, or that the government can enforce the equilibrium through an administered planning process. Upton is still our representative producer. If property rights can be established for externalities, Downley is our representative recipient. Otherwise he is a single recipient whose losses from any level of externality production are the sum of the losses of all recipients. Both Upton and Downley act as price takers.

Our problems, alas, are not over. If Upton and his fellow polluters dump enough, it is absurd to expect that Downley will remain in the business of providing swimming facilities. Perhaps, if he had rights, he might stand around idle in order to collect his due from Upton. But from a resource allocation standpoint he would be out of business. This possibility was ruled out in Section 1 of this paper, where full convexity was assumed. Downley's quite appropriate shutdown decision must be associated with a nonconvexity in his production set.

To see that there is indeed such a nonconvexity, consider a single-product competitive firm that suffers from an external diseconomy. In the absence of externality markets or taxes, it could make a short-run,

---

*If the recipients were acting in concert, they would act as a monopolist and sell an amount which equates marginal revenue (derived in the usual manner) and marginal cost (the sum of their marginal cost curves). In the strategic situation described, unless the sellers can achieve some degree of coordination they will sell even less than they would if they were acting in this monopolistic fashion.

5. TREATING EXTERNAL DISECONOMIES

profit-maximizing decision for each potential level of externality. Plot its level of maximum profits as a function of the level of externality it receives. We know a lot about the shape of this function. It must be downward sloping, since z is a diseconomy. If, as assumed, the firm's production function is concave, this curve must be concave as well. This implies that it must reach any arbitrarily great negative value, since z is unbounded.* But the firm could always choose to produce nothing and thus incur at most fixed costs. This implies that the maximum-profits vs externality curve must be bounded below. The maximum loss floor that is established by the shutdown possibility rules out global concavity of the production function.

Figure 1 shows two possible shapes for the profits function when there is a nonconvexity in the production set.

A nonconvexity will obviously have no effect on any outcomes so long as N, the level at which the externality would be produced in the absence of any compensation arrangements, is less than K, the point at which the nonconvexity begins.

In the analysis that follows, we are worried about cases where nonconvexities matter, where N > K. Such a location reflects a situation

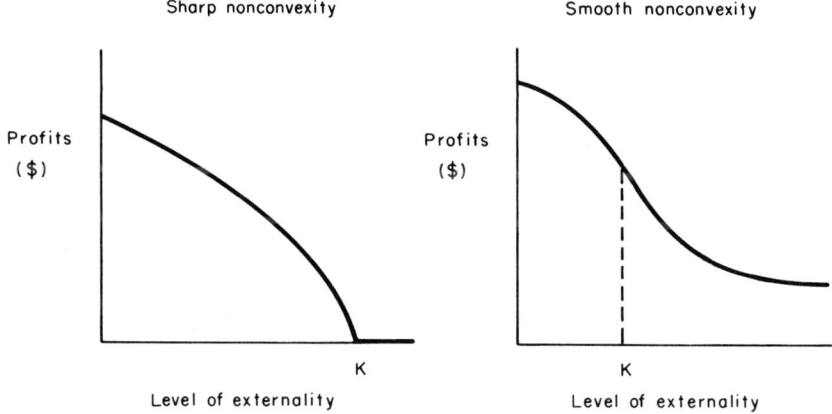

Fig. 1. Externality recipient's profits.

---

* A concave function always lies below its tangent line. If the tangent is ever downward sloping, it must cross the z axis, and so, therefore, would the profit function if it were everywhere concave.

where the potential harm is serious enough that it may induce a firm to shut down operations.

Drawing on historical precedent, we turn from swimming to agriculture and consider as our example the classic farmer-cowman illustration developed by Coase [4]. Coase observes that wandering cattle trample the neighbor farmer's crops. Consider two possible schedules of damage, representing, respectively, a sharp and a smooth nonconvexity (Table 1).

For full convexity the cows must do increasing marginal damage; something like Schedule A must apply. The difficulty is that the addition of a fifth cow to the herd induces the farmer to discontinue operations. Beyond four cows profits are identically zero,* and the profit function is not convex.

TABLE 1

Two Schedules of Damage to Farmer

Size of herd	Schedule A: sharp nonconvexity		Schedule B: smooth nonconvexity	
	Total profits	Marginal loss	Total profits	Marginal loss
0	8		8	
1	$7\frac{1}{2}$	$\frac{1}{2}$	6	2
2	6	$1\frac{1}{2}$	5	1
3	$3\frac{1}{2}$	$2\frac{1}{2}$	$4\frac{1}{2}$	$\frac{1}{2}$
4	0	$3\frac{1}{2}$	$4\frac{1}{4}$	$\frac{1}{4}$
5	$-4\frac{1}{2}$	$4\frac{1}{2}$	$4\frac{1}{8}$	$\frac{1}{8}$
	0^a	0^a		

[a] This figure applies if the farmer stops farming.

*Profit here should be thought of as return above and beyond variable cost, since in the short run the firm will quit only when it cannot cover variable cost. Returns to the fixed factor (land) are ignored in making short-run decisions.

5. TREATING EXTERNAL DISECONOMIES

Schedule B presents what we call a smooth nonconvexity. It illustrates a situation in which the marginal damage per cow is decreasing. Casual consideration of technological factors might lead us to suspect that this would be the case, that two cows would trample less apiece than one, and that a dozen would not multiply by six the damage done by but a pair.

Decreasing marginal damage over some range is not the only cause of the types of nonconvexities we have been considering. Nonconvexities will crop up if there are economies of scale in treating external diseconomies. For twice the price of a fence that keeps out one cow, the farmer could probably construct one that could keep out eight. Even if there were increasing marginal trampling damage over the one-to-eight range, once there are enough cows to make a fence worthwhile, a nonconvexity presents itself; extra cows will impose decreasing marginal costs (losses of profit) on the farmer. If, as seems not unlikely, the fence can keep out ten or one hundred, the nonconvexity will be of the sharp variety.

The results we are about to discuss hold equally whether nonconvexities are smooth or sharp. However, the exposition is somewhat easier when nonconvexities are of the sharp type; therefore that is the case we shall discuss.*

2.3. Artificial Markets

Consider Schedule A and the operation of an artificial market. Give the farmer, the recipient, rights. Say that the price, p^*, of the externality (cows that trample) is $2. The farmer, adopting traditional marginal optimization procedures, decides to accept two cows, the number for which his marginal loss does not exceed the price. Assume that the cowman's profit schedule leads him to demand rights to graze two cows at this price. Do we have an equilibrium? No! Surely, with the present price, this position is not optimal for the farmer. He would make greater total profits (infinite) if he accepted an infinite number of trampling cows at this price, or indeed at any positive price.

Thus, unless there is some artificial restriction on the number of rights which can be sold, equilibrium cannot exist for this artificial-market scheme. The farmer who is offered a positive price for accepting cows which can cause him at most a finite amount of damage (his farming profit), will surely want to supply an unlimited number of rights; but if he is offered a zero price, he supplies no rights, so equilibrium is impossible.

---

*The two types of nonconvexities may be combined in a single instance if the total profits curve of a smooth nonconvexity cuts the horizontal axis, the axis on which the externality is measured. But either nonconvexity, by itself, is sufficient to hinder the effectiveness of the corrective measures outlined above.

Nonexistence is somewhat surprising, since the presence of arbitrary nonconvexities does not ordinarily rule out the possibility of equilibrium.

There may still exist equilibria if we modify the rules by introducing an upper limit on the number of rights which can be sold. Indeed, an equilibrium will surely exist if we set the upper limit at the level K in Fig. 1, since then the problem is convex and we can appeal to a standard existence theorem. However, such an equilibrium no longer has any efficiency properties; for example, if it happens that the best solution is to have the farmer out of business and pollution levels above K, any equilibrium which excludes this possibility is surely inefficient. To be sure that the upper limit does not exclude the best outcome, we would have to set that limit outside the "feasibility" region. In our example, we would set the limit at N, the maximum number of cows our rancher would put on the land if the farmer were absent.

Having taken this precaution, special circumstances are required in order that an equilibrium exist. Let us return to our example. When the externality market is introduced, the farmer receives an externality payment, p^*c, where c gives the number of cows grazed. His aggregate profits, π^a, are equal to his profits from farming, π, plus the externality payment. The line through the point $(2,6)$ in Fig. 2 is an iso-aggregate-profits line; it offers a value of 10. The point $(2,6)$ is a local optimum for the farmer, but he could increase his profits by stopping all production and selling N externality rights. In the case shown, where $N = 8$, the farmer could secure aggregate profits of 16 at the present externality price by ceasing farming operations. This means that $(2,6)$ cannot be a global equilibrium; there is no price that would lead the farmer to choose to operate at that point.

No other price can be an equilibrium either. At prices above $\hat{p}$, the farmer will always want to sell N externality rights, while the cowman will not want to have as many as N cows.* At prices below $\hat{p}$, the farmer will want fewer than two cows (the number he would choose when the price is p^*), while the cowman would want at least the two he had when the price was higher. In terms of supply-and-demand analysis, the farmer's supply curve of cattle externalities is discontinuous and the cowman's demand

---

*In the numerical example, the externality level can only take on integer values; $\hat{p}$ turns out to be 1 1/14, the price that makes the farmer's profits equal whether he accepts eight cows or one, the latter being his optimal inferior point.

5. TREATING EXTERNAL DISECONOMIES

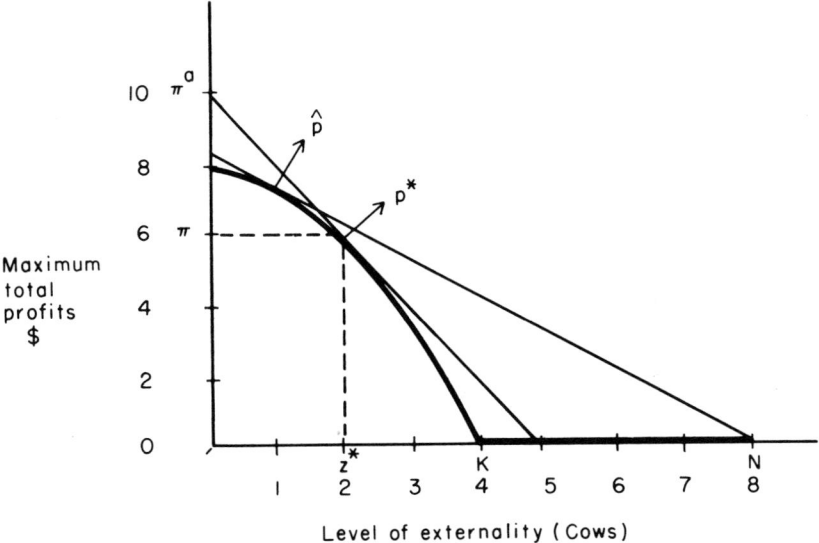

Fig. 2. Externality recipient's profits - sharp nonconvexity. The p's in the diagram represent the gradients of the total profits function for their respective price vectors.

curve passes through the gap of the discontinuity. Equilibrium will exist only if N is below 5 (so that at price p^* our farmer cannot offer enough rights to improve on the position $(2, 6)$).

The instability in any proposed equilibrium which derives from the farmer's threat to quit may seem somewhat silly, but, as we shall see below, it is a signal that an efficient allocation of resources may require that the farmer get out of farming.[*] We conclude that in cases where externalities are potentially strong enough to drive out firms, situations where N is large, the artificial-markets scheme will not work at all.[†]

---

[*] Note that instability is even more likely in the long run, since then the farmer will quit whenever he cannot cover his fixed as well as his variable cost.

[†] Our conclusion that equilibria will generally fail to exist seems to be related to some results of Shapley and Shubik [9]. They showed that in games with external diseconomies, the core frequently fails to exist. However, Starrett [10] has shown elsewhere that the two results are unrelated. The Shapley-Shubik result depends on a particular definition of property right, whereas ours does not.

2.4. Taxes

Can we expect a tax scheme to lead to the optimum equilibrium? Returning to our diagram, it appears that not one, but two distinct points are candidates for such an equilibrium. The point (2,6) is a possibility. If the government tells the farmer that there are going to be two cows, the farmer's quoted marginal losses will be $2, and at a tax (price) of $2, the cowman will want to have exactly two cows. But the point (N,0) is also a contender. At that point, with the farmer out of business, marginal losses (and hence taxes) are zero and the cowman will want to have N cows, his profit-maximizing number of cows assuming zero externality payments.

The presence of multiple equilibria should not surprise us. Pigou [8] points out that nonconvexities will lead to multiple tax equilibria. The point that we want to make here is that since external diseconomies logically imply nonconvexities, multiple equilibria are going to be the rule rather than the exception.

Should the farmer operate or not? To find out, we need to know whether society is a net gainer or loser from having the farmer operate at this location. Assuming that prices of private goods correctly reflect social opportunity costs, the best point is the point of maximum combined profit.* In Fig. 3, we plot the marginal profit loss of the farmer, ss, and the marginal profit gain, dd, for the cowman (derived from figures not given here), as functions of the number of cows.

Note that there are two points, A and N, where the marginal gain is equal to the marginal loss. These naturally correspond to the two potential tax equilibria. Whether we would prefer A or N depends on whether the total area under the ss curve between A and N (the farmer's profit loss moving from A to N) is greater or less than the total area under the dd curve over that range (the cowman's gain for the corresponding move).

We can conceive of situations in which the balance might be tipped either way. For example, if the soil is particularly rich, so that the farmer could earn considerable differential rent on it, and if it is the case that if the cowman intensified operations he would not secure much of a profits gain, then the farmer should continue farming. If the land is poor for farming, then the situation would probably be reversed, and the farmer should cease operations.

---

*We are assuming that our firms are small enough that their behavior does not affect final demand prices. Furthermore, the statement that everyone is better off at the point of maximum combined profit may require transfers among owners of the farm and the ranch.

5. TREATING EXTERNAL DISECONOMIES

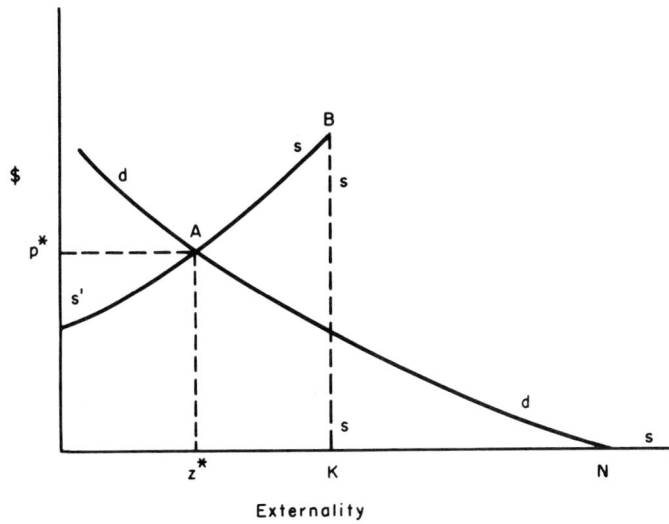

Fig. 3. Marginal profit and loss curves.

Unfortunately, if we set up the adjustment mechanism for taxes described earlier, there can be no assurance that the preferred equilibrium will be found. Assuming that we follow the conventional marginal prescriptions, we will end up at A or N depending on whether we start with externality levels less than or greater than K. Where we start, not the merits of the situation, will determine the outcome. The only sure way to find the optimum when there is the possibility that it will be appropriate for the externality recipient to cease operation, is to engage in the type of macro cost-benefit analysis described above. And in a general equilibrium framework with a highly interdependent externality structure, the required analysis would be incredibly complicated. Searching for shortcuts in such a procedure will provide an important, and hopefully productive, area for future research.

The above analysis suggests at least one important implication for current policy far beyond the narrow example on which it was based. Until recently, externality producers have not been forced to bear many of the costs they impose on others. Thus, current pollution levels are probably beyond optimal levels. What taxes should be charged? The above analysis suggests that using marginal losses may understate the relevant costs and lead to less than desirable adjustments. Firms, or households (see below), which have quit or moved will not be counted at all, even though they would benefit from substantially lower pollution levels. Furthermore, agents who have already absorbed the bulk of their potential damage from pollution (whose loss-of-profit functions show a smooth nonconvexity) may have small marginal losses now. Under these circumstances a nonoptimal equilibrium may be established with relatively low taxes and relatively high pollution levels.

2.5. Consumers as Externality Recipients

What about consumers? They too suffer from external diseconomies. In fact, many would argue that the most important externalities are in consumption. There would seem to be two significant ways in which consumers as externality recipients are unlike firms. First, consumers do not present a convenient cardinal measure of their welfare which would correspond to profits for a firm. In a variety of capacities, the consumer's willingness to pay to reduce the level of an externality may play a parallel role to the marginal loss in profits concept when the externality recipient is a firm. But if we are interested, as many economists may be, in the question of whether an artificial-markets scheme can be counted on to lead to an efficient equilibrium, we must look deeper and see whether the consumer's indifference surfaces display nonconvexities.

Second, consumers, unlike producers, cannot shut themselves down. For them, going out of business is not an approved enterprise. But the consumer does have an equivalent to the shutdown option; he can cease consuming a particular good. Such an action would merely indicate a situation where an indifference curve met an axis. If the increased presence of an externality induces a consumer to stop consuming an affected good, we might expect to find a nonconvexity in the consumer's indifference map. We will find one if the consumer does not care about the level of externality once he ceases consumption, if in this sense his after-shutdown feelings about the externality parallel those of a firm.

That a nonconvexity will be present in such an instance can best be illustrated with reference to our polluted-river example. Consider, for instance, a consumer who swims not in the Downley Baths, but directly in the river. Beyond a certain level of filth, the consumer will swim elsewhere or take up another form of recreation. He will cease consuming a particular commodity.

A full geometric representation of the nonconvexity involved would require a complex three-dimensional diagram displaying the three arguments: pollution, swimming, and other goods (money). Fortunately, there is a simpler, though indirect, way in which we can demonstrate the presence of the nonconvexity. Look at a typical consumer's indifference

5. TREATING EXTERNAL DISECONOMIES

curve on one swimming-pollution plane (Fig. 4). Every point on the curve represents some positive level of swimming and offers satisfaction level I. We assume that by giving the consumer a sufficiently large increment in his consumption of other goods we could just compensate him for a loss of all his swimming; that is, we can find a point (in general, there will be many) in commodity space that has zero swimming and offers satisfaction level I.

We will now argue that, given these assumptions, there is no point (such as Q) on the indifference curve displayed that can be a competitive equilibrium in the sense that there are prices (on real and artificial markets) for which this point is the least-cost way of achieving satisfaction level I. Since the consumer prefers less pollution, its price must be negative.

Now find the point in commodity space mentioned above, one that has zero swimming and offers satisfaction level I. Call it P. To purchase the goods bundle represented by P would cost a finite amount. But with zero swimming, the consumer no longer cares about the pollution level. With a negative price on pollution, the consumer would demand an arbitrarily large amount of it and would attain point P and satisfaction level I at as low a total cost as he wished. Thus, no point such as Q can represent a competitive equilibrium. This means that the consumer's indifference

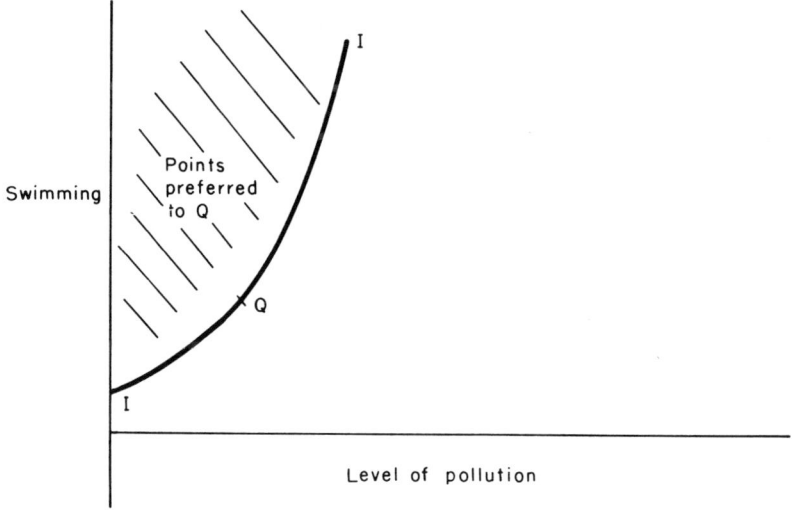

Fig. 4. Swimming-pollution indifference curve.

curves must present a nonconvexity; with full convexity of preferences, every commodity bundle is the least-cost way of achieving its own satisfaction level for an appropriate choice of prices.

With the consumer an externality recipient, there is no more assurance that an artificial-markets scheme will lead to an equilibrium than there is when the recipient is a firm. In effect, the two types of recipients display the same pathologies in these circumstances. It can be shown equivalently that if a tax scheme were employed in a consumer-recipient situation, the same form of macro cost-benefit analysis required with firms as recipients would be needed to find the efficient equilibrium among the multiple contenders.

To summarize, a consumer externality recipient may cease consuming a commodity and thus like a firm may present a nonconvexity that renders irrelevant conventional analyses about the existence and efficiency of competitive equilibria.

Despite these similarities, the consumer nonconvexity is less general than that for the firm. A consumer cannot always escape the effects of an externality. Suppose in our example that the consumer felt badly about river pollution even if he were not swimming in it. Then, although it is still likely that there will be a nonconvexity (since his dislike for pollution probably diminishes when he quits swimming), we can no longer demonstrate it logically. Furthermore, if an externality affects an individual through his consumption of an essential commodity, there will be no shutdown possibility and hence no essential nonconvexity. Air over New York City would not be such an essential commodity; one could always move away. But one cannot escape from radioactive contamination of our atmosphere. For this type of consumer externality there may not be a nonconvexity.

Even if full convexity is assured, the results for consumers differ from those for producers. With producers, the shuttling of rents that result from transfers of rights or switches from tax to market schemes do not affect resource allocations. But with consumers, these rents will exert income effects; these in turn will alter the efficient allocation of resources. Although tax and market schemes will both lead to Pareto-efficient outcomes when consumers are involved with externalities, they will lead to different efficient levels for the production of the externality. *

---

*The income effect on the purchase of externality production may well be negative. Rich people can swim in their own pools and will not have river pollution affect their natatory pleasures.

5. TREATING EXTERNAL DISECONOMIES

3. CONCLUSION

Ever since we left our opening section with its very restrictive assumptions, the problem of external diseconomies has looked more complex. In the real world we should expect to find externalities functioning as public bads for a variety of firms and consumers all of whom might exercise a shutdown option. The entire array of problems considered in the second half of this paper will be encountered.

Our conclusions, therefore, must be rather negative. We have shown that none of the decentralized schemes, even if they can be administered properly, can be expected to work in any straightforward manner when externalities are serious. An equilibrium may not even exist with artificial-markets setups. Even with tax schemes, our procedures of greatest promise, there may be multiple equilibria, only one of which will be efficient. To find and enforce the optimal solution may be a costly and complex procedure.

Despite these pessimistic findings, it would be wrong to conclude that a laissez-faire approach should be chosen by default. That mode of non-treatment has led to very serious present symptoms. Even rudimentary remedies, such as effluent charges or quota schemes, might be better than leaving matters untended. Surely we should conduct extensive research, both experimental and theoretical, to find positive treatments that might ameliorate existing conditions.

ACKNOWLEDGMENTS

An earlier version of this paper was presented at the Second World Econometrics Congress, September, 1970. This research was supported by NSF Grant GS-2797 (Starrett) and NSF Grant GR-88 to the Analytic Methods in Public Policy Faculty Seminar (Zeckhauser). That paper was distributed as Starrett and Zeckhauser [12].

REFERENCES

1. M. Aoki, Two planning processes for an economy with production externalities, Harvard Institute of Economic Research Discussion Paper No. 157, Harvard University, Cambridge, Mass., 1970.

2. K. J. Arrow, The organization of economic activity: issues pertinent to the choice of market versus non-market allocation, in The Analysis

and Evaluation of Public Expenditures: The PPB System, U. S. Congress, Joint Economic Committee, U. S. Government Printing Office, Washington, 1969, pp. 47-64.

3. J. Buchanan, External diseconomies, corrective taxes and market structure, Amer. Econ. Rev., 59, 174-177 (1969).

4. R. H. Coase, The problem of social cost, J. Law Econ., 3, 1-44 (1960).

5. F. Dolbear, On the theory of optimum externality, Amer. Econ. Rev., 57, 90-103 (1967).

6. E. Lindahl, Just taxation - a possible solution, in Classics in the Theory of Public Finance (R. A. Musgrave and A. T. Peacock, eds.), Macmillan, London, 1958, pp. 168-176.

7. J. Meade, External economies and diseconomies in a competitive situation, Econ. J., 62, 54-67 (1952); reprinted in Readings in Welfare Economics (K. J. Arrow and T. Scitovsky, eds.), Irwin, Homewood, Illinois, 1969, pp. 185-198.

8. A. C. Pigou, The Economics of Welfare, Macmillan, New York, 1920 (1st ed.) and 1932 (4th ed.).

9. L. Shapley and M. Shubik, On the core of an economic system with externalities, Amer. Econ. Rev., 59, 678-684 (1969).

10. D. Starrett, A note on externalities and the core, Harvard Institute of Economic Research Discussion Paper No. 198, Harvard University, Cambridge, Mass., 1971.

11. D. Starrett, Fundamental non-convexities in the theory of externalities, J. Econ. Theory, 4, 180-196 (1972).

12. D. Starrett and R. Zeckhauser, Treating external diseconomies - markets or taxes? Public Policy Program Discussion Paper No. 3, Harvard University, Cambridge, Mass., 1971.

6

ENFORCEMENT MEASURES IN WATER QUALITY CONTROL

Murray Stein

Office of Water Enforcement
Environmental Protection Agency
Washington, D. C.

1. INTRODUCTION

The Environmental Protection Agency (EPA), was established in the Executive Branch of the United States Government on December 2, 1970, with the assent of Congress to the President's Reorganization Plan No. 3 of 1970.* EPA is charged with the administration of Federal laws and programs in water quality control, air quality control, solid waste management, pesticide control, generally applicable environmental radiation standards, and noise abatement. Its mission is to mount a coordinated attack on pollutants which debase water, air, and land resources. Its principal functions are establishment and enforcement of standards; conduct of research and gathering and use of pollution information; grant, technical, and other assistance to others in arresting environmental pollution; and assistance to the Council on Environmental Quality in the development and recommendation to the President of new policies for the protection of the environment. The Council (CEQ), established pursuant to the National Environmental Policy Act, approved January 1, 1970, is a high-level advisory body in the Executive Office of the President. CEQ and

---

*The Federal Water Pollution Control Act Amendments of 1972, enacted October 18, 1972, supplant the former enforcement procedures of the Act with a new regulatory system based primarily on permits issued to municipal, industrial, and certain other dischargers. This paper reflects Federal water quality law as of the date of delivery, September 2, 1971.

EPA, a regulatory agency, play complementary roles in the campaign against environmental degradation.

2. PRESENT ENFORCEMENT AUTHORITIES

The Office of Water Enforcement is placed in the Environmental Protection Agency in the Office of the Assistant Administrator for Enforcement and General Counsel, under the supervision of the Deputy Assistant Administrator for Water Enforcement. The basic statutory authority for the enforcement of water quality requirements is the Federal Water Pollution Control Act, which provides for the abatement of pollution of interstate or navigable waters which endangers the health or welfare of persons, under authority first provided in 1956 and since expanded, and for the abatement of pollution which lowers the quality of interstate waters below the water quality standards established for those waters under the Act, as amended in 1965. The Act was amended in 1970 to provide for the abatement of pollution by oil, authority which EPA shares with the Coast Guard and other Federal agencies.

A new mechanism for the abatement of water pollution is the Refuse Act of 1899, administered through the years by the Army Corps of Engineers primarily in the interest of navigation. The old law, which prohibits the discharge of refuse into navigable waters without a permit, or in violation of the conditions of a permit, was the basis for ten suits brought in 1970 at the request of the Secretary of the Interior (then responsible for the Federal water quality program) to halt the discharge of mercury into navigable waters. The Attorney General has since been requested to bring water pollution abatement actions under the 1899 law in other cases, including the case of an industry which had failed to comply with the remedial action requirements of an enforcement conference convened under the older enforcement authority of the Federal Water Pollution Control Act, and had failed to comply with water quality standards after the expiration of the requisite period, at least 180 days, following notice of violation.

The President, by executive order issued December 23, 1970, established the Refuse Act Permit Program, which took effect July 1, 1971. The Permit Program requires that a Refuse Act permit must be issued by the Army Corps of Engineers for all discharges or deposits into navigable waters or tributaries, or into a waste treatment system other than municipal from which the matter will flow into a navigable waterway or tributary. The law does not apply to discharges or deposits of matter "flowing from streets and sewers and passing therefrom in a liquid state." The Permit Program accordingly excepts such discharges,

6. WATER QUALITY ENFORCEMENT MEASURES 87

and further excepts certain other discharges, although the Act itself is still applicable to such other discharges. The deadline for filing applications for permits for existing facilities was July 1, 1971. New dischargers are required to file for permits at least 120 days prior to discharging. The Permit Program is keyed to water quality standards, as well as navigation and fish and wildlife values. Refuse Act permits in turn are subject to the requirement of a 1970 amendment to the Federal Water Pollution Control Act that a Federal permit or license to conduct an activity which may result in a discharge into navigable waters (or may otherwise affect the quality of navigable waters) cannot be issued unless the State water quality agency certifies that there is reasonable assurance that the activity will not violate applicable water quality standards.

3. ADMINISTRATION LEGISLATIVE PROPOSALS

The President on February 8, 1971, recommended to Congress a far-reaching environmental protection program, building on the program which he set out in 1970. Among the Administration legislative proposals now before Congress are bills to regulate dumping in the oceans, estuaries, and the Great Lakes; to regulate the use and distribution of any substance which the EPA Administrator finds hazardous to human health or to the environment; to tighten the regulation of pesticides; and to strengthen and expand the water quality standards and enforcement authorities of the Federal Water Pollution Control Act.

The proposed amendments to Section 10 of the Act would (1) authorize the establishment, revision, and enforcement of water quality standards, now applicable to interstate waters only, for navigable waters, ground waters, and certain other waters; (2) include as elements of water quality standards (which now consist of water quality criteria and a plan for their implementation and enforcement) water use designations and effluent limitations; (3) streamline the procedure for the revision of water quality standards; (4) provide for the issuance by the EPA Administrator of administrative orders and the assessment by the Administrator of administrative penalties; (5) provide new authority for the institution of civil actions to enforce water quality standards and other requirements; (6) provide authority for the EPA Administrator to seek court relief in emergency situations; (7) confer on the EPA Administrator the subpoena power, the right of entry, and authority to require monitoring by dischargers and submission of reports; (8) regulate hazardous substances; and (9) provide for citizen suits to enforce water quality requirements.

4. STATISTICS IN WATER QUALITY CONTROL

Statistics, the science of assembling, classifying, and tabulating facts or data of a numerical kind, so as to present significant information about a given subject, are a valuable tool in water quality control. In enforcement work, the development of reliable economic, financial, scientific, and technical data is an important activity.

Statistical information bears on decisions as to whether an enforcement action should be brought in a given case. Statistics are critical to the successful outcome of an enforcement action.

The following questions arise:

1. Do Federally approved water quality standards apply to these waters?

2. If so, what are the applicable water quality criteria, including numerical criteria, to protect the assigned uses of these waters?

3. Are the criteria being met?

4. Do the standards include, in the State implementation and enforcement plan, which is an integral part of the standards, a phased schedule for the provision of needed waste treatment facilities or other control measures by municipal, industrial, and other dischargers?

5. What is the record of compliance with the schedule?

6. What is the estimated cost of needed waste treatment facilities or other control measures?

7. In the case of eligible public entities, what is the expected Federal share, State share, and local share of the project costs?

8. What is the present and expected future population of the area?

The Refuse Act Permit Program requires that the industry, or other nonmunicipal discharger, state in the permit application information which will fully identify the character of the discharges, including data pertaining to chemical content, water temperature differential, toxins, sewage, amount and frequency of discharge, and type and quantity of solids. Proposed changes in the Federal Water Pollution Control Act would apply precise effluent requirements, as part of water quality standards, to municipal, industrial, and other dischargers into all interstate, navigable, and certain other waters. The complexities involved in the establishment

6. WATER QUALITY ENFORCEMENT MEASURES

of specific quantitative effluent limitations are reflected in the recent EPA determination that it is not possible at this time to issue formal guidelines setting forth a method of computation whereby a specific discharge might be determined for general application in the Refuse Act Permit Program. Regional Administrators are directed to select cases for the establishment of effluent limitations on the basis of carefully identified priorities, with major emphasis on the most seriously polluted waters and on major sources of pollution on those waters.

Water quality requirements imposed in enforcement cases are frequently expressed in statistical terms. Percentage reduction of pollutants is sometimes applied, such as the requirement that the aggregate of phosphate discharges originating in each State to waters in the conference area must be reduced by a stated large percentage by a given date. Waste load allocations are sometimes applied, a variant of total percentage reduction. Thermal discharges became an issue in the Lake Michigan enforcement conference. The requirements approved by the EPA Administrator for the control of excessive waste heat discharges to the lake include limitations on the rise above existing natural temperature and the stated maximum temperature, whichever is lower, as well as other requirements relating to closed-cycle systems.

Stipulations were entered in nine of the ten Federal court cases taken under the Refuse Act in 1970 to abate mercury discharges into the Nation's water environment, and the tenth plant closed down. Analyses show a 98.4% reduction in mercury discharges in the ten cases. The overall reduction in mercury discharges by fifty industries identified as of approximately one year ago is 97%. The inclusion of mercury dischargers more recently identified gives a total reduction of 91%. Pounds per day of mercury discharge is the measurement.

5. EPA BUDGET

The Department of Agriculture-Environmental and Consumer Protection Appropriation Act, 1972, provides $441 million for EPA operations, research, and facilities, increased from $299.9 million for the EPA programs then conducted in other Federal agencies for the previous year. The appropriation for Federal grants for municipal waste treatment projects was doubled from $1 billion to $2 billion (available within limits of amounts authorized). These statistics bear brief mention as indications of the importance assigned to the mission of the new agency.

6. THE ECONOMY AND THE ENVIRONMENT

The Council on Environmental Quality, in its second annual report, released August 6, 1971, gave prominent attention to the relationship of environmental problems and their solution to the national economy. Media coverage of the report was extensive, but its interest is such that I want to take brief note of its discussion of the economics of pollution.

The Council reiterates the lack or inadequacy of data in presenting some of the cost estimates. The report also observes that environmental values are not easily reflected in our usual accounting systems, and that degradation is a real cost not subtracted from past calculations of the value of national output, and that when enhancement of the environment is uncounted the value of total output is understated. The Council gives an EPA estimate of the total annual toll of air pollution at over $16 billion. The costs of water pollution damage are even less well documented, the Council reports, than those of air pollution. Heavy losses of ocean and coastal fisheries and shellfish production by estuarine pollution are noted, as are salinity costs, and losses of recreational opportunities, which rise in value as demand for recreation and scenic amenities grows.

To estimate current and prospective costs to meet environmental objectives is described as a hazardous exercise. Summary statistics, based on EPA data, are staggering (a total of $105.2 billion through 1975), but, the Council states, "appear to be well within the capacity of the American economy to absorb." The temporary dislocation of workers who lose their employment because of plant closings is put as the problem which will require the most difficult adjustments in its solution.

The total cost estimate for environmental control of $105.2 billion breaks down into $38 billion for water, $23.7 billion for air, and $43.5 billion for solid wastes. For water the figure represents $24.5 in public costs and $13.5 in private costs; and for air $1.6 in public costs and $22.1 in private costs. Less than 25% of the total solid waste expenditures fall directly on the private industrial sector. Nearly two-thirds of the water costs will be paid by the public sector, but most air costs will be paid by the private sector.

7. CONCLUSION

The Environmental Protection Agency is dedicated to the restoration and enhancement of the natural environment, to the betterment of the quality of life. In the furtherance of that end, the refinement of knowledge of quantity in many areas, and the correct interpretation of statistical data,

6. WATER QUALITY ENFORCEMENT MEASURES

are indispensable tools. A splendid scientist, F. W. Kittrell, with whom I was long associated in the national water quality program, cautioned against "data interpretation as strictly an exercise in mathematics rather than an exercise in evaluation of an actual situation with mathematics used as a tool to assist in the formation of judgments."

Thoreau, more than a century and a quarter ago, lamented that the Yankee race found the Merrimack River offering its ages-old privileges of lakes for millponds and of natural dams and saw fit to "improve" them. The restoration of the environment does not require that we undo all of the "improvements" made by man in his attempts to conquer nature and turn natural resources to his use. It does require a realistic assessment of damage done, and of remedial measures required, and a determination to leave the planet Earth at least a little fairer than we found it as a habitat for the generations to come.

7

PROGRESS IN ENVIRONMENTAL QUALITY*

Thomas C. Winter, Jr.

Council on Environmental Quality
Washington, D. C.

0. SUMMARY

This paper reviews some of the progress made to preserve and improve our environment. It includes the new Federal institutional arrangements; the establishment of national air quality standards for certain substances, of emission standards for automobiles, and of steps to control certain categories of new facilities and certain hazardous substances; the permit program for industrial discharges into navigable waters and the Federal Water Pollution Control Act, also covering municipal discharges; and the progress in Federal agencies and facilities, particularly the environmental impact statement mechanism. It is emphasized that establishing environmental standards and making environmental considerations a part of decision making are not enough. We do not have a blank check to implement everything we want to do to achieve environmental quality. We must carefully balance the environmental considerations in our decisions against social, economic, technical, and other considerations, and choose the solution which will best meet our goals at the least cost, both to natural resources and in dollars. To make these decisions, good data are required to detail possible courses of action and their effects in a manner understood by decision makers and by the public. Statistical analysis plays an extremely important role in this process. Some examples are given to illustrate this.

---

*This paper was adapted, with the author's kind permission, from an after-dinner address he delivered at the symposium which led to this book. The lively questions and answers which followed unfortunately could not be reflected here. [Ed.]

1. INTRODUCTION: FEDERAL INSTITUTIONAL ARRANGEMENTS

Historians will probably look back on 1970 as the year of the environment. Many individuals had been concerned about our state of environmental quality prior to that, but it was not until 1970 that the general public and Federal decision makers became aware of environmental quality as a national issue. What was done at the Federal level in the first year after its identification as a national issue?

Before answering this question, let me review some of the institutional arrangements made by the Federal government to deal with this problem. In the Executive Office of the President, the Council on Environmental Quality was established in early 1970. Its duties are to advise the President on environmental matters and to make national policy recommendations to him in this area.

The operating agencies of the government all have a strong role to play in preserving and enhancing our environment. A provision of the National Environmental Policy Act of 1970 established the requirement that all Federal agencies must prepare environmental impact statements on all actions they take which significantly affect the environment. Until recently the chief factors in decisions were the mission and the cost. The purpose of this new mechanism is to make environmental considerations a major part of every decision. Another objective of this mechanism is to carry out this process in a "goldfish bowl" type of visibility so that the public and certain nongovernmental groups with interest in the project can have a chance to feed their input into the decision-making process at the initial stages. The intent is to cause these decisions to be made in a much more objective fashion.

In December, 1970, a new agency was created to act as a pollution control policeman - the Environmental Protection Agency. EPA has the job of establishing criteria and, when necessary, of setting standards. It must see that regulations to enforce these standards are promulgated and that programs are implemented to enforce them. When it was initially formed, EPA absorbed a number of agencies organized along categorical lines, such as air, water, and solid wastes, from other Federal departments. Examples are the National Air Pollution Control Administration from the Department of Health, Education, and Welfare, and the Federal Water Quality Administration from the Department of Interior. However, insults to the environment usually do not come from one particular pollutant alone or through one medium. They usually result from the combined effects of a number of pollutants acting through several media. This was one of the reasons why EPA was quickly reorganized along functional lines. For example, it now has an Office of Standards and Enforcement,

7. PROGRESS IN ENVIRONMENTAL QUALITY

which looks at enforcing standards for all pollutants and all media. Similarly, all research, development, and demonstration activities are now combined into one entity, the Office of Research and Monitoring.

2. OTHER FEDERAL ACTIONS

Let me summarize some other Federal actions taken - under categorical headings, since most of the authorizing legislation is still along categorical lines.

Let's first look at air. The Clean Air Amendments of December, 1970, authorize some sweeping improvements in our programs dealing with air pollution. In the area of auto emissions, they decree that a 90% reduction will occur in carbon monoxide and hydrocarbon emissions from 1975 model cars as compared with the 1970 models. Further, they decree the same reduction in the emissions of the oxides of nitrogen in 1976 model cars as compared with 1971 models. This authority was promulgated by EPA regulations published on June 30, 1971.

This is quite a radical move in legislation. It sets standards by law for which technology is not now available. By technology, I am referring to the means to mass produce cars which will meet these standards during 50 to 100,000 miles of the type of driving which you and I and our wives do, with the type of maintenance which the corner gas station mechanics perform on vehicles.

The Clean Air Amendments of 1970 also address national ambient air quality standards, specifying a sequence of steps, with time limits, to implement these on a national scale. For example, EPA was given 4 months to promulgate standards. Within 9 months of promulgation, plans to meet the standards had to be adopted by the states involved in each air quality region, after public hearings, and submitted to EPA for approval. EPA then has four months to approve or disapprove, etc.

These standards apply to ambient air - the receiving medium. They are necessary, but in some cases are not sufficient. There are emission sources whose effects have such magnitude that they must be separately controlled by Federal regulations in order to protect the general public. A sequence with time limits has been established by law for EPA to implement national controls on certain categories of new facilities. EPA's initial list of these categories includes contact sulfuric acid plants, nitric acid plants, portland cement plants, large incinerators, and fossil fuel steam generators.

The Clean Air Amendments of 1970 also specify a sequence which must be followed to control hazardous substances, the first three identified by EPA being mercury, beryllium, and asbestos.

I should emphasize that the categories for new stationary sources and the hazardous air pollutant substances which have been identified are just a start. EPA is charged with the responsibility of adding to this list as sufficient information becomes available to show that other stationary sources and hazardous air pollutants are a serious threat to the air quality of our country. The substances which have been identified thus far are not necessarily the worst, but are those for which the most definitive information existed at the time that EPA so designated them.

In addition to the legislation, Executive Order 11602 bars facilities involved in air quality convictions from procurement and other contracts with the Federal government and from receiving Federal financial aid.

In the case of water, the most powerful existing authority is the Refuse Act of 1899. It applies only to industrial discharges, however. The more recent Federal Water Pollution Control Act covers municipal discharges. Although much more difficult to enforce than the Refuse Act, it has been put to good use. For example, in December, 1970, the Administrator of EPA issued notices to the cities of Atlanta, Cleveland, and Detroit that they were violating water quality standards. Under the framework of this authority, officials from the cities and EPA have since worked out agreements for these cities to eventually meet the water quality standards, with some assistance from the Federal government. Stein's paper, immediately preceding in this book, gives additional information about these two acts.

It was recognized from the start that the Federal government would have to set the example by cleaning up its own facilities in order to get the other sectors of the country to clean up theirs. Military facilities are the most numerous of the Federal facilities. It was estimated in fiscal year 1970 that it would take $359 million over the next three years to enable military facilities to meet pollution standards or at least to have remedial action under way by the end of 1972. In fiscal year 1971, $131 million was appropriated.

3. NEEDED LEGISLATION

In order for the Federal government to fulfill its role in enhancing and preserving the environment, it must have the necessary authority. Some applicable laws, examples of which I have already described, are on the books, but many were ineffective. Other environmental areas were

7. PROGRESS IN ENVIRONMENTAL QUALITY

not covered by Federal authority. In February, 1971, the President proposed to Congress eighteen bills to fulfill this need. Stein's paper outlines some proposals for water. Let me talk briefly about a few others.

Chemical pesticide control has been effected essentially by labeling requirements. However, there is no guarantee that the customer will read the label or follow its instructions. A new approach is needed. To fill this the President proposed a new pesticides control act. Basically it categorizes chemical pesticides into three groups: (1) for general use; (2) for restricted use, to be applied by trained applicators; (3) for use by permit only, to be approved for use by trained consultants and then applied by trained applicators. These consultants and applicators would be licensed by the State.

Another new bill, the Toxic Substances Control Act of 1971, would give the Administrator of EPA the authority to prevent the distribution of new substances until they have been shown to be compatible with the environment, and to ask the courts to restrict the use of existing products if there are indications that they are imminently hazardous to human health. Hopefully it will shift the burden of proof of determining the environmental compatibility of a product from public and conservation groups to the manufacturer, prior to their actual distribution.

Another bill deals with noise. It will provide the Administrator of EPA with the authority to establish criteria and set standards for noise sources which are particularly obnoxious, except aircraft, where authority would remain with the Federal Aviation Administration.

To reduce the amounts of lead and sulfur oxides which get into our air, the President will resubmit to Congress a proposed tax on the lead added to automobile fuels which did not pass last year, and will also submit a tax on the sulfur content of fossil fuels used in combustion and distillation.

4. TRENDS

What are the trends which have emerged from our activities relating to the environment? First, it is obvious that the public is becoming better informed concerning environmental issues and is demanding more comprehensive controls. Second, the Federal government has a major role in this effort. Part of it is to establish criteria, which I define as data describing quantitatively the effects of a pollutant on a variety of receptors. Another function is to establish standards, which I define as regulated levels of quality for a natural resource or emission rates of pollutants prescribed for specific control purposes.

The implementation and enforcement of these standards must be accomplished primarily by the private sector and by State and local governments if we are to be successful. However, the Federal government has a role in accomplishing this in regard to its own facilities and in stepping into the breach if State and local governments cannot or will not accomplish this function.

Another trend concerns the burden of proof as to whether a product has detrimental effects on the environment. This is gradually shifting to the producer. There are growing pressures on him to show that there is no significant environmental harm before the widespread distribution of the product.

Another trend is reevaluation of our galloping technology. More and more we are looking at the environmental assessment of new technology before it goes into effect rather than after it goes into effect. Not only do we consider the primary side effects to the environment, but also the secondary and tertiary effects.

Some extremists are contending that we must do away with technology. This is nonsense. We need technology to provide the solutions to our present environmental problems and to devise ways to prevent further environmental insults in the future. We must control technology, but do it in a manner which will still encourage initiative on the part of individuals to develop new technological processes and to develop incentives on the part of industry to implement these new processes.

5. COSTS

What will environmental preservation and enhancement cost us? The estimated cost over 1970-75 to implement pollution control is $105.2 billion. I must add that this is based on very rough data. Good data with which to assess our environmental situation and to make the necessary decisions to improve it are just not available at this time.

One hundred five billion dollars is a lot of money. Pollution control will be costly. However, it is less than 2% of the estimated gross national product over 1970-75, $6.7 trillion. This is less than the increase in other social costs which we have grown to expect over the same time period (an example is the increased cost of labor).

We can also look on this not as cost, but as investment. For example, one estimate indicates that the cost of damages from air pollution is $16 billion each year, including doctor bills incurred in counteracting the

7. PROGRESS IN ENVIRONMENTAL QUALITY

degradation of health, e.g., by emphysema; more frequent paint jobs to structures whose paint has been damaged by air pollutants; and the like. The estimated costs to control air pollution come to around $4 billion per year. So in effect we are investing $4 billion to save $16 billion each year. Of course, this is a general statement, and there are specific instances where it might cost us as individuals more in the short run than the benefits which we derive.

6. INFORMATION REQUIREMENTS

We need comprehensive and reliable data and facts on which to make our decisions. Right now we are in very poor shape as far as this is concerned. In our Council's second annual report we indicated that we don't know if pollution was better or worse in 1971 than in 1970. In most cases the data are simply not available in usable form to show this. That which is available is not reliable or comprehensive enough to provide definitive answers.

Here is a statement which discusses the data which was included in the report [1] :

"The tables and graphs are not the real world. They are but the final outcome of selecting a few dimensions from an almost unlimited number of environmental variables, selecting a few sample sites from an almost unlimited number, collecting and analyzing the data on the basis of often very inaccurate techniques, and then aggregating and statistically analyzing the collected data on the basis of a number of crude assumptions. How close the results of this process come to reflecting actual conditions is uncertain."

On the other hand, there are some areas in which we have too much data available in unmanageable form. An example of this is the data which some satellites are continuously transmitting to the ground and which are accumulating on thousands of magnetic tapes and being stored without anyone even looking at them. We are producing a new type of pollution - information pollution!

There are two major tasks facing statistical analysts in the environmental field.

1. We must make sense out of the available data and present it in a form which can be used.

2. We must guide future efforts to get data so that we obtain it in the most efficient manner to provide the most useful results, and are not burdened with data that we cannot use.

However, we cannot wait until these techniques are developed. We must make decisions and provide environmental controls now. We are, and these decisions are by necessity quite often based on insufficient data and on "guesstimates."

Let me provide an example, in reference to the control of toxic substances, of the attempts to grapple with these complex issues when insufficient data are available. In the past couple of years we have had national scares with such substances as mercury and polychlorinated biphenyls, more commonly known as PCBs. There are now rumbles about cadmium, arsenic, and others. EPA is attacking the problem of how to identify such potentially harmful substances through the concept of a toxic substances early warning center. How would the people in this center screen the over 2 million substances which man is dispensing? They would work with two basic sources of data. One is the production figures of industry. The second is the levels of the buildup of the substances in certain critical species in the biota. Examples are the level of mercury in rainbow trout and the level of PCBs in eagles. By putting these two categories of information together, it may be hoped that the analyzers will be able to identify a group of substances on which we should place priority for devoting our limited resources and expertise to try to determine what the effects on the environment are and what controls should be instituted.

We live in a complex world. We are running into limitations on the space in which we operate and the natural resources which we need in order to live, particularly at the standard of living to which we have become accustomed. We must make decisions on how to manage this space and these resources; how to minimize pollutants. There is an extremely important role for the statistical aspects of pollution control in the formulations of these decisions. We need sufficient and reliable data. These data will be used as a basis for decisions on environmental policies and for regulations and other implementation measures. These data will be used for enforcement. They must stand up in courts of law. Finally, these data are needed to assess how well we are doing and whether some of our present policies must be changed or modified.

The results of these environmental statistical analyses must be understandable. They must be in a form which the scientists and engineers can use to implement environmental controls. They must be in a form that decision makers can understand in order to fulfill their function. And they must be in a form which is meaningful to the general public, so that

7. PROGRESS IN ENVIRONMENTAL QUALITY

they can assess what is being done, what has to be done, and provide the political support which the decision makers need.

In 1970-71 we defined our game plan at the Federal level on how to achieve environmental quality. Of course we have more to do on this game plan; we must work out many of the details and fill in the gaps. However, the Federal focus is now shifting toward how to implement these procedures, particularly in view of the constraints that we are under.

We have a lot of hard questions to answer. We need the results of good statistical analyses of accurate and reliable data in order to provide these answers.

REFERENCE

1. Second annual report, Council on Environmental Quality, August 1971, p. 263.

8

DEVELOPMENT OF ENVIRONMENTAL INDICES

Robert Pikul

The MITRE Corporation
McLean, Virginia

0. SUMMARY

This paper presents an initial formulation of environmental indices categorized into the following classes: air pollution, water pollution, hazardous substances, land management, waste disposal, recycling, resources, natural phenomena, social aesthetic conditions, population statistics, human health, biological health and ecological balance, economic loss, and control measures. The intended uses of these indices are to describe environmental status and trends, guide environmental policy and legislation, and evaluate effects of environmental policies and programs.

Over 100 indices have been proposed in the 14 environmental classes. This paper discusses a method of ranking indices, taking cost and value considerations into account. It also presents as an example the definition and calculation of an index of air quality. (For another example, see Ref. 4.)

1. INDICES AND INDICATORS OF ENVIRONMENTAL QUALITY

1.1 Introduction

Indices and indicators are useful means of observing trends, analyzing programs, and informing the public of important concepts in a simple, understandable manner. A perfect set of measurements of a large number of parameters which are reported in a timely and effective manner merely

provides a decision maker with large amounts of data. To be useful for
evaluation and assessment, these data must be aggregated in a meaningful
way to show magnitudes and trends. Evaluating the state of the environment as well as assessing the impact of programs on the environment will
be facilitated through analyzing and tracking the behavior of indices and
indicators over time.

In the context of this discussion, an _index_ is considered to be a mathematical combination of two or more parameters which has utility at least
in an interpretive sense. An unemployment index may be as simple as the
ratio of unemployed persons to the total number of employable persons,
where the unemployed persons constitute a subset of the total set of employable persons. A cost of living index, on the other hand, may be a linear
combination of costs of some set of essential goods and services, each
weighted by the portion of the total dollar amount (for essential goods and
services) a "typical" family may spend (for the particular good or service
purchased).

An _indicator_ phenomenon serves as an alerting device to call attention
to a potential change in a related phenomenon. Changes in the prime interest
rate indicate similar changes in the interest rates consumers must pay for
home mortgages. An indicator measurement may be an index (e.g., a
continuous rise in the wholesale commodity price index indicates a rise in
the cost of living index) or a basic parameter measurement (e.g., a
significant reduction in the work force in a one-industry town indicates a
reduction in the general economy of the town).

Economic and social indicators and indices, such as the cost of living
index and cars per household index, have been utilized by governmental
decision makers and the general public for years. More recently, environmentally related indices have been developed to assess the quality of
our air [1], the quality of our water [2], and the environment in general
[3]. These environmental indices serve a useful purpose but they are
somewhat limited in scope, failing to consider many important factors.
The last of the three, the EQ (Environmental Quality) Index, published
annually by the National Wildlife Federation, is strikingly illustrated and
contains a variety of interesting, often alarming, facts, but the Federation
does not claim to be analytically rigorous in its derivation of quantitative
results.

Because of the broad range of environmental factors which must be
addressed, a single index, no matter how desirable from a standpoint of
reporting simplicity, would not be adequate to highlight significant trends
for each environmental area.

8. DEVELOPMENT OF ENVIRONMENTAL INDICES 105

1.2. An Analogy from Economics

To illustrate this point, let us consider various economic indices. The Gross National Product, which is the total national output of goods and services valued at market prices, is regarded by many as the best single indicator of the state of the economy. The GNP, reported since 1929, consists of (1) private and Federal purchase of goods and services, (2) gross private domestic investment, and (3) net exports of goods and services [5]. Even though the GNP is a useful index, it is informative to examine other economic indices, such as those for wholesale prices, retail prices, unemployment, and cost of living. Such supplementary indices will always be useful, especially since many experts disagree as to the meaning and relative importance of the various indices.

In contrast to the highly developed economic indices, environmental indices are just now being defined. Current knowledge about the environment is so rudimentary that identification and measurement of parameters for specific indices and indicators is extremely difficult. It would be premature to attempt to develop an overall index for the environment in any rigorous fashion, although subjective judgments of the "state of the environment" may be useful. Consequently, the first step in the development of environmental indices must be to focus upon specific areas of concern.

Illustrations have been borrowed from the field of economics for two primary purposes.

1. Economic indices and indicators, having been utilized and reported for many years, are fairly familiar to decision makers and provide a good analogy to clarify basic concepts which will be described in subsequent sections of this paper.

2. Economic indices, in spite of their refined state of development and use (relative to environmental indices) contain many conceptual difficulties, such as the following: (a) What is a "typical" family and how does one measure the distribution of its expenditures for essential goods and services? (b) How does one determine the components of the basic commodity group and the essential goods and services to be included in the cost of living index? (c) What is the rationale for assigning weights to terms in an index? (d) How does one obtain representative samples?

These difficulties have been resolved on a practical basis, and in spite of many other problems related to definition, interpretation, and value judgments, economic indices have proven useful in assisting other processes of assessment, evaluation, and control. Similar difficulties will be encountered with environmental indices, a fact which emphasizes the necessity for continuing development and refinement.

The following sections provide a point of departure for this effort and suggest potential avenues for further development, some of which can be more definitely selected once data are available and operational experience in measurement and use has been obtained.

2. COMPOSITE GROUPING OF INDICES

Fourteen composite groupings which focus on critical environmental areas are shown in Table 1. Over 100 indices/indicators are included in this grouping (some entries refer to more than one index, as indicated by the code assigned to each entry). While it would be desirable to have a single index for each group composed of an aggregate of indices within each group, our concentration on the component indices has precluded an attempt at such a formulation.

A brief description of each group is presented in Section 2.1. The basis for developing the importance rank shown in Table 1 is discussed in Section 2.2.

2.1 General Description of Index Groups

Air pollution and water pollution are the most obvious examples of man's impact on the environment, and the results of effective cleanup measures would presumably be noticeable soon after enactment. Both emissions (or effluents) and concentration levels are included in the indices, so that all of the pollution factors for a single medium are grouped together. It is recognized, however, that control must be directed toward emission or effluent sources in order to meet standards for air and water quality. Fish kills and drinking water acceptability are grouped under biological health and water resources, respectively, although they of course relate to water pollution. Such grouping problems will inevitably arise. In these composite categories, each specific index is included once only, but a strong argument can be made for including certain indices in more than one category.

Hazardous substances, such as mercury, cadmium, lead, and persistent pesticides, are of increasing concern and danger. Radioactivity levels may become a serious problem as nuclear reactors become a widely used source for electric power or as other uses of radioactivity increase. For

8. DEVELOPMENT OF ENVIRONMENTAL INDICES

many substances, "normal" or "background" levels are not known. As more and more substances are shown to be damaging to human or other life, dumping or use of dangerous or unstudied substances will decrease. Because these substances are potentially fatal, they have been classed separately from air and water pollutants. Both amounts used and concentration levels should be measured.

Land management reflects not only soil conservation practices, but also land usage shifts brought about as a result of growth of urban populations. Irreversible land use shifts (such as the development of wetlands) are of concern as "idle land" becomes more scarce and as the ecological value of areas previously thought worthless becomes known. In the future, land use practices will probably be regulated by increased zoning regulations (which are included under "control measures") if the public good is to be served. Categories such as wildlife habitat are of course related to endangered species, but wildlife habitat has been considered here as a land use. By focusing upon land use and soil conservation practices in a land management index, the choices made by an increasingly urban population become more visible, and alternative practices may be more carefully considered.

Waste disposal relates to water pollution (overflow of untreated sewage), air pollution (incineration), land use (sanitary land fill, open dumps), and recycling. Most air and water pollution is a type of waste disposal, but this index is intended to cover sewage treatment facilities (which could be included under water pollution) and solid waste disposal. This index may be considered a quality index and computed relatively easily once standards are set. It has been chosen as a separate category because it is in the disposal of sewage and solid waste that many environmental problems and solutions lie.

Recycling is a partial solution to problems of waste disposal and to exhaustion of natural resources. Wood, paper products, glass and metal cans are particularly amenable to this approach. Material balance studies are useful to trace the flow of expensive, dangerous, or otherwise important substances. For a given material, the amount produced plus the amount recycled less the amount utilized will represent additional waste. The amount of waste that cannot be accounted for by effluent and emission studies is referred to as the mass gap. Material balance studies are a check upon the completeness and accuracy of emission and effluent measurement.

The supply of resources is of concern as population size grows and per capita consumption increases. Living resources, if managed properly, need not be depleted. Nonliving resources, especially minerals,

TABLE 1

Major Categories and Ranking of Indices

Category	Index name	Code	Importance rank
Air pollution	aggregate pollution	02	2
	individual pollutant	01	22
	episode days	38	23
	dust and smoke	108	38
	turbidity	63	54
	upper atmosphere pollution	03	
	pollutant dosage	39	
	sentinel indicators	31	
Water pollution	aggregate quality	14	4
	aggregate pollution	05	5
	oil spills	43	6
	oceanic pollution	06	16
	percent acceptable drinking water	95	19
	episode days	42	27
	individual pollutant	04	28
	stream mile quality	96	
Hazardous substances	radioactivity levels	17	9
	pesticide usage	21	11
	radioactivity spills	09	12
	pesticide residues	08	14
	food residues	27	33
	tissue residues	26, 33	34

8. DEVELOPMENT OF ENVIRONMENTAL INDICES

Land management	land shift	55	8
	wetland habitat	50	17
	closed shellfish areas	105	29
	erosion potential	61	32
	dam siltage	107	43
	urban green	90	
Waste disposal	solid waste disposal practices	15	13
	solid waste per capita	07	47
Recycling	solid waste recycled	20	
	mass gap	11	
Resources	water table depth	53	15
	timber	56	21
	agriculture	57	24
	aggregate water use	49	30
	fish catch	51	40
	electric power per capita	65	58
	minerals	54	
	potential water use	97	
	river navigability	52	
Natural phenomena	climatological trends	47	25
	solar radiation amount	62	35
	floods	68, 69	41
	severe storms	66, 67	51
	precipitation	19, 110	52
	earthquakes	70, 71	
	runoff	106	
	salt spray	111	

TABLE 1 (cont'd)

Major Categories and Ranking of Indices

Category	Index name	Code	Importance rank
Social-aesthetic conditions	rat infestation	81	31
	outdoor recreation	77	36
	housing shortages	102	45
	residential amenities	89	46
	urban parks/playgrounds	103	49
	noise	10, 18, 80, 84, 85, 86	
	urban sprawl	91	
	traffic congestion	94	
	substandard housing	88	
	refuse collection	82	
	street maintenance	87	
	odor	79, 83	
	work/leisure time	98	
	cultural facilities, events	99, 78	
	billboards/junkyards	100	
	roadside litter	101	
	roadside rests	104	
Population statistics	vital statistics	25	55
Human health	respiratory disease	22	10
	eye irritation	23	
	hearing loss	24	
	gastrointestinal disease	112	

8. DEVELOPMENT OF ENVIRONMENTAL INDICES

Biological health and ecological balance	pest species	60	18
	endangered species	59	20
	fish kills	40	26
	algal blooms	41	39
	troubled wildlife	58	
	population changes	28, 32	
	plant growth suppression	30	
	biogeochemical cycles	64	
Economic loss	farm losses	76	50
	soiling costs	35	
	crop loss	29, 109	
	pest control expenditures	75	
	corrosion damage	34, 36	
	natural disasters	72, 73, 74	
Control measures	pollution research funds	37	42
	adequately zoned communities	93	48
	states with water quality standards	45	56
	regions with air pollution standards	44	57
	cease and desist orders issued	46	

may become scarce if wasteful practices are encouraged and if alternative energy sources and materials are not found. The drain on our resources should be made evident so that intelligent choices can be made.

Natural phenomena include those events which are relatively unaffected by man's actions but which either are critical to survival (weather, climate) or cause injury, death, or severe property damage. Some of these events may be affected by man's actions, e.g., forest fires and climatic trends, but the magnitude of such effects may show up only in long-term measurements.

In contrast to natural phenomena, social-aesthetic conditions are entirely the results of human actions. Social conditions relate to interactions among individuals and groups, whereas aesthetic conditions relate to sense perception, usually to the perception of beauty. Social-aesthetic conditions reflect the state of the man-made environment. Recreational and cultural facilities and residential conditions are included in the social part of the index, and noise, odor, and visual insults such as "urban blight" are included under aesthetic conditions. Rat infestation and noise levels are included in this index; rat bites and hearing loss are included under human health.

Population statistics reflect not only such measures as infant mortality and life expectancy, but, more important, population growth trends. If we are to continue to lower the death rate by advances in medical care, then a choice must be made between lowering the birth rate more rapidly or coping with the demands of a rapidly growing population.

The impact of pollution on human health is of increasing concern because of its implications relative to human survival. Respiratory and cardiovascular symptoms may be correlated with air pollution episodes. Factors such as the possibility of enzymatic alterations and the long-range effect of the pollutant burden (the tissue concentration of various pollutants) are being studied by the Community Health and Environmental Surveillance Studies (CHESS) program of OAP.

The biological health and ecological balance index reflects not only pollution effects (eutrophication) but also effects of poor management (pest species). The tendency of man to simplify natural ecosystems (e.g., by monoculture) has resulted in the reduction of species diversity in particular ecosystems, and in outbreaks of pest species. The combined effects of pollution and habitat reduction contribute to the number of endangered species. Long-term effects of man's activities may also include effects on the biogeochemical cycles vital to all life.

8. DEVELOPMENT OF ENVIRONMENTAL INDICES

Both natural and human-caused effects (pollution, poor management) are included in the economic loss index. In many cases, it may be possible to separate the cost of pollution (soiling, corrosion) from other economic losses. Agricultural costs include control expenditures, which are presumably offset by reduced losses from crop damage. The economic loss category represents, in addition to those costs man can change relatively little (natural disasters), a category that would be affected by pollution control effectiveness and by more ecologically sound agricultural practices. Comparison of trends in this index with trends in control measures would be interesting.

The composite index for control measures provides a "report card" for efforts to reduce pollution and increase the number of ecologically sensible practices.

These fourteen composite indices are thought to cover most areas of concern without unduly splitting related subjects. The eventual computation of these indices will require agreement on certain standards for the computation of "quality" and the assignment of weights to various components.

2.2 Cost-Value Considerations in Ranking of Indices

All 107 indices associated with Table 1 are not equally important. Perhaps some should be discarded and others added. A priority ranking will be subjective but the value judgments made must be explicit. A preliminary analysis has been based upon the following factors: impact, utility, value, cost, and importance. Impact evaluates the effect on various human activities or environmental factors. Utility refers to the purpose (scientific, administrative, or CEQ-use) for which the index may be used. Value is understood as the product of impact and utility. Cost involves a relative cost estimate as opposed to an actual dollar amount. Importance is defined in terms of a scheme for analyzing a cost vs value graph.

The first step in the analysis consisted in rating each index as to its relevance to an impact area. The relevance scale and impact weights are shown in Table 2. These impact weights represent the subjective judgment of several analysts. Other weights or tables of impacts may readily be applied. Each index was then assigned a total impact by summing weighted ratings over all impact areas. In the same way a utility was assigned each index using the utility weights listed in Table 2. The total value was calculated for each index by multiplying impact by utility.

TABLE 2 [a]

Impact/Utility Index Analysis

Impact		Utility		Relevance scale	
Human death	0.20	Scientific	0.25	None	0.0
Ecosystem disaster	0.18	Administrative	0.25	Slight	0.1
Human health	0.12	Direct CEQ	0.50	Moderate	0.5
Species threat	0.10		1.00	High	1.0
Economic loss	0.10				
Food crop loss	0.10				
Plant damage	0.06				
Discomfort	0.06				
Recreation	0.04				
Aesthetics	0.04				
Total	1.00				

[a] Impact x Utility = Value.

8. DEVELOPMENT OF ENVIRONMENTAL INDICES

The index cost analysis is illustrated in Table 3. Each index was evaluated as to the condition of the system needed to acquire the data for the index calculation.

Perhaps a more meaningful consideration than mere rank ordering is to evaluate the indices on a cost vs value basis. This approach lends itself very well to graphical analysis, as in Fig. 1. The graph is divided into zones as an aid to providing a relative importance ranking. Detailed development and implementation should concentrate initially, for example, where the cost/value ratio is less than 2.5 (the slope of the line separating Zones II and III). Zone II contains indices where the cost/value ratio ranges from 2.5 to 5.5. Although many of the indices in this zone may be described as marginal, there are some important indices included: urban green, mass gap, and ecocycle normality. These latter indices and their seemingly anomalous position on the graph only serve to point out the tentative nature of this "first-cut" analysis. Zone I contains indices whose cost/value ratio is greater than 5.5. Based on the present value analysis, it is doubtful whether significant resources should be devoted to indices in this zone. They should be reappraised periodically, however, to determine if priorities need to be upgraded.

The areas in Zone III of Fig. 1 may be subdivided to evaluate the higher priority indices according to their importance. Importance cells are numbered in circles from 1 to 25 in Zone III. The ranking procedure is lowest cost within highest value, starting first with the highest value (cell 1). In Table 1 only the indices in the most important group, Zone III, were assigned importance rankings; others have a ranking greater than 58.

As a final note, it should be pointed out that any analysis like the one presented here relies heavily on value judgments throughout the process. The procedure itself is relatively straightforward and useful. The final rankings and zonings will, of course, require the knowledge, judgment, and, most important, cooperation of other experts.

3. EXAMPLE: AIR QUALITY INDEX

This Section illustrates computational algorithms and data tabulation in the case of the air quality index.

3.1 General Development

The air quality index is defined as

$$Q_A = 1 - P_A, \tag{1}$$

where P_A is an index of air pollution. The air pollution index is defined without regard to synergistic effects, which occur as a result of reactions between two or more pollutants. The keys to determining P_A are the index standards for each pollutant, which need not correspond to legal standards.

One method of determining index standards is to relate them to points on a cumulative distribution curve for each pollutant. For example, let

S_{i1} = standard concentration at 50th percentile for pollutant i

S_{i2} = standard concentration at 84th percentile (one standard deviation) for pollutant i

S_{i3} = standard concentration at 99th percentile (two standard deviations) for pollutant i

represent three index standard points. (The best percentile points to select depend on the specific distribution.)

If m_{iK}, $K = 1, 2, 3$, correspond to measured concentration values of pollutant i for the Kth-percentile index, then a pollution index for pollutant i is computed as

$$P_{Ai} = \frac{1}{M} \sum_{K=1}^{3} b_{iK} \frac{m_{iK}}{S_{iK}}, \qquad (2)$$

TABLE 3

Index Cost Analysis

Data acquisition system condition	Cost estimate (relative)
Exists in desired form	0.05
Exists but needs reformatting	0.1
Exists but needs expansion	
Slight	0.2
Moderate	0.3
Major	0.5
Does not exist	
Minor system	0.6
Medium-size system	0.8
Major implementation	1.0

8. DEVELOPMENT OF ENVIRONMENTAL INDICES

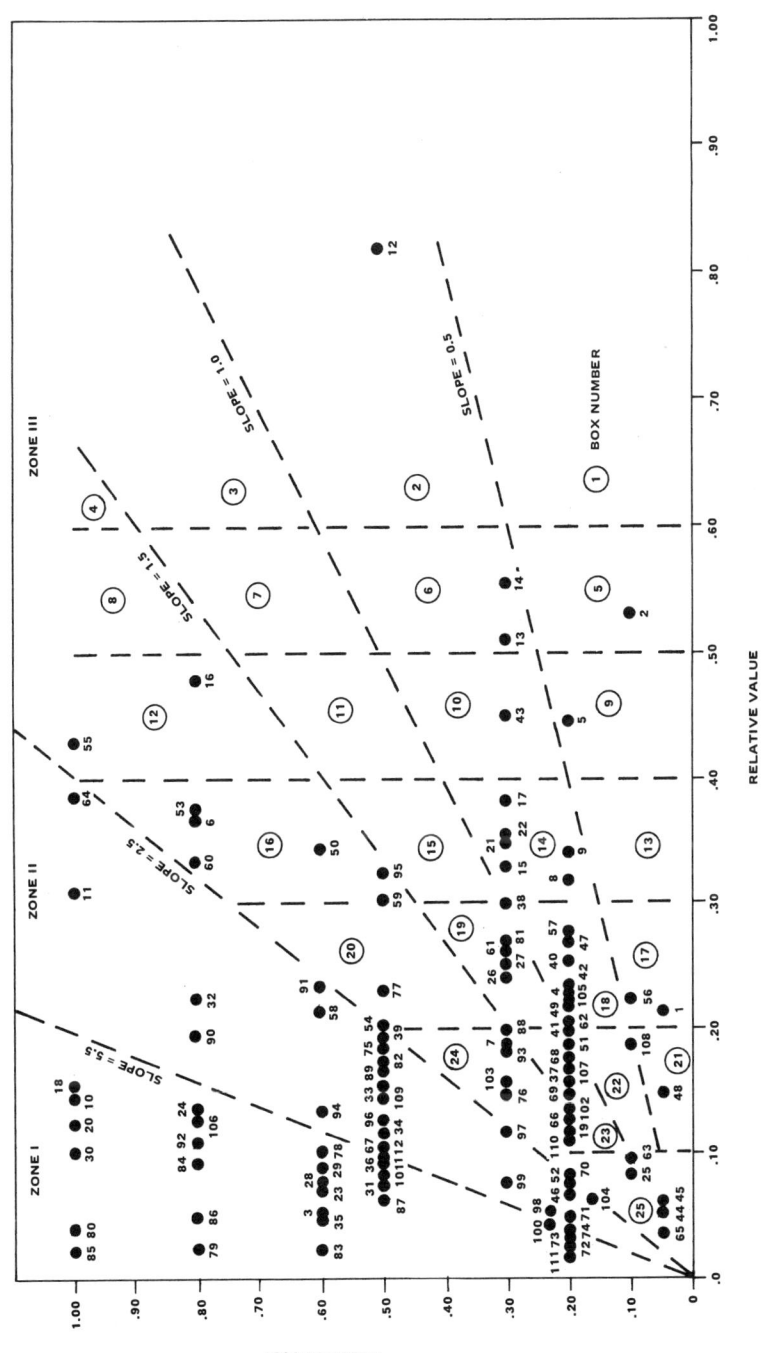

Fig. 1. Cost vs value.

where the b_{iK} are relative weights assigned to each percentile value ($\Sigma_K\, b_{iK} = 1$), and M is a factor that ensures that P_{Ai} does not exceed unity. If each measured value were equal to the corresponding standard value ($m_{iK} = S_{iK}$), then the weighted sum $\Sigma_K b_{iK} m_{iK}/S_{iK}$, would equal unity and the value of P_{Ai} would be $1/M$. In general, however, a value of $1/M$ for P_{Ai} simply indicates that the weighted ratios ($b_{iK} m_{iK}/S_{iK}$) add up to unity, and not necessarily that each measure coincides with the standard. For P_{Ai} in excess of $1/M$ at least one m_{iK} exceeds the standard, and this tendency dominates the index value. Similarly, if P_{Ai} is less than $1/M$, at least one m_{iK} is less than the standard, and this tendency dominates the index value.

The value of P_{Ai} computed according to Eq. (2) provides an index of pollutant i at a particular location. Values at different locations may be weighted by relative population and added to provide an index of pollutant i over a geographical area.

To obtain an overall index of pollution (P_A), one could simply average P_{Ai} over all pollutants:

$$P_A = \frac{1}{n} \sum_{i=1}^{n} P_{Ai}. \quad (3)$$

The air quality index would be computed from Eq. (1) by substitution from Eq. (3).

3.2 Sample Calculation

Index standards have been arbitrarily defined for the six primary pollutants shown in Table 4. The values do not correspond to the present standards announced by the Environmental Protection Administration (EPA), but the proposed methodology can be modified easily to include any set of standards.

Table 5 shows results of applying the methodology of the previous paragraph to three cities for which annual data were available. Two cases show the results of equal weighting ($b_{iK} = 1/3$) and of an unequal weighting where greater emphasis is placed on the higher percentile values, corresponding to acute or episodic conditions ($b_{i1} = 0.2$, $b_{i2} = .3$, $b_{i3} = .5$). The normalization factor M employed in these computations has been set at 10, so that the overall pollution index is 0.1 and the overall quality index is 0.9 when all measurements equal standard values.

8. DEVELOPMENT OF ENVIRONMENTAL INDICES

TABLE 4

Illustrative Index Standard Values

i	Pollutant	Index standards		
		S_{i1}	S_{i2}	S_{i3}
1	SO_2 (ppm)	0.02	0.04	0.10
2	CO (ppm)	2.0	3.3	10.0
3	oxidants (ppm)	0.02	0.03	0.08
4	NO_2 (ppm)	0.05	0.10	0.20
5	hydrocarbons (ppm)	0.10	0.14	0.30
6	particulate ($\mu g/m^3$)	70	110	180

The following statements may be summarized from the results of Table 5.

Pollution severity P within a city differs for each pollutant.

Comparisons between cities depend on pollutant. For example, City B has a greater SO_2 problem than either of Cities A or C. If CO is used, however, City C is worse than both A and B.

The numerical value of the aggregate quality index depends on the component pollutant indices included in the aggregation. If a consistent pollutant set (with four or five pollutants) is selected, however, the rank ordering among different locations is relatively less sensitive to different weightings.

Particulates were excluded in one computation for Q_A because measured values were available for only one city, and a consistent set of components should be retained for comparative purposes among cities. Hydrocarbons were excluded because some specialists have suggested that these are important only insofar as they contribute to photochemical creation of oxidants and less significant from the standpoint of direct contribution to a health hazard.

More experimentation and experience is required with values resulting from a specific form for the quality index before generalized statements (similar to those made for Table 5) can be made with a high degree of confidence.

TABLE 5

Comparison of Air Quality for Individual Cities

Indices	Equal weight [a]			Unequal weight [b]		
	CITY A	CITY B	CITY C	CITY A	CITY B	CITY C
SO_2 (P_{A1})	0.009	0.260	0.080	0.01	0.289	0.086
CO (P_{A2})	0.339	0.233	0.423	0.406	0.229	0.420
Oxidants (P_{A3})	0.150	0.143	0.250	0.161	0.159	0.300
NO_2 (P_{A4})	0.035	0.136	0.209	0.034	0.149	0.384
Hydrocarbons (P_{A5})	0.643	0.480	0.800	0.786	0.524	0.860
Particulate (P_{A6})		0.160			0.163	
Q_A	0.76	0.76	0.65	0.72	0.75	0.59
Q_A [c]	0.76	0.75	0.65	0.72	0.73	0.59
Q_A [d]	0.87	0.81	0.76	0.85	0.79	0.70

[a] $b_{iK} = 1/3$.
[b] $b_{i1} = 0.2$; $b_{i2} = 0.3$; $b_{i3} = 0.5$.
[c] Exclusive of particulates.
[d] Exclusive of particulates and hydrocarbons.

8. DEVELOPMENT OF ENVIRONMENTAL INDICES

ACKNOWLEDGMENTS

The material in this paper is based on a study performed by The MITRE Corporation for the Council on Environmental Quality (CEQ), relating to development of a system design concept for monitoring the environment of the nation. The author supervised the technical effort. The study reflects a cooperative contribution of many MITRE staff members, and contributions in particular areas are difficult to assign to specific individuals. Some of the concepts relating to environmental indices presented in this paper reflect not only the author's work but also that of Dr. Charles A. Bisselle, Steven N. Goldstein, Dr. Richard S. Greeley, Martha J. Lilienthal, and William D. Rowe, all members of the MITRE Technical Staff. Any conceptual deficiencies resulting from the summarization of their original work in this paper are the sole responsibility of the author.

REFERENCES

1. L. R. Babcock, PINDEX, J. Air Pollut. Contr. Ass., 10, 653-659 (1970).

2. R. N. Brown et al., A water quality index--do we dare?, paper delivered at Natl. Symp. on Data and Instrumentation for Water Quality Mgt., Univ. Wisconsin, July 1970.

3. 1970 National Environmental Quality Index, Natl. Wildlife Mag., October-November 1970.

4. R. P. Pikul, C. A. Bisselle, and M. Lilienthal, Development of environmental indices: outdoor recreational resources and land use shift, in Indicators of Environmental Quality (W. A. Thomas, ed.), Plenum Press, New York, 1970.

5. Statistical Abstract, 1970, U. S. Dept. of Commerce, Bureau of the Census.

9

A STATISTICAL SYSTEM FOR ESTIMATING THE DETERIORATION
OF THE HUMAN ENVIRONMENT*

R. Hueting

Netherlands Central Bureau of Statistics
The Hague, Netherlands

1. ENVIRONMENTAL GOODS: FUNCTIONS BY COMPONENT

The development of statistics on the deterioration of the human environment at the Netherlands Central Bureau of Statistics (C.B.S.) is based on the classic division of the environment into the components (media) water, air, and soil. Each of these components has many uses. A use of an environmental component that has utility for man is called a function. As a result of social development (population growth, increase in the amount of goods and services produced per capita), each of these components is falling increasingly short of supplying existing needs.

Initially a rough division of the functions of the environmental components has been made. Thus the functions of water are divided into drinking, processing, cooling, agriculture, recreation, water in the natural environment, navigation, water as an element in the social environment, construction, and water as a dumping ground for waste products. The functions are split up in more detail as the system is elaborated. Each of the functions mentioned fulfills different, specific needs and conversely makes its own specific demands regarding quality of the water. The functions of air

---

*This paper is a summary of part of the author's thesis, which is to be published in Dutch at the end of 1973. For the greater part, this thesis refers to an investigation currently being carried out at the Environment Section of the Netherlands Central Bureau of Statistics. It is also intended to publish an English version of the thesis.

are, among others, as a factor in human life (where loss of function may take many forms, for example irritation of the respiratory organs through inhaling polluted air), as a factor in agriculture, and as a medium for storing matter (where loss of function may consist of accelerated corrosion).

Deterioration of the environment is defined as <u>decreased availability of functions of an environmental component</u>, or, in other words, as <u>loss of function</u>. Loss of function has apparently arisen because the increasing burden on the environment has placed such a great strain on the different environmental functions that they have been thrown into competition with one another. Thus the strain placed on a certain function by any human activity is at the expense of another function. It is therefore obvious that by checking where <u>competition of functions</u> takes place in an environmental component, one can trace the losses of function that occur.

2. QUANTITATIVE, SPATIAL, AND QUALITATIVE COMPETITION OF FUNCTIONS

An environmental component always has three aspects: quantitative (the amount of the component), spatial, and qualitative (the degree of pollution).* The decreased availability of a function may relate to each of these three aspects. Therefore the competitive use of functions is divided into quantitative, spatial, and qualitative competition.

A situation where the quantity of a component is inadequate for its use or its intended use is known as <u>quantitative competition of functions.</u> The functions here are placed in direct confrontation with each other. Quantitative competition is absolute; if a drain is placed on one particular function this completely prevents the use of any other function. It takes place in the component water by actual withdrawal of the water from the environment.

When there is not enough space for the use or the proposed use of functions, one refers to <u>spatial competition of functions</u>. This is especially prevalent with the component soil. The investigation at C.B.S. has not yet covered this component.

When a qualitative change is brought about in water, air, or soil, resulting in the component becoming less suited for many different uses, <u>qualitative competition of functions</u> occurs. Qualitative changes may be brought about in two ways: first, by disposing of refuse directly into the

---

*Note that, in this paper, "qualitative" always refers to a quantitative measure of quality, not, as often, simply to a difference in kind. [Ed.]

9. ESTIMATING ENVIRONMENTAL DETERIORATION

soil, draining off waste matter directly into the water, emitting materials directly into the air, or causing excessive noise; second, as a result of unintentional or "unavoidable" side effects when utilizing other functions. Both cases of qualitative competition are classified together as one separate function. In water this function is called water as a dumping ground for waste matter. In air it is called air as a medium in which to release waste matter (this includes noise nuisance as well as emissions of harmful substances).

The utilization of this function decreases the availability of other functions. In other words, the function of dumping ground for waste matter competes with the remaining functions. This competition is indirect, and occurs by way of qualitative change, which is the vector of the competition. Qualitative competition is gradual; loss of function depends on the degree of change. Both change and loss of function may be described in natural, scientific, or technical terms, as may measures for both elimination and compensation, i.e., for both recovery and substitution of functions. While tracing the losses of function it became evident that the utilization of water and air to accommodate the waste products of human activities leads to a decreased availability of all their other functions.

In order to present a complete picture, the competition of functions has been represented in tabular form (see Fig. 1) by arraying the different functions of an environmental component against one another in a table with identical row and column classifications. It was then investigated and indicated in the cells of the table where utilization of a certain function reduces the availability of any other function (in other words, where losses of function occur). Such losses can result from quantitative, spatial, or qualitative competition. As already stated, the first two types result directly from the confrontation of functions. All cases of qualitative competition result from the utilization of the function of dumping ground for waste matter in any form, which is indicated on the last line of Fig. 1a.

The differences between the three kinds of competition make it desirable to derive from this table, for each component, three tables or sets of tables for quantitative, spatial, and qualitative competition, respectively. The losses of function resulting from both quantitative and spatial competition are classified in tables having the form of Fig. 1b. In each such table the functions are in direct competition with each other. It can therefore be immediately ascertained which losses of function occur and which activities cause the losses. The next stage is to investigate what measures can compensate for these losses of function and what sums of money they involve. One example is the withdrawal of large quantities of fresh groundwater for the drinking-water supply, which will lower the groundwater level. This in turn will reduce agricultural yield, which will have to be compensated for by extra irrigation.

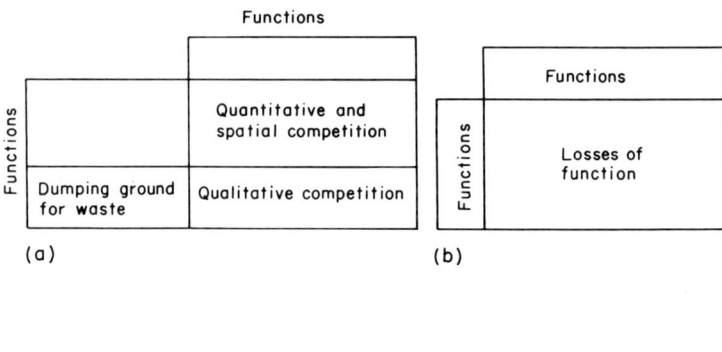
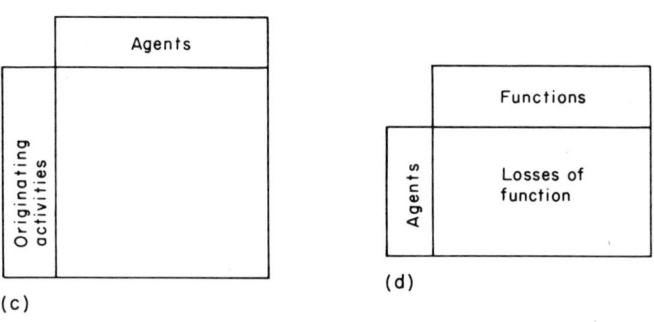

Fig. 1. Schematic tables of interactions between functions of an environmental component (a and b), between environmental agents (physical, chemical, or biological) and their originating activities (c), and between functions and agents (d).

In the third set of tables, which show the losses of function due to qualitative competition, the functions do not compete directly. There is an intermediate stage: a human activity may introduce an agent into the environment. By agent, in this context, is meant an element (chemical substance, plant, animal, heat, etc.) that may cause loss of function, either by its addition to or withdrawal from the environment. The agent can trigger a change in the environment which results in loss of function.*
In the tables, agents and changes are viewed collectively and not separately. The agent is mainly used for nomenclature.

The agents are classified as biological, chemical, and physical. In water, for example, we find pathogenic organisms, animals and plants, as biological agents. The terms "animals" and "plants" refer here to

---

* The effect of the agent will not be eliminated by the environment itself after a short while (short in relation to the length of the process of disturbance brought about by the change).

9. ESTIMATING ENVIRONMENTAL DETERIORATION

either the addition of organisms to the environment or their withdrawal from it. Among the chemical agents are biodegradable organic matter, inorganic toxic substances, saline substances, and eutrophicating matter. Physical agents include heat (added or withdrawn), radioactive materials, and silt. The different originating activities are arrayed against the agents in the Fig. 1c. With the help of this table the C.B.S. investigation has traced where a given activity introduces a certain agent into the environment.

In the Fig. 1d the different agents are arrayed against the functions of a component.

This method of classification (the separation of a number of attacks on the environment, with many interrelated connections, into separate cases of environmental pollution) in fact stretches the truth. The method is needed if one is to arrive at some kind of orderly quantification, yet the results cannot fully indicate the whole deterioration of the environment: the sum of all the cells in the tables is not equal to the total deterioration. This is particularly apparent when one calculates the costs of elimination. Elimination of an agent often infringes on other functions, either qualitatively, quantitatively, or spatially. The method stated here must therefore be completed with material balances. Furthermore, the C.B.S. investigation only examines that which has been observed or can be derived from observations. An example which may clarify this limitation is the case of pesticides brought into the environment. Pesticides will be absorbed by different organisms and will be passed on via the food chain to other organisms, where they will reach higher concentrations. In birds of prey, which are at the end of a food chain, this has lead to mortality; the consequences for man, who is also at the end of a food chain, are still unknown.

A final example of loss of function as a result of qualitative competition should be mentioned here. The depositing of the agent biodegradable organic substances into water (originators being industry and population, among others) leads, via an increased concentration of this matter, to a scarcity of oxygen in the water. Losses of function here are, for example, increased mortality among fish and, in very serious cases, putrefaction ("dead water").

3. INDICATORS

Progress of the qualitative deterioration of the environment manifests itself in many ways. The following indicators can be used:

1. emission of waste matter over a given period;

2. concentration of pollution in the environment (immission);

3. effects of pollution (e.g., increased mortality among fish, disappearance of oxygen-loving water organisms).

Indicator No. 1. This indicator's statistical reflection is the total emission per period. The level of emission caused by man can be raised, for example, by increasing the activities of production and consumption or by less careful behavior. Emission can be reduced by taking all kinds of measures to remove the agent from the source (including reduction of the polluting activity itself).

Indicator No. 2. The emitted materials dispersed in the environment can sometimes undergo changes and will reach a certain level of concentration. This concentration can be measured or calculated. One can measure the total concentration, whether natural or induced by man. The part caused by man, to which the C.B.S. investigation refers, must be estimated. The total concentration in the air of, among other things, hydrogen sulfide, ozone, and methane, and the total concentration in water of, say, salt, biodegradable organic materials, and phosphates, are partly caused by natural processes. When no measurements are available, the concentration caused by man can be calculated from emissions, using a rough model. Also, the existing data on concentrations can be gathered and reproduced in statistical form.

Indicator No. 3. Concentration of a pollutant disturbs utilization of an environmental component: it causes loss of function. But the degree of concentration is only one of the variables that determines the extent of this loss; other variables are temperature, humidity of the air, etc. Because of this fact, the connection between concentration and loss of function is not easily seen. It is therefore useful to deal with a third indicator: the effects of pollutants in an environmental component. This third indicator is to be seen as the last link in the chain of indicators. It gives directly the quality of the environment and may also serve as a means of tracing emitted matter, the harmful effects of which are still unknown.

4. SHADOW PRICES OF ENVIRONMENTAL FUNCTIONS

When data on materials are known a cost-benefit analysis can be attempted. Ideally the aim should be to provide an analysis for each function. In practice this is seldom feasible. More feasible is a cost-benefit analysis for one agent and some of the functions affected by the

9. ESTIMATING ENVIRONMENTAL DETERIORATION

agent. For simplicity the following supposition is made: a given function is affected by one and only one agent. That function is not influenced by other agents and the agent is not detrimental to other functions. Naturally this situation will seldom or never occur in practice. The supposition is made only to simplify the theoretical explanation.

Loss of function as a result of qualitative competition can be restored by elimination measures. These may consist of making various provisions or reducing the polluting activity. Elimination is defined here as the removal of the agent from the source, resulting in total or partial recovery of the function.

The costs of purification are, of course, offset by the benefits gained by recovering the function. In a number of cases these benefits can be partially measured by the market. The benefits include reduction both in the cost of compensatory measures taken in response to the loss of function and in the financial damage resulting from the pollution. Compensation is defined here as the making of alternative provisions for total or partial recovery of the function. Compensation is nearly always far less effective than elimination; it results mostly in the creation of a substitute for the original function. Financial damage is defined here as the deterioration of the position of income and property in private households, industries, or government resulting from loss of function.

By adopting measures to eliminate the agent, compensatory measures that have already been put into practice will become totally or partially superfluous, and financial damage will be reduced.

The costs and benefits of recovery of function can be displayed graphically (Fig. 2). A scale is adopted for the abscissa, giving the degree of purity of the environmental component. No particular assumptions are made about this scale except that the numerical value increases with increasing purity, i.e., with reduced concentration of the agent in the environmental component. The ordinate stands for both the elimination costs (curve E) and the sum of compensation costs and financial damage (curve C + D).

Under the extreme assumption - which will seldom hold in practice - that the C + D curve reproduces the total of the preferences for the function, the optimum degree of purification occurs where the sum of the curves C + D and E shows a minimum (curve T).

This information can also be displayed in marginal curves (Fig. 3). The ordinate then stands for the annual cost per extra unit of purity. The point of intersection G of the marginal elimination curve e and the marginal

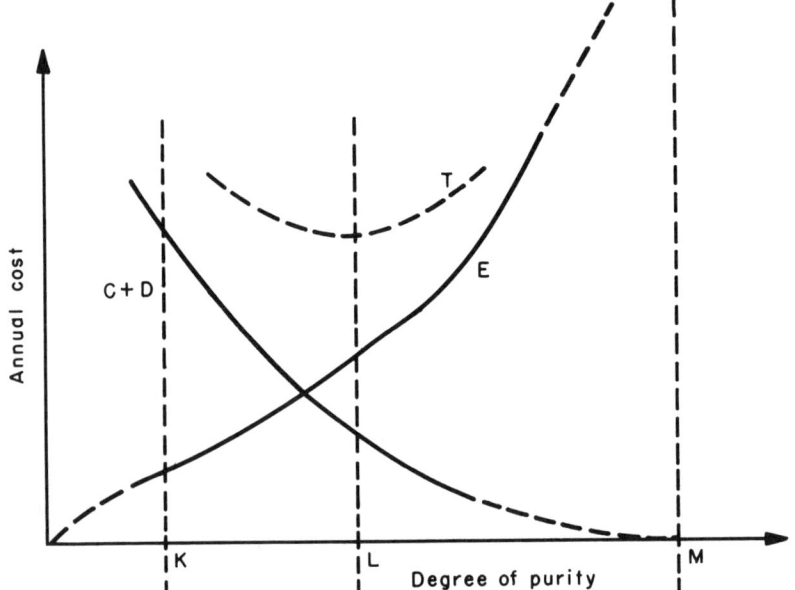

Fig. 2. Cost of recovery of function.

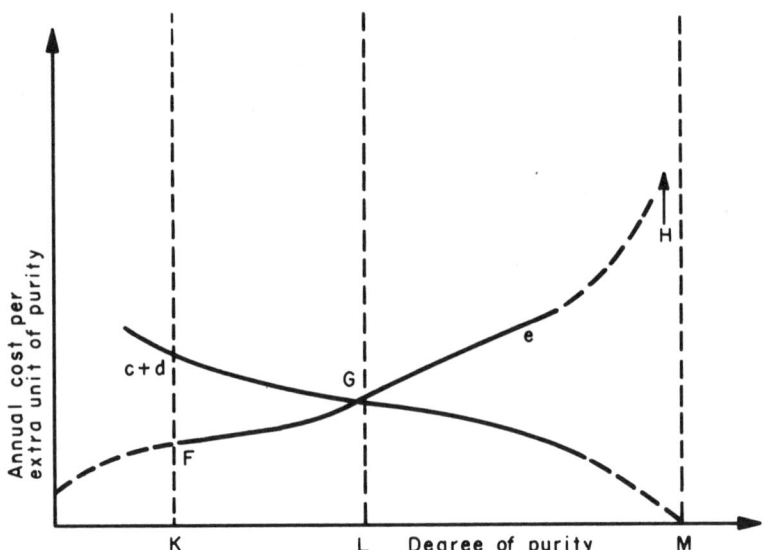

Fig. 3. Marginal cost of recovery of function.

9. ESTIMATING ENVIRONMENTAL DETERIORATION

compensation-plus-damage curve c + d corresponds, of course, to the minimum point of curve T in Fig. 2. (To make c + d positive, we define it as the negative of the derivative of C + D). The marginal costs and benefits of function recovery are equal at this point. The point shows also the shadow price per unit of purification at the optimum degree of purification. The shadow price of an environmental function is equal to the marginal elimination cost at the optimum recovery of function.

Figure 3 once again draws attention to the fact that pollutant production and environmental functions can be regarded as each other's opportunity costs. Purification from K to L involves sacrificing alternative uses of the resource KLGF for the sake of the recovery of the function. The remaining loss of function LM is sacrificed for the sake of the remaining polluting activities; area LMHG represents the sacrifices still required to recover the function completely.

It is clear that only to a slight extent are losses of uses of the human environment compensated for or experienced as financial damage. Double-glazed windows may reduce traffic noise inside houses, but not in their surroundings. Stench is almost impossible to avoid. Damage from noise nuisance and air pollution is very poorly reflected in property prices. The loss of clean beaches can scarcely be compensated for by artificial swimming pools. The loss of flora, fauna, and landscapes and the gloomy promise for the future cannot be measured in terms of compensation or in terms of financial damage.

This means that the shadow price of most environmental functions is higher, in some cases much higher, than the point of intersection of curve e and curve c + d (Fig. 3). Additional preferences x must be added to curve c + d. The point of intersection of curve e and the curve c + d + x represents the real shadow price of the functions.

The following three approaches to evaluating this x are open in principle.

1. With the aid of market data an attempt can be made to derive the preference for a function from the behavior pattern regarding an unaffected function.

2. One may attempt to express the preference for a function in monetary terms by asking subjects what they would spend for recovery of the function, what they would give in order to be spared loss of the function, or what they would consider adequate compensation for an existing loss of function.

3. The utility of a function or the consequences of loss of the function can be described. To this end the government - bearing in mind the elimination costs of a given degree of purification - can fix a point on curve e. Lack of space precludes going into these methods further here.

Since methods 1 and 2 in particular have great limitations, in practice the government will usually have to fix the amount of x. When doing so there are many reasons to regard environmental functions as merit goods. Again further detail cannot be given here.

REFERENCES

The following publications resulting from the C. B. S. investigation have been completed:

1. on elimination costs: "Water pollution by biodegradable organic and eutrophicating materials," 120 pages, 1972.

2. on elimination costs and compensation-plus-damage costs: "Water pollution caused by salinity," 30 pages, 1973.

Other publications are planned. Among the topics are:

1. elimination costs and compensation-plus-damage costs: air pollution by fossil fuels;

2. emission quantities: data regarding cattle dung polluting the water;

3. general statistics on indicators in the environment.

Part III

MATHEMATICAL MODELS

10

PENALIZATION OF THE ENVIRONMENT DUE TO STENCH:

A study of the subjective experience of the population

L. A. Clarenburg

Rijnmond Authority
Rotterdam, The Netherlands

0. SUMMARY

In this paper a mathematical model is developed with the aid of which the subjective appreciation of the population living in the vicinity of chemical plants can be described quantitatively. The study is centered around the relation between odorous air pollution and health, in particular mental well-being. To express the latter concept in an operational way, a new notion is introduced, namely, penalization of the environment due to stench, which is defined as the percentage of the population capable of perceiving odorous pollution.

A mathematical model is derived based on the assumption that the great number of emissions of odorous compounds, even those seeming completely unimportant, on a chemical industrial terrain, determines the appreciation of the population for the quality of the air. The model is applied to the situation in five areas. Results of calculation are compared with results of public polls and with the complaint pattern as registered by the Central Complaint Office of the Rijnmond Authority. The agreement between calculation and observation is highly significant in the statistical sense. This permits the drawing of some policy-related conclusions concerning a directed abatement of odorous pollution, improved regional planning, and a standard for odorous pollution.

1. INTRODUCTION

The Rijnmond area is a densely populated area, having in its heart one of the most impressive industrial agglomerations in the world. With the population living in the immediate proximity of the petrochemical and chemical industry, it is not surprising to find an air pollution problem here.

Some years ago the Rijnmond Authority installed a central telephone, where the inhabitants of the area can deliver their complaints about air pollution and noise. During 1970 over 17,000 complaints were received, the majority related to odorous pollution. Thus it makes sense to consider the problem of air pollution as a subjective experience of the consumers of the air, the population; the first task is to diagnose the origin of the displeasure signalled.

Approaching the problem of air pollution from the consumers' point of view corresponds closely with the definition of health given by the World Health Organization: a condition of physical, mental, and social well-being. In the light of recent sociological studies stench appears to be "mentally irritating." As the presence of stench violates mental well-being, an approach based on the subjective experience of the population may contribute to achieving new insights useful for regional planning. Hence the second task is to provide materials for conducting regional planning more than hitherto in tune with the notion of health. If these tasks are carried out successfully it will prove relatively easy to formulate air quality standards for odorous pollutants.

2. PENALIZATION

In an important industrial area, stench is a composite name for an unknown multitude of odorous compounds. Many of these compounds cause nuisance even in unmeasurably low concentrations. In "normal" situations it is hard to distinguish which odor is perceived predominantly. Odors are always mixed, as evidenced by the wide variety of complaints about the same situation. Therefore, in order to relate any one analysis of the complaint pattern to actual perception, one can profitably model a situation by developing a description of the simultaneous action of all odorous pollutants.

To do so a new notion is introduced, penalization of the environment due to stench, defined as follows:

10. PENALIZATION DUE TO STENCH

<u>The penalization of the environment due to stench in an area is equal to the percentage of the population living in that area capable of perceiving it.</u>

This definition requires some clarification:

1. The word "stench" is used in a composite sense, meaning all odorous pollutants;

2. As the concentration and the composition of the joint pollutants may vary greatly from place to place (and from time to time), the "area" is thought of as having (very) limited dimensions. If "penalization of a residential district (or a town)" is used, the average penalization is meant, with possibly appreciable deviations from this mean within the area considered;

3. In this paper "penalization of the environment" applies to the situation in the area considered, averaged over a longer period of time (e.g., one year). Depending on the meteorological conditions, the penalization during shorter periods of time (hours, up to one day) may deviate greatly from the mean situation;

4. The penalization, as used in this paper, should thus be considered as the mean of all conceivable penalizations, if the wind were to blow 100% of the time from the industry; consequently, penalization as used here has in the first instance an arithmetic significance;

5. The frequency with which a given penalization occurs at a given site depends entirely on the distribution of the wind direction;

6. The penalization of the environment due to stench depends only on the ability of the population to perceive an odor; however, the number of complaints connected with a given degree of penalization is variable from year to year, as it depends on the tolerance level of the population.

3. MATHEMATICAL MODEL

Consider an industrial terrain of dimensions X parallel to the mean wind direction and 2Y laterally. On such a terrain odorous compounds escape into the free atmosphere at tens of thousands of locations, through leakage, spillage, evaporation, etc. Each separate emission might be completely harmless, but what is the effect of the very great number of such emissions? The first and most fundamental assumption, hereafter referred to as the basic assumption, is that the great number of emissions of odorous compounds, even those seeming completely unimportant, on a chemical industrial terrain, determines the appreciation of the population for the quality of the air. The other assumptions are as follows:

1. The sources of odorous air contaminants are in a statistical sense homogeneously spread over the entire surface area of the terrain considered. This means that in the direction of the wind a mean distance a can be assigned to the mutual distance of sources, and a mean distance b laterally;

2. The emission rates of the individual sources follow a statistical distribution, and the mean source strength $\overline{Q}$ can be defined;

3. The emissions take place at or near ground level;

4. Although the perception of various pollutants differs, it is assumed that one overall perception function can describe the sensitivity of the population to the mixture of pollutants "normally" present in the air in the vicinity of large industrial complexes.

Let the origin of the coordinate system be O (see Fig. 1). Assuming the Gaussian plume model for the dispersion of the odorous components in the atmosphere [1,2], then the joint contribution of all sources to the concentration at a point ($\hat{x}$, 0, 0) is found by integrating over the sources in the y direction and by subsequent integration over the rows of sources in the x direction. The approximate result for $\hat{x} = X + x$, is

$$C_{tot}(X+x, 0, 0) \approx \frac{2}{\sqrt{\pi}} \frac{\overline{Q}}{n_z \overline{U} C_z ab} [(X+x)^{n_z} - x^{n_z}], \tag{1}$$

where n_z is a parameter indicating the vertical stability of the atmosphere, C_z is a measure of the dispersivity of the atmosphere in the vertical direction, $\overline{U}$ is the mean wind velocity, and x is the distance from the downwind edge of the terrain considered.

The concentration downwind from a point source is:

$$C(x, 0, 0) = \frac{2\overline{Q}}{\pi \overline{U} C_y C_z x^{2-n_z-n_y}} \tag{2}$$

According to the literature [3-6] the probability of perception of any one odor by a group of people is given by a log normal distribution function for the just perceptible concentration C,

10. PENALIZATION DUE TO STENCH

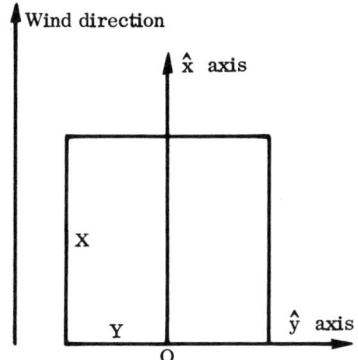

Fig. 1. Schematic diagram of an industrial terrain.

$$P_p(C)\, d \ln C = \frac{1}{\sqrt{2\pi} \ln \sigma_g} \exp \left[-\frac{(\ln C - \ln C_{th})^2}{2 \ln^2 \sigma_g} \right] d \ln C. \qquad (3)$$

Here C_{th} is defined as the threshold concentration perceived by 50% of a group of people, in other words, C_{th} is the median of the perception distribution. Hence σ_g is the geometric standard deviation of the distribution, which should be regarded as an odor characteristic, assuming different values - as does C_{th} - for different odorous compounds. It will be assumed here, however, that the way in which the mixture of odorous compounds emerging from a large industrial terrain is experienced by the population can be described by one overall perception function, an assumption not contradictory to the results obtained by Lindvall [6].

Equation (3) was established in a laboratory, and is based on experimental results with a very limited group of persons. The tail of the distribution function, related to the most sensitive part of the population, deviates from log normality; therefore Eq. (3) does not hold beyond distances downwind where C is less than some value C_{lim}, in which case $P_p(C) \approx$ constant.

An estimate can now be made of the percentage of the population living somewhere downwind of an industrial terrain and capable of perceiving odorous pollution. Let $C(x)$ represent the concentration of odorous compounds at distance x from the downwind edge of the terrain considered. Then the percentage of the population living at distance x capable of perceiving the odor (the penalization), denoted by f(x), is found by integrating Eq. (3) between the limits C = 0 and C(x). Thus

$$f(x) = \int_{-\infty}^{\ln C(x)} P_p(C) \, d \ln C = \Phi \left\{ \frac{\ln [C(x)/C_{th}]}{\ln \sigma_g} \right\}, \qquad (4)$$

where $\Phi(u) = [1 + \text{erf}(u/\sqrt{2})]/2$ is the standard normal cumulative distribution function. Substitution of Eq. (2) yields, for a point source,

$$f(x) = \Phi \left(\frac{2-n}{\ln \sigma_g} \ln x_{th}/x \right), \quad x \leq x_{lim}, \qquad (5)$$

where $n = n_z + n_y$. After substitution of Equation (1) the penalization function downwind of an industrial terrain is found to be

$$f(x) = \Phi \left\{ \frac{1}{\ln \sigma_g} \ln \left[\frac{(X+x)^{n_z} - x^{n_z}}{(X+x_{th})^{n_z} - x_{th}^{n_z}} \right] \right\}, \quad x \leq x_{lim}. \qquad (6)$$

In both equations, x_{th} represents the downwind distance at which C_{th} is found; beyond the distance x_{lim}, at which C_{lim} is found, the equations are invalid. Using the penalization functions (5) and (6) it is already possible to predict the result of a public poll, since it is possible to estimate the percentage of the population capable of receiving odorous air pollution, if the values of the various parameters are known.

From Eq. (6) it follows that on the downwind edge (x = 0) of an industrial terrain, the normal deviate p corresponding to the penalization satisfies

$$\frac{1}{\ln \sigma_g} \ln \left[\frac{X^{n_z}}{(X+x_{th})^{n_z} - x_{th}^{n_z}} \right] \geq p. \qquad (7)$$

Here the value of p can be approximated. Recognizing that on the downwind edge of the terrain considered certainly not less than 95% of a group of persons would perceive the odor, one obtains the value 2 for the lower limit of p. If 100% of the group of persons could perceive the odor, then the upper limit of p would theoretically be infinity. However, it is very unlikely to find p > 5, indicating that more than 99.9999% of the group of persons would perceive the odor. Moreover, the tail of the perception

10. PENALIZATION DUE TO STENCH

distribution deviates from log normality. Therefore the upper limit of p will not be more than 5. Arbitrarily the value p = 3.5 was adopted throughout this study.

It is known [3, 4, 5] that values of σ_g for many chemicals lie between 1.6 and 2.0. Again an arbitrary choice was made; the value σ_g = 1.88 was used throughout this work. In agreement with Gifford and Hanna [7] the value n_z = 0.2 was selected as a long-term average value. Accordingly, considering the symmetry of turbulence at neutral conditions, n = 0.4.

With p, σ_g, and n_z selected in the way described above, and with X known, the values of x_{th} for the various cases studied could be determined with the aid of Eq. (7). Although these cases are quite different, no attempt was made to fit the model to observation by adjusting the parameter values.

4. TESTING OF THE MODEL

4.1. The Available Data

Since a number of political decisions can be based on the penalization functions (5) and (6), it is important that their functional form be well tested. The only direct application of the penalization functions is the prediction of the outcome of a public poll, in which the population of a town is asked whether or not they are bothered by air pollution.

The result of only one such investigation was available in the Netherlands; it is described in Section 4.2. The bulk of the available "experimental" data is formed by telephone complaints received by the Central Office of the Rijnmond Authority.

To illustrate how the complaints about odorous pollution are analyzed, let us consider a living area situated opposite an industrial area (Fig. 2). Complaints may be received when the living area is downwind of the industrial area. Besides many other data, a record is kept of the complainant's address. After about one year the complaints originating from one city are sorted out.

The city is subdivided into residential districts, and both the number of complaints and the number of inhabitants per residential district are determined, so that the number of complaints per 1000 inhabitants (N) can be calculated. Next, lines are drawn of equal distance from the downwind edge of the industrial area. Thus a number of bands are generated. Within each

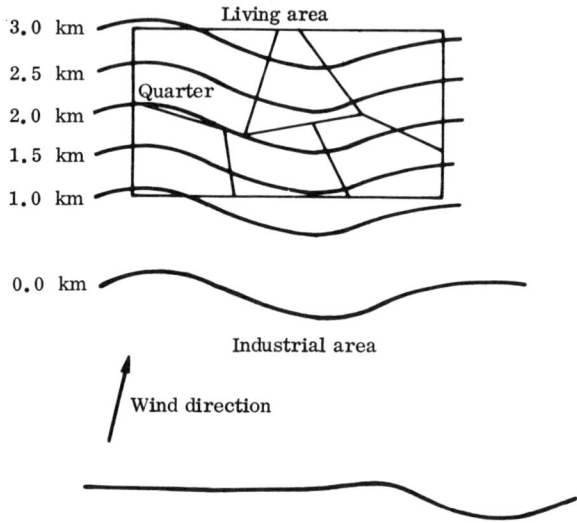

Fig. 2. Schematic diagram of a living area adjacent to an industrial area.

band the number of complaints per 1000 inhabitants is averaged ($N(x)$), taking into account as a weight factor the fraction g of the residential district falling within the distance band considered ($N(x) = \Sigma g N/\Sigma g$). Thus a relation is found between N and the downwind distance x. These relations form the "experimental" material to be analyzed.

However, these "experimental" relations cannot be compared directly to Eqs. (5) and (6); they must be modified. Let the probability of response (i.e., telephoning) after someone has perceived an odor be denoted P_r. It will be assumed here that, provided the period of time considered is long enough, the probability of response is linearly proportional to the probability of perception:

$$P_p(C) = KP_r(C), \qquad (8)$$

where K is a social constant representing the behavior of the population.

Consider a group of 1000 persons. Of these a fraction F_t has the disposal of a telephone. After a period of time of D days, of which the fraction F_w had a wind blowing from the industry, the maximum response

10. PENALIZATION DUE TO STENCH

($P_p = 1$) is clearly $1000 F_t F_w D/K$. However, on the average after a period of D days a fraction $f(x)$ of the population has perceived an odor. As a result, the number of complaints of a group of 1000 persons living at distance x from the downwind edge of a chemical industrial terrain will be, after a period of D days,

$$N(x) = \frac{1000 F_t F_w D}{K} f(x). \tag{9}$$

As has already been outlined, since the penalization function $f(x)$ depends only on the ability of the population to perceive an odor, it is independent of time. The number of complaints, however, depends on the tolerance level of the population, represented by the social constant K in Eq. (9). For a fixed period of time, K can be estimated using Eq. (9):

$$K = 1000 F_t F_w D \, \overline{f(x)}/\overline{N(x)}. \tag{10}$$

Here $\overline{f(x)}$ is the penalization averaged over a town or a part of a town; if the number of inhabitants in a distance band is $I(x)$ and the penalization within that band is $f(x)$, then

$$\overline{f(x)} = \frac{\Sigma \, I(x) f(x)}{I_{tot}}$$

where I_{tot} is the total number of inhabitants in the town considered. Similarly, $\overline{N(x)}$ stands for the mean number of complaints per 1000 inhabitants averaged over a town or a part of a town. If N_{tot} is the total number of complaints from the town considered, then

$$\overline{N(x)} = \frac{N_{tot}}{I_{tot}} \times 1000.$$

It can easily be verified that Eq. (9) can be written as

$$N(x) = \frac{1000 N_{tot}}{\Sigma I(x) f(x)} f(x). \tag{11}$$

It should be emphasized that Eq. (11) was derived only because the bulk of the "empirical" data consist of telephone complaints. The fundamental relations to be verified are Eqs. (5) and (6).

4.2. The Western Mine Area

In the Western Mine Area (Province of Limburg) an important chemical industry is developing, surrounded by residential districts. From Fig. 3 it can be seen that three centers of activity can be indicated: Complex A, where nutrients for plants, animals, and man are manufactured, as well as starting materials for synthetic yarns and fibers, covering an area roughly 1 x 1 km^2; Complex B, where plastics and rubbers and basic materials for synthetic resins are produced, covering an area roughly 0.5 x 1 km^2; Complex C, quarry and stone works, to be regarded as a point source.

In the town of Geleen (36,125 inhabitants) a public poll was taken, including among others the question: "Are you bothered by air pollution?" The sample of the population, drawn at random, consisted of 2000 addresses; of the questionnaires delivered, 537 were recovered. The percentage of the population responding alternatively in the various districts is given in Table 1 (last column). These percentages were also calculated using the Eqs. (5) and (6). To do so, the penalization functions downwind of each complex were determined, always taking into account only the highest contribution of one of the complexes. For example, the calculation of the mean penalization of residential district 3 is

$$f(x)_{calc} = \frac{3}{4}\left[\frac{0.99+0.52+0.99+0.60}{4}\right] + \frac{1}{4}\left[\frac{1.00+0.81}{2}\right] = 0.805.$$

All relevant data are given in Table 1; it appears that the model is quite capable of predicting the results of a public poll well within its confidence limits.

4.3. Living Area around the Gekro Incinerator

The Gekro incinerator is situated in an open space surrounded by relatively wealthy residential districts, starting at about 500 m (Fig. 4). As this incinerator is relatively small, it was treated as a point source. Consequently the bands of equal distance are circles, taken at intervals of 250 m. Within each band the number of complaints received during the period Oct. 1, 1968 to Sept. 25, 1969 and the number of inhabitants were determined. The data are summarized in Table 2. It was assumed that the telephone density is constant over the whole area, not too bad an assumption in view of the social homogeneity.

10. PENALIZATION DUE TO STENCH

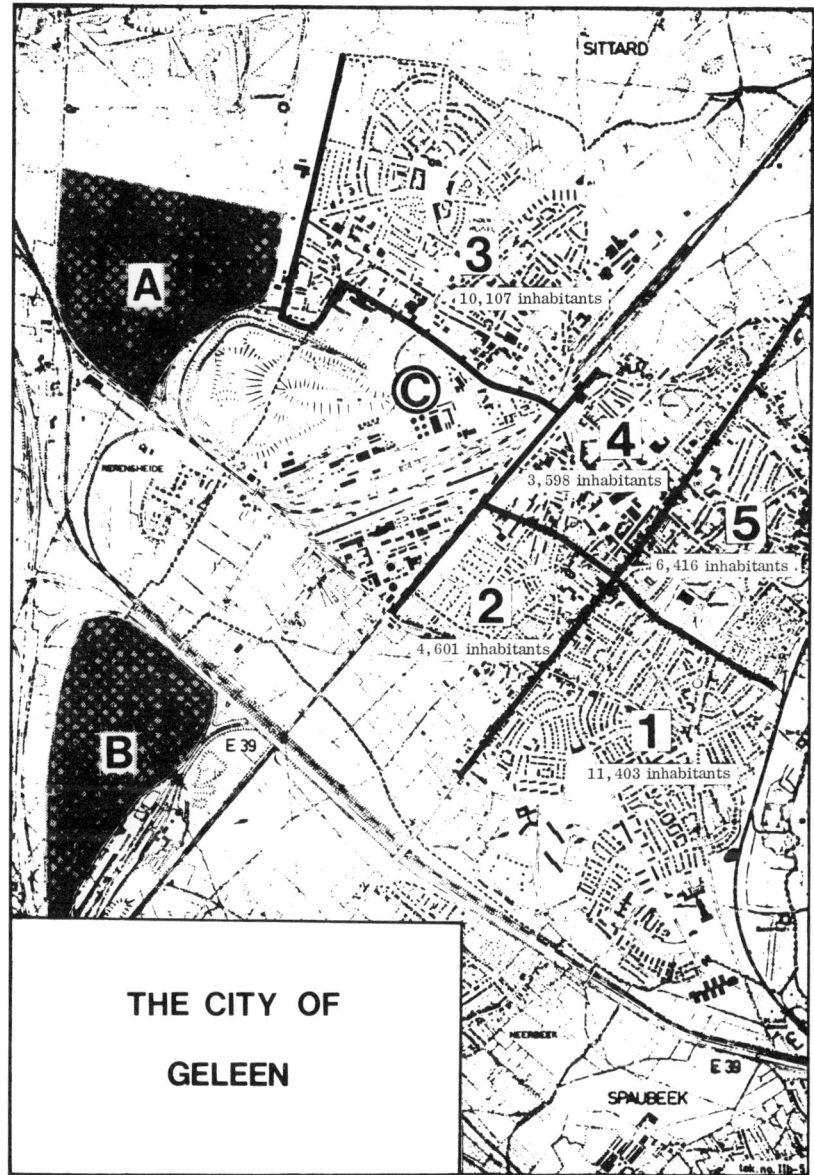

Fig. 3. Map of Geleen, showing industrial concentrations and residential districts.

TABLE 1

City of Geleen: Comparison of Calculation and Public Poll

District	Distance from complex (km)			f(x)			Mean f(x) (%)	
	A	B	C	A	B	C	Calc.	Poll[a]
3	0.1-1.5 0.1-1.2 g=3/4		0.25-0.6 g=1/4	0.99-0.52 0.99-0.60		1.00-0.81	80.5	83 ± 11.5
2	1.5-2.2 g=3/4		0.6-0.85 g=1/4	0.52-0.38		0.81-0.51	50.5	51 ± 13
5	2.2-2.9 2.3-3.1 g=1			0.38-0.27 0.37-0.24			32	35 ± 10
4	1.7-2.4 1.8-2.6 2.0-3.0 g=2/3		0.6-0.95 g=1/3	0.48-0.35 0.46-0.32 0.41-0.26		0.81-0.40	41.5	34 ± 9.5
1	2.2-3.3 2.3-3.5 g=9/10	1.1-1.3 g=1/10		0.38-0.20 0.37-0.17	0.37-0.32		27	26 ± 5
					Weighted average:		47	47.5 ± 4

[a] Two-sided 95% confidence limits indicated.

10. PENALIZATION DUE TO STENCH

TABLE 2

GEKRO: Relevant Data [a]

Distance band (km)	I(x)	Complaints	N(x)	f(x) [b]	I(x) × f(x)
0.5 - 0.75	1871	309	164	0.785	1470
0.75 - 1.0	3743	335	94	0.49	1835
1.0 - 1.25	4912	171	34	0.24	1180
1.25 - 1.5	8103	100	13	0.105	850
1.5 - 1.75	6615	79	12	0.043	285
1.75 - 2.0	3259	6	2	0.018	59
2.0 - 2.25 [c]	(10,030)	(14)	1.4	0.007	(70)
Total:	28,503	1000			5679

[a] $\overline{N(x)} = 1000 \times 1000/28,503 = 35$, $\overline{f(x)} = 5679/28503 = 0.20$.
[b] Mean value of f(x) over the band indicated.
[c] The figures in parentheses are not included in the sum.

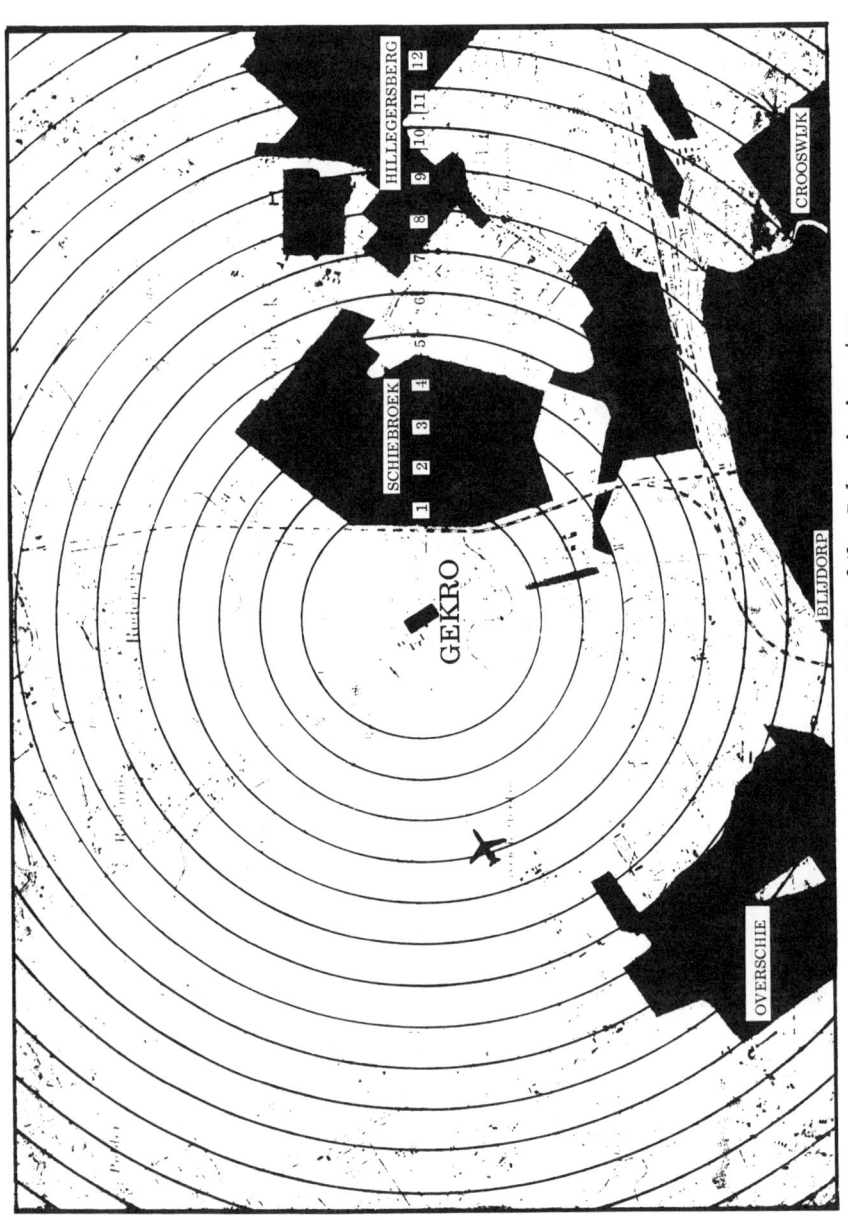

Fig. 4. Map of the vicinity of the Gekro incinerator.

10. PENALIZATION DUE TO STENCH

From Eq. (5) it follows that somewhere downwind the penalization falls under 1%; let this distance be denoted by x_{lim}. Then

$$\frac{2-n}{\ln \sigma_g} \ln \frac{x_{th}}{x_{lim}} = -2.33.$$

From Table 2 it may be inferred that, according to the complaint pattern, $x_{lim} \approx 2125$ m, yielding $x_{th} = 850$ m. Using Eq. (5) the penalization function was calculated; it is also listed in Table 2. With the aid of Eq. (11) the number of complaints as a function of distance was calculated, the result being shown in Fig. 5. The agreement between model calculation and observation is good; it could even have been improved by selecting a somewhat smaller value for σ_g.

4.4. City of Vlaardingen

The city of Vlaardingen is situated north of a very important complex of refineries and chemical plants, covering an area roughly 1.75 km by several kilometers (Fig. 6). The industrial area is cut by wide inlets; this seems to upset the assumption of statistical homogeneity of emissions on the terrain considered. However, during the loading and unloading of vessels there are numerous spillages onto the water surface, which thus becomes a not negligible source. Possible inhomogeneities cannot be taken into account without complicating the model considerably; moreover, at this stage it has not been proved warranted to introduce these complicating refinements. As a consequence, the results should be considered first-order approximations.

The town was subdivided into districts, as shown in Fig. 6. In each district the number of complaints received during the period Jan. 1, 1968 to July 31, 1969 and the number of inhabitants were determined, as listed in Table 3. At 500-m intervals, lines were drawn parallel to the downwind edge of the industry considered. Estimates by eye were made of the fraction g of the residential district falling within each band. Subsequently the number of complaints per 1000 inhabitants was averaged (Table 3). It had to be assumed that the telephone density was equal in all residential districts, not too good an assumption in view of the social inhomogeneity. Though the effect of this assumption will be smoothed by the procedure of averaging over the various residential districts, one should anticipate some spreading of the experimental data.

The calculation of the penalization function was made using Eq. (6), the complaint pattern with the aid of Eq. (11). The results, depicted in Fig. 7, again show a good agreement between model calculation and observation.

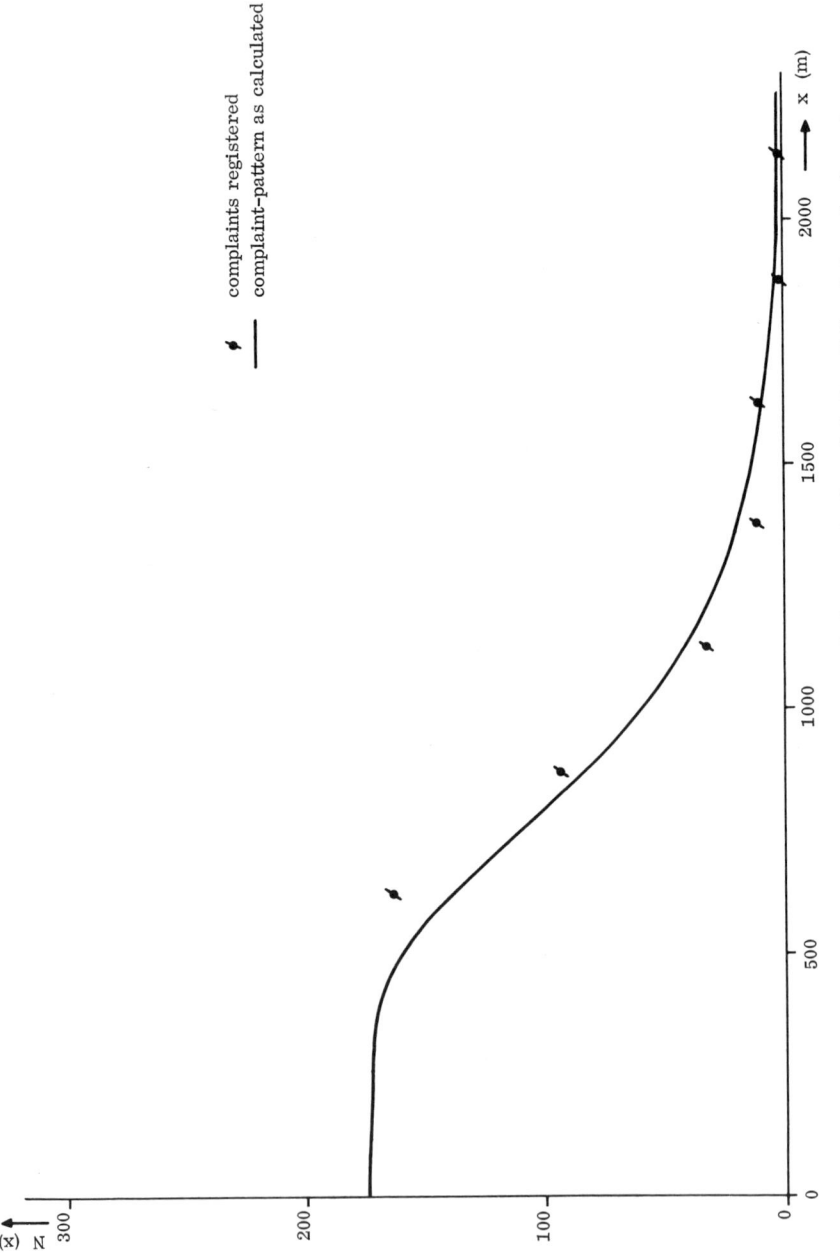

Fig. 5. Number of complaints per 1000 inhabitants in the vicinity of the Gekro incinerator.

10. PENALIZATION DUE TO STENCH

Fig. 6. Map of Vlaardingen. The residential districts listed in Table 3 are shaded.

TABLE 3

City of Vlaardingen: Relevant Data [a]

Distance band (km)	District	Inhabitants	Complaints	N	g	N(x)	I(x)	f(x) [b]	I(x) · f(x)
1 – 1.5	Vettenoord	8,719	398	46	2/3	37	5,820	0.75	5,900
	Oost	8,236	103	13	1/4		2,055		
1.5 – 2	Zuid-West	17,801	875	49	1/3	26.5	5,934	0.66	12,900
	Vettenoord	8,719	398	46	1/3		2,906		
	Ind. Buurt	5,700	176	31	1/3		1,900		
	Centrum	9,321	82	9	1/2		4,660		
	Oost	8,236	103	13	1/2		4,118		
2 – 2.5	Zuid-West	17,801	875	49	1/3	19	5,934	0.59	12,860
	Ind. Buurt	5,700	176	31	1/2		2,850		
	Centrum	9,321	82	9	1/2		4,660		
	Oost	8,236	103	13	1/4		2,059		
	Babberspolder	12,583	116	9	1/2		6,291		
2.5 – 3	Zuid-West	17,801	875	49	1/3	25	5,934	0.51	6,860
	Ind. Buurt	5,700	176	31	1/6		950		

10. PENALIZATION DUE TO STENCH

	Ambacht	5,144	87	17 2/3		3,420	
	Babberspolder	12,583	116	9 1/4		3,146	
3 – 3.5	Ambacht	5,144	87	17 1/3	14.5	1,715	0.435 3,220
	Holy	10,246	159	16 1/4		2,561	
	Babberspolder	12,583	116	9 1/4		3,146	
3.5 – 4	Holy	10,246	159	16 3/4	16	7,680	0.39 2,995
Total	Vlaardingen:	77,750	1,996				44,745

^{a} $\overline{N(x)} = 1996 \cdot 1000/77,750 = 25.7$; $\overline{f(x)} = 44,745/77,750 = 0.575$.
^{b} Mean value of $f(x)$ over the distance band indicated.

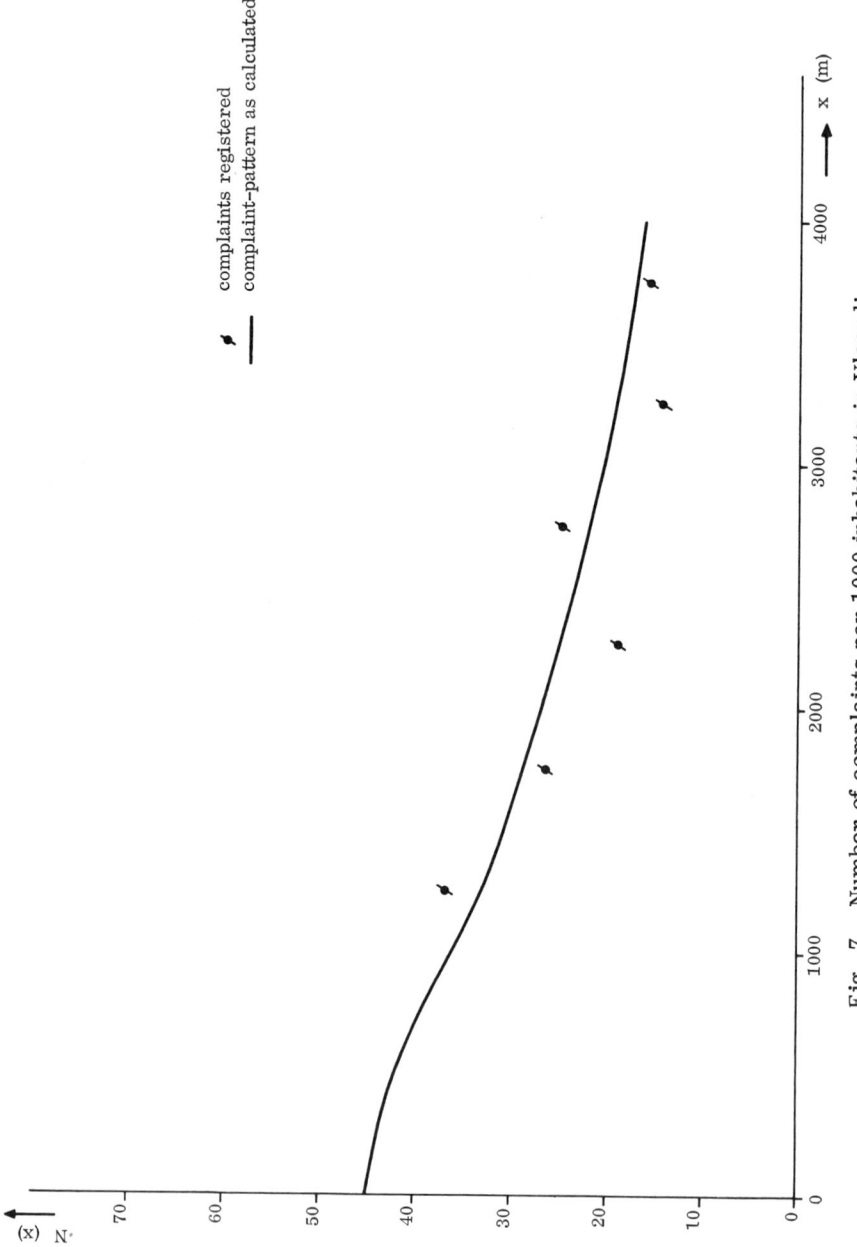

Fig. 7. Number of complaints per 1000 inhabitants in Vlaardingen.

10. PENALIZATION DUE TO STENCH

Following the same procedure, the complaint patterns were calculated in Hoogvliet and Spijkenisse, two cities in the Rijnmond area which are situated much as Vlaardingen with respect to industry. In all cases good agreement was obtained between calculation and observation. In the light of a recent sociological study in the Rijnmond area it appears that 56% of the population perceives air pollution from time to time. Combining the mean penalization values of Vlaardingen (57.5%), Hoogvliet (62%), and Spijkenisse (51%), using the total number of inhabitants (77,750, 37,390, and 28,327, respectively) as weight factors, the overall value of 57.5% is obtained. This compares favorably with the experimental value.

4.5. Statistical Tests

The question arises as to how significant the results presented in the previous sections are. The answer to this question is of great importance, as the results could well form the basis for far-reaching political decisions.

The investigations in Geleen and in the Rijnmond area can be regarded as replications; a similar calculation was carried out five times. The results were compared with observed data, obtained independently. The correlation coefficients and their significance levels are given in Table 4. In four of the five cases the significance is very clear, even at the 99% level.

Doing replication research one can draw up a composite null hypothesis. For Vlaardingen, Hoogvliet, and Spijkenisse, for which the complaint patterns were calculated in a completely similar way, but for which the observed complaint patterns are completely independent, the composite significance level is certainly smaller than 0.000002, as can be derived from Table 4. Consequently, one cannot doubt the efficiency of the mathematical model.

In Fig. 8 all calculated values are plotted vs observed data. Calculated and experimental results were tested against each other in three different ways:

1. using the Spearman rank correlation method; $r_s = 0.96$ ($P << 0.0005$);

2. using Robinson's [8] intraclass correlation coefficient; $r_i = 0.93$ (at the 99% significance level, $0.81 < r_i < 0.98$);

3. using the F test: the least-squares fit of the straight line through the points shown in Fig. 8 is $y = 1.08x - 2.63$; this line was tested against the line $y = x$; $F(2,25) = 1.56$ ($P >> 0.10$).

A difference between calculated and experimental values cannot be shown with the available means. Consequently, the basic assumption cannot be rejected.

5. CONCLUSIONS

From the foregoing, the following conclusions can be drawn:

a. The long-term average of the subjective appreciation of the population for the quality of the air in the vicinity of chemical industries can be estimated with the aid of a mathematical model. By statistical testing the efficiency of the model has been demonstrated.

b. This mathematical model is based on the following fundamental assumption: The great number of emissions of odorous compounds, even those seeming completely unimportant, on a chemical industrial terrain, determines the appreciation of the population for the quality of the air. Based on highly significant tests it was proved that this hypothesis cannot be rejected. Consequently, it offers a clear clue to the abatement of odorous pollutants; among other measures, to strive for a drastic improvement of the "leak-tightness" of chemical plants.

c. Apparently it is possible to estimate a priori the value of the mixed odor characteristic emerging from chemical industry. This offers the possibility, albeit in a careful way, to predict the effect on the environment of new industrial settlements, or expansions of existing industries, an important starting point for air pollution management.

d. Much work remains to be done to study the day-to-day (hour-to-hour) variation of the penalization under varying meteorological conditions.

6. APPLICATIONS

As penalization of the environment due to stench is uniquely dependent on human perception, it is possible to uniquely define the maximum allowable penalization level. This is a standard for odorous pollutants. As such a standard cannot be based on dosage-effect relations, fixing a standard requires a political decision. As the tolerance level of the population will change with time, the number of complaints associated with a given penalization level will also change with time; this means that any standard for odorous pollutants is intrinsically temporary.

10. PENALIZATION DUE TO STENCH

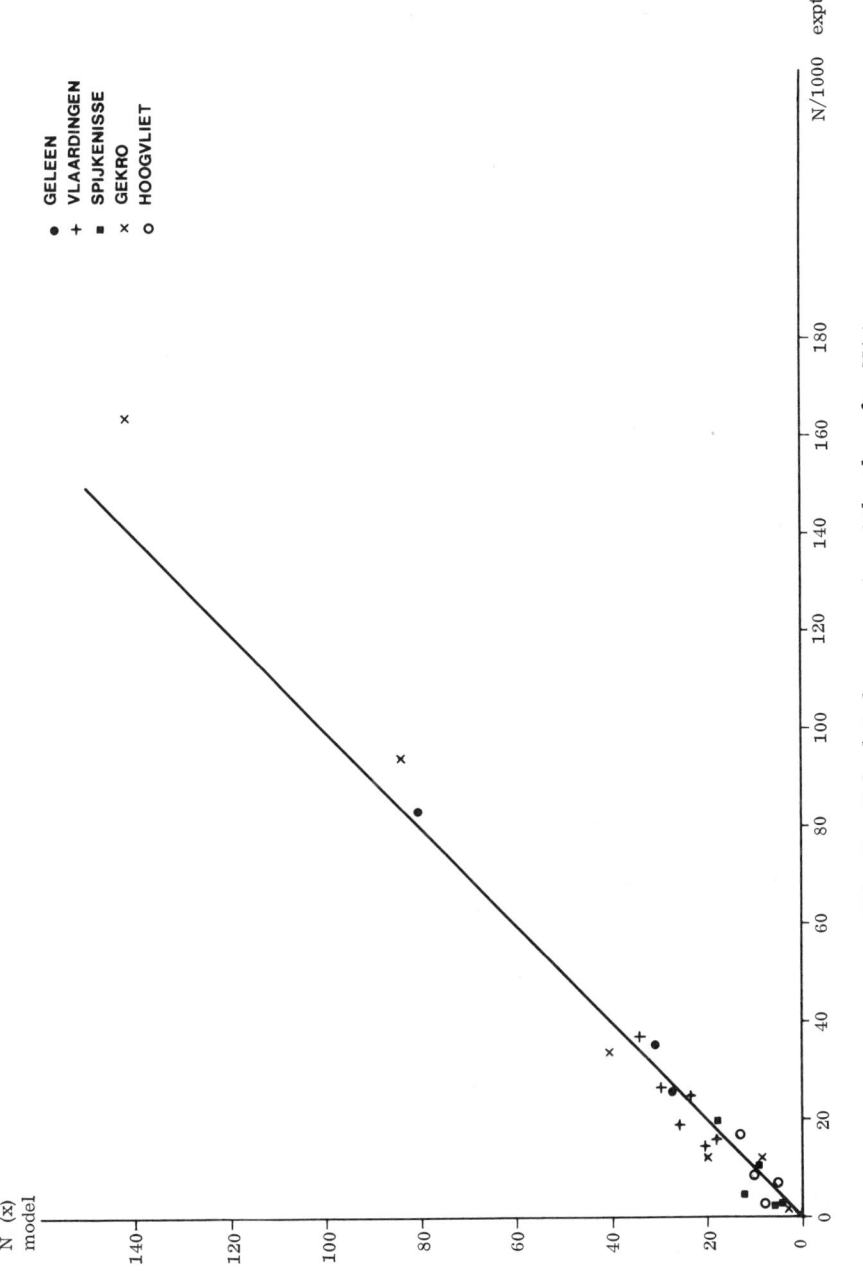

Fig. 8. Calculated vs experimental values for N(x).

TABLE 4

Correlation Coefficients between Calculated and Observed Data and
Their Significance Levels

City	r	p[a]
Geleen	0.98	< 0.001
Vlaardingen	0.90	< 0.01
Hoogvliet	0.69	< 0.20
Spijkenisse	0.97	< 0.001
Gekro	0.996	< < 0.0005

[a] Tested against independence.

In Fig. 9 two penalization curves are shown. Suppose the maximum allowable penalization is fixed at the 50% level; it then follows from the figure that new house building should preferably not be initiated within a distance of 2750 m from an existing chemical complex of a depth of 2 km. Or, conversely, no new chemical plant of any proposed dimensions should be erected closer to an existing residential area than some predictable distance. Especially in densely populated areas this can contribute to better regional planning, more in tune with the notion "health" than has hitherto been the case.

If the result of an abatement action against odorous pollution has to be monitored, one cannot rely on the number of complaints received in a given period of time. If the tolerance level of the population decreases, the number of complaints may increase, while in reality the situation has improved. The penalization function, plotted on log probability paper, yields a straight line. As penalization is independent of the number of complaints, comparison of the slopes of two penalization functions, for two nonoverlapping time periods, reveals whether or not the situation has improved. In this respect also the penalization function may prove to be a useful tool.

By Eq. (7) a direct relation exists between p and the distance x_{th}, at which the threshold concentration occurs: the greater p, the larger x_{th}. Let the index 1 indicate the present situation, and 2 the situation to be achieved. Then on substitution of Eq. (1) in Eq. (7) the following relation

10. PENALIZATION DUE TO STENCH

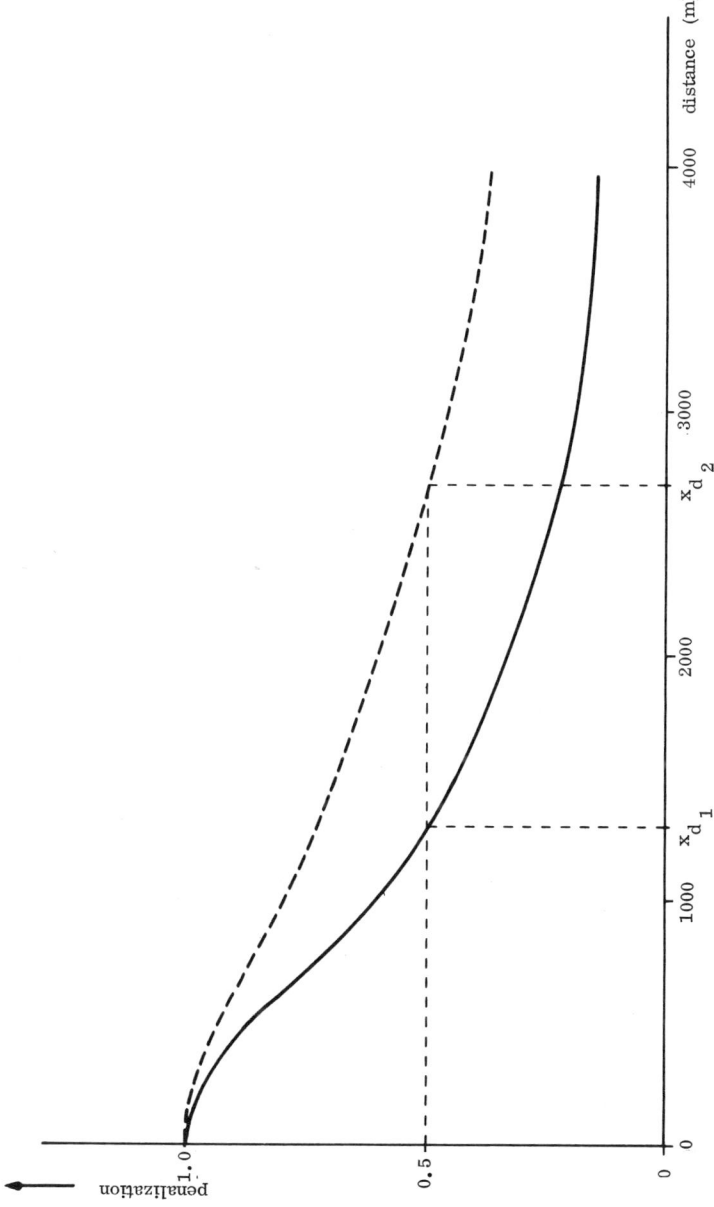

Fig. 9. Penalization of the environment: --- downwind from an industrial terrain of a depth of 2 km; ——— downwind from an industrial terrain of a depth of 1 km.

can be derived, assuming $\overline{Q}$ remains approximately constant:

$$p_2 = p_1 - \frac{\log (a_2 b_2 / a_1 b_1)}{\log \sigma_g} \; . \tag{12}$$

With the aid of Eq. (12) the per cent emission reduction to be effectuated at any one industrial site to meet the penalization standard can be calculated.

REFERENCES

1. D. H. Lucas, Int. J. Air Poll., 1, 71 (1958).

2. M. E. Miller and G. C. Holzworth, J. Air Poll. Control Assoc., 17(1), 46 (1967).

3. G. Leonardos, D. Kendall, and N. Barnard, J. Air Poll. Control Assoc., 19, 91 (1969).

4. F. V. Wilby, J. Air Poll. Control Assoc., 19, 96 (1969).

5. B. K. Krotoszynski and A. Dravnicks, Odor-relevant measurement of odorous pollutants in air, Paper 68 - 17, 61st A.P.C.A. Meeting, Saint Paul, Minn., 23-27 June 1968.

6. T. Lindvall, On Sensory Evaluation of Odorous Air Pollutant Intensities, Uno S. Andersons Tryckeri AB, Stockholm, 1970.

7. F. A. Gifford and S. R. Hanna, Urban air pollution modelling, Paper ME-32D, 2nd Int. Clean Air Congr. I.U.A.P.P.A., Washington, D. C., 1970; J. Air Poll. Control Assoc., 20(11) (1970).

8. W. S. Robinson, The statistical measurement of agreement, Amer. Sociol. Rev. (1957), p. 17.

DISCUSSION

Gerald van Belle

Florida State University
Tallahassee, Florida

In the brief span of these four days we have come to know Dr. Clarenburg as a perceptive contributor to the discussion of papers, bringing balanced comments matched by technical competence. This paper further confirms that impression and a vote of thanks is extended to the speaker.

A major problem for the reader is that this paper is, apparently, a condensation of a much larger investigation, so that at times the argument is not immediately clear and calculations cannot be reproduced. As a result the comments cannot be as constructive as one would hope. Within this limitation this discussion will focus on three aspects:

1. The meteorological argument
2. Statistical aspects
3. Some comments about the model.

1. THE METEOROLOGICAL ARGUMENT

It is not clear how the meteorological argument is relevant, especially in situations where the complaint pattern is the summary of an annual complaint log. Meteorological conditions vary considerably over the year; winds shift directions, seasonal variations of all kinds take place, and whatever values for meteorological parameters are assumed they cannot be too meaningful in relation to the complaint pattern as summarized over the year.

With a point source, Gekro data, Table 2, a contradiction would seem to be involved in that the bands are concentric rings around the point source and no distinction is made between down-wind and up-wind sections of the same band.

In view of the cumulative effects of stench (see the argument in Section 4.2) it is not clear how the bands can be (or are) derived from meteorological considerations when multiple sources (either point or area) are present. Furthermore, there is a major contribution to stench from nonstationary sources such as buses and cars.

Finally, if penalization varies from year to year one would not want to advance the argument that the meteorological situation had varied because that would then preclude a logical basis for action.

2. STATISTICAL ASPECTS

If the meteorological basis for the derivation of the model is dubious the question may be asked: Why does the model fit so well? There are three answers to that question. First of all, the model does not fit too well. Secondly, values of some parameters have been estimated from the data (Geleen study). Finally, in the two other studies, Gekro and Vlaardingen, a very small portion of the "dose-response" curve is estimated - and this portion is near the tail of the distribution.

Before discussing these points some remarks should be made about the calculations as presented. Consider Table 1 (Geleen data). There are several entries that are not clear: (a) Why are there two or more measurements in column 2, distance of residential districts from Complex A, for each residential district? (b) On what basis was one complex selected over another in calculating $f(x)$, the penalization function? (c) What are the weights g in this Table? They do not seem to fill the same role as in Table 3.

The model does not fit too well. It is misleading to compare $N(x)$, the number of complaints per 1000 population, and $\hat{N}(x)$, the predicted number of complaints per 1000 population. What should be compared are the original number of complaints and the predicted number. Using the data as presented in Tables 1, 2, 3, Table A can be constructed. For each of the three sets, Geleen, Gekro, and Vlaardingen, the observed number of complaints per band (or residential district) is compared with the expected number of complaints. No χ^2 goodness-of-fit test is carried out, but clearly results indicate significant departures from the model. A goodness-of-fit test should include the observed and expected frequencies of noncomplaints, of course. In the Geleen study observed and expected numbers of

DISCUSSION

complaints were calculated by multiplying each district's population (as given in Figure 3) by f(x), as calculated, and as obtained from the poll results.

A second comment about the statistical aspects is that the choice of parameter values may have been arbitrary for the Gekro and Vlaardingen data, but this does not appear to be the case for the first set of data, the Geleen study. Values of p and σ_g have to be assumed before X_{th} and $f(x)$ can be calculated. By Eq. (15), X_{th} can be calculated for an area source if values of p, σ_g, and X are assumed. Assuming values of $\sigma_g = 1.88$ and $X = 1$ (for complex A) the threshold distance X_{th} varies drastically with change in p. The following table indicates some values:

p	X_{th} in kilometers (Eq. 15)
2	0.26
2.5	0.54
3	0.98
3.5	1.65
4	2.65
5	6.5

Only a value of $p = 3.5$ is consistent with an observed threshold distance of approximately 1650 meters (Table 1). A similar argument can be made about the choice of $\sigma_g = 1.88$.

It is true that p and σ_g are maintained in the following two studies, Gekro and Vlaardingen, and the question can be asked why the predicted values "fit" as well as they do. There are three reasons; (a) p is not used in estimating X_{th} for the Gekro data; Eq. (9) is used. (b) In the case of the Vlaardingen study the socialization constant K is in effect used to raise or lower the response curve. This explains the reason for the correct level of the curve (essentially determined by p or K). (c) What about the slope of the curve, which is essentially fixed by σ_g? The reason that σ_g "fits" is that a minute portion of the dose-response curve is being estimated. The range of proportion of complaints for the Gekro study is 0.001 to 0.16; for the Vlaardingen study 0.009 to 0.049. In terms of a dose-response curve this is a very small region and considerable variation in σ_g will give the same "good fit."

A final statistical comment relates to statistical tests carried out by the author to "confirm" the model. Correlating rates per 1000 does not take into consideration the accuracy of the data (as measured by the number of people involved). Secondly, since correlation coefficients are invariant under relocation and rescaling, a regression line is more informative. This the author calculates and tests, but it is clear that the proper error term was not used in the regression analysis, i.e., the proper error term is based on the band population sample sizes (using the binomial distribution) and not on deviations of frequencies from a line at 45° through the origin.

3. SOME COMMENTS ABOUT THE MODEL

The basic assumption in the model is that there is a monotone gradient in contamination from an area source (or point source). It has not been verified that the form of the monotonicity is as indicated. To test the form one would have to (a) estimate σ_g and x_{th} for each situation (b) get responses in the form of a poll - not telephonic complaints (under specified meteorological conditions).

If σ_g and x_{th} vary markedly from situation to situation the model is useful for evaluative purposes in general and for predictive or prescriptive purposes for specific locales when σ_g and x_{th} have been estimated.

TABLE A

Observed and Expected Number of Complaints for Geleen, Gekro, and Vlaardingen Data

Band or residential district	No. of complaints					
	Geleen		Gekro		Vlaardingen	
	obs.	exp.	obs.	exp.	obs.	exp.
1	2965	3261	309	259	233	263
2	2347	2310	335	323	641	575
3	8389	8156	171	208	549	578
4	1223	1637	100	150	409	306
5	2246	2021	79	50	96	144
6			6	10	119	134

DISCUSSION

In view of the model as postulated it would seem to be more reasonable to assume the socialization constant to be proportional to the "probit" of the response. The postulated linear relationship

$$P_p(C) = KPr(C)$$

results in response curves that vary in intercept as well as slope with varying K.

ACKNOWLEDGMENT

Research supported by Grant AP00880-02, National Air Pollution Control Administration, Department of Health, Education and Welfare, USA.

11

ON THE ESTIMATION OF POLLUTION-CAUSED GROWTH REDUCTION IN FOREST TREES

Rolf Sundberg[*]

Stockholm University
Stockholm, Sweden

0. SUMMARY

This paper contains some models, methods, problems, and experiences associated with the estimation of growth reduction in forests. The main part of the paper concerns a situation with a local growth reduction caused by smoke emission, but the case of regional acidification is also considered.

1. GENERAL INTRODUCTION

For more than a century it has been known that air pollution can have deleterious effects on vegetation. It has been found that chronic exposure to smoke containing, e.g., SO_2 or HF causes damage, a damage which normally leads to a reduction of growth. This paper is concerned with estimation of the reducing effect of pollution on the growth of forest trees. For several reasons this is a worthwhile task. Forests are of high economic value and a reduced growth rate means reduced yields, a reduction which can be converted into monetary units. Another reason is that trees store the history of their growth development in the form of distinguishable annual rings which can be measured from increment (boring) cores. This provides an almost unique opportunity to study past changes, and it makes continuous measurements unnecessary. Also, the actual populations of forest trees are often pleasantly large. However, drawing inferences from measurements of annual rings poses several difficult

---
[*]Present address: Royal Institute of Technology, Stockholm, Sweden.

problems. The present study is an attempt to base the attack on those problems on a statistical model of growth.

The primary concern of the paper, taken up in Section 2, is the estimation of the local effects of smoke emission on the forests in the vicinity of an emission source. Among other papers on this subject we would mention a recent account of Austrian methods and experiences by J. Pollanschütz [1]. Our main background is a case study, financed by Mosjøen Aluminum Works, Norway, in the assessment of possible effects of fluorine-containing smoke on growth development in forests of Norway spruce [2].

Local damage is usually caused by gas absorption. More recently it has been discovered that large-scale smoke emission can have deleterious effects on large regions hundreds of miles away from the sources. Emitted sulfur may be deposited with precipitation in the form of sulfuric acid, which in large amounts may cause an acidification and impoverishment of sensitive regions. This may in turn be expected to result in reduction of growth. In Section 3 we touch on the problem of determining such a growth reduction. The results of an analysis carried out in cooperation with B. Jonsson, The Royal College of Forestry (Sweden), appear in the Swedish contribution to the U. N. Conference on the Human Environment, Stockholm 1972, and a detailed report is available [3].

2. THE LOCAL PROBLEM

2.1. Introduction

We make the following assumptions, which are strongly influenced by a practical case. In a region there is one emission source from which emmision started at some year t_0. We possess data on the annual rings representing some period about t_0 from a sample of trees of a single variety, the sample covering such a large region around the source that the effects on growth in the outer parts are certainly negligible. For practical and economic reasons the sampling is usually made in two stages: first a choice of a set of plots, and then a choice of a subsample of trees within each plot. Such a procedure is assumed here.

It might be desired to answer the following questions. Has the smoke emission caused any growth reduction? If so, what is the magnitude and what is the extent of the reduction, and what is its development in time? As the conditions are not experimental, the first question about causes

11. GROWTH REDUCTION IN FOREST TREES 169

cannot in principle be settled only from an observed growth reduction. All possible explanations must be examined and compared as to their plausibility, and to that end we should use supplementary data, e.g., chemical analyses of concentration, wind direction statistics, and observations of visible damage on trees as well as on other vegetation. Such data are of value also when determining the spatial distribution of growth reduction, but will not be further considered in this paper.

In what follows we shall mainly consider the problem of estimating the growth reduction for an individual (single tree) or an area. This is a basic problem, for the resulting estimates should be used in statistical tests of significance to answer the question of whether any growth reduction has occurred, and also for classifying areas according to existence and magnitude of growth reduction. But we do not pretend that the latter problems are more trivial or less important than the one treated here.

2.2. A Model of Growth

From the economic point of view growth should naturally be measured in volume units, but radial growth is preferable for obvious practical reasons. Fortunately, experience indicates that the relative growth reduction should be approximately the same for both radius and height. Provided that the annual growth is small in comparison with the tree itself, an "infinitesimal" argument then shows that the two measures should lead to approximately the same result. We will therefore be satisfied with measurements on the annual rings.

We now construct a model describing how the annual radial growth depends on exterior and individual elements. We measure time in discrete years and denote it by t. The radius and radial growth will be denoted by $r(t)$ and $\Delta r(t)$, respectively.

To each tree we associate a growth level function $\lambda(t)$, determined by such variables as the age and size of and ground area covered by the tree, and the soil and climate at the locality. The values of most of these variables are normally unknown, and in any case the form of their influence on the annual ring is not satisfactorily known. The growth level for a not very young tree normally decreases with age. Thinning a forest raises the growth level. The level $\lambda(t)$ should be thought of as normally smooth and slowly varying with t for each individual, but the variation between individuals at a plot and between plots may be considerable.

If desirable a parametric assumption on the form of the function $\lambda(t)$ may be imposed. But in that way we cannot obtain more than a numerical fit, not allowing on its own merits any prediction from a period before t_0 to the years after t_0. This is the principal reason why we will avoid any such assumption here whenever possible. For a case where a parametric assumption was considered necessary, see Section 3. The annual growth of a tree also depends on the annual variation of weather conditions, mainly temperature and precipitation. Theoretical arguments as well as empirical knowledge indicate that many conditions, including the weather conditions, have a multiplicative influence on growth. Some tests in the present case study support this hypothesis; see the end of this subsection. The introduction of the growth level $\lambda(t)$ should be regarded in this light, and the effect of weather variations is now logically introduced as a climatic factor $\varkappa(t)$, multiplying $\lambda(t)$. In the present case of a local investigation, the value of $\varkappa(t)$ for fixed t can probably be regarded as common for all the trees in the region, or at most as having a slight and smooth spatial trend.

It is well known that the weather conditions in one year have a significant influence on growth not only that year but several years afterward. The weather influence cannot be satisfactorily calculated. Hence $\varkappa(t)$ is naturally assumed to be the realization of some random process, and without loss of generality this process may be normalized in some way so as to vary around the level 1.

These two factors are not sufficient to describe the annual growth of a tree. In addition to the process $\varkappa(t)$ we must permit individual variations and fluctuations of an apparently more purely random kind, included as a multiplicative random process $\xi(t)$. With satisfactory accuracy we may probably usually assume that $\log \xi(t)$ is a stationary Gaussian process centered to have expectation zero, with independence between the realizations for not very closely adjacent trees.

For the unaffected radial growth of a tree in the year t we have thus arrived at the following statistical description:

$$\Delta r(t) = \lambda(t) \times \varkappa(t) \times \xi(t),$$

where λ is an individual function, $\varkappa$ a common random process, and ξ an individual random process, all of which have been discussed above.

Now suppose that a growth reduction occurs. In the present case the growth reduction caused by smoke emission is of primary interest, but temporary reductions occur for other reasons, e.g., because of insect attacks. It should also be noted that growth-reducing insect or fungus attacks may be a secondary effect of the smoke emission. The relative growth reduction will be included in the model in the form of a function $\theta(t)$, where $\theta(t) \leq 1$ (negative values are permitted) and $\theta(t) = 0$ for

11. GROWTH REDUCTION IN FOREST TREES

$t < t_0$, so that the resulting observable growth of a tree is described by

$$\Delta r(t) = \lambda(t) \times \varkappa(t) \times [1 - \theta(t)] \times \xi(t).$$

The individual trees are unequally susceptible to the smoke, so there is a random element in θ. It might be realistic to consider $\theta(t)$ for given t as the product of a smooth ordinary function in the plane depending on the exposure to smoke and an individual random disturbance factor describing the effect of smoke on the growth of that tree. It is then of great importance both for inference and for economic conclusions to know whether or not the magnitude of this random factor is independent of the growth level λ. If independence holds, the growth reduction factor θ is the natural measure of the damage to a tree, and the arithmetic mean of θ over all the trees in a plot is the natural and appropriate measure of the growth reduction in the plot. In estimating this plot reduction from a sample it is then not necessary that the growth levels λ of the sampled trees be representative of the plot population, and the same may be the case in estimating the total annual yield loss for a subregion. Actually, tests performed in the case study on data from sample plots with established growth reduction gave no indication of any nonnegligible dependence.

Other tests of implications of the growth model have also been carried out in the case study, indicating satisfactory fit between the data and the model. We mention briefly two such implications. For the trees in a plot consider the amounts of growth in two years not too remote from each other. An approximately proportional functional relationship should then be expected from the multiplicative model. Furthermore, the quotient of the two growth quantities should be approximately independent of the expected growth in any of the years.

2.3. Estimation of Growth Reduction

To determine the growth reduction in year t for a tree, a plot, or some larger area, we should eliminate as far as possible the nuisance factors $\lambda(t)$, $\varkappa(t)$, and $\xi(t)$ in the growth model. The influence of the random factor ξ and other individual elements is automatically reduced when we estimate the growth reduction for an area from the estimates of θ for the individual sampled trees. As concerns the two other factors, our growth measurements must be referred to growth free from smoke effects, and we will say that we need some reference growth.

To remove the individual growth level $\lambda(t)$, we should divide the growth for $t \geq t_0$ by some reference growth originating from the same tree. A growth that is certainly unaffected for all trees is found only before the year t_0, when the emission started. In view of the multiplicative model, it is natural to choose as reference growth a weighted geometric mean of the growths before t_0. (However, arithmetic means are simpler and have not appeared to behave worse.) In the resulting growth quotient the individual growth level λ is eliminated. Instead, the growth quotient is influenced by spatial variations in the rate of relative change with time of the growth level, although usually to a much lower degree. (In fact, experience indicates that this procedure should perform better than a prediction of the growth level function by conventional regression techniques.) The reference growth brings to the growth quotient also a random factor corresponding to $\xi(t)$. The contribution of the latter to the total random quantity depends on the autocorrelation of the process $\xi(t)$, the time difference between t_0 and the year $t \geq t_0$ under consideration, and the rate of decrease of the weights in the reference growth.

How should the weights in the reference growth be chosen in order to minimize the spatial variation in growth quotient over an area with equal climatic process and equal growth reduction? To minimize the spatial variation in the growth level quotient it is likely that the last year before t_0 should be given weight 1. To minimize the variance of the random quotient it is likely that all weights should be equal. These considerations conflict. A determination of the optimal weights is a difficult matter, and the optimal solution will depend on the year $t \geq t_0$ in question. This would be reason enough to recommend parallel usage of several sets of weights; also, comparisons may give valuable information about the factors involved. For instance, if a plot is extremely sensitive to the choice of weights, that plot must have had an abnormal growth history in the years before t_0, a fact which must be considered.

The procedure for eliminating the effects of the climatic process $\varkappa(t)$ relies on the assumption that the climatic process is approximately the same all over the region. Provided this is true, we can leave the values of the climatic factor out of consideration if we restrict ourselves to comparisons and differences between subregions. And in fact, remembering the assumption that the outer parts of the region should be unaffected by smoke, comparisons with those regions are sufficient. Thus the growth in the outer parts of the region serves as reference growth in the elimination of the effects of the climatic process. If the latter assumption is not valid, a general underestimation of the growth reduction must be expected. If the spatial variation of the climatic factor (or, slightly weaker, of the climatic factor quotient appearing in the growth quotient) is not negligible and care is not taken, the estimates may be seriously erroneous. Spatial trends in the climatic factor may be detected and corrected for by, for example, comparison of different outer subregions.

11. GROWTH REDUCTION IN FOREST TREES

2.4. Some Further Remarks

If great damage has been caused, dead trees will occur. For such trees we may talk of a 100% growth reduction. These trees present the problem that the annual rings may be destroyed within a few years, and after that it might not even be possible to determine the year of death. Of course, attention must be paid to the frequency of dead trees, but we leave open the question of how that should best be accomplished.

Thinning a part of a forest has a positive effect on growth. This effect should be kept in mind when one is performing a growth reduction analysis, and taken into account as far as possible. However, this poses difficult problems. In the ideal situation none of the sampled plots has been thinned soon before t_0 or after t_0, but unfortunately such a situation usually cannot be expected.

3. THE REGIONAL PROBLEM

In the preceding section a growth model was constructed so as to be appropriate for the specified case of local damage. In other situations it might be necessary to use a modified version of that growth model. As one such example we briefly consider the regional problem mentioned in Section 1.

For a large region with varying growth conditions, the annual precipitation is assumed to have been unnaturally acid from some year t_0 onwards, with a steadily decreasing pH value. The acidification may have caused a cumulative growth reduction in sensitive parts of the region, a reduction which we wish to estimate and to test statistically. It would seem to be impossible to proceed by simply correlating measurements on growth and acidity, because the annual amount of deposited acid depends strongly on the simultaneous precipitation, and this precipitation also has a strong direct influence on growth (which depends on other climatic factors as well), and none of these biological relationships is quantitatively well known. We therefore proceed by analogy with the local case.

Reference growths may be obtained by the same principles as in Section 2, namely, by comparing the growth development before t_0 with that after t_0 and by making comparisons between subregions. To be of use in the latter comparisons, two classes of subregion may be chosen, on the basis of soil and acidification data, so as to consist of areas which are a priori most likely and least likely, respectively, to have been subject to a growth reduction. As in Section 2, these comparisons should take care of the growth level λ and the climatic factor $\varkappa$, respectively.

However, the climatic factor $\varkappa$ cannot be considered constant over a large region. This makes the present situation less satisfactory than the local one. In the case study mentioned in Section 1, further restrictions and assumptions were introduced in order to get around the difficulties. It was assumed that the climatic process could be split into two factors, one common to all the subregions and another characteristic of the individual subregion, with approximate statistical independence between subregions. To what extent this assumption is acceptable should be judged with due consideration to the location of the regions. No attempt was made to determine the growth reduction in an individual subregion, but, the procedure being based on the aforementioned independence assumption, the (geometric) mean process for the sampled trees in a subregion was considered as one of a set of independent realizations of random processes with expectations depending on the common climatic factor introduced above as well as on the (geometric) mean growth reduction in the actual subregion.

The spatial variation of $\varkappa$ has consequences also for the use of the growth before t_0 to remove the growth level λ. When some substitute is introduced for λ, we automatically get a contribution from $\varkappa$ to the spatial variation. To make that contribution small, large samples are not sufficient. Rather, we should estimate λ from a long time period before t_0. But if we consider a long time period we must take the development of the growth level $\lambda(t)$ into account. This may be accomplished by introducing a parametric description of the growth level $\lambda(t)$ as a function of time. In the local case this was avoided, but here it might be necessary. A natural and simple function to describe the decrease in growth due to aging is the exponential

$$\lambda(t) = \lambda e^{-\mu t}, \quad \mu > 0.$$

Perhaps this expression is only a crude approximation, but it may be hoped that the deviations from the true function do not differ systematically between the two classes. This hope can be made more realistic by, for example, making the age distributions of the samples similar for all subregions. The hope may also be tested on data from the time before t_0, which is not disturbed by the acidification. Other statistical tests may clarify whether μ should best be chosen as a common parameter or as an individual one (like λ).

A parametric attempt may also be advisable for the growth reduction factor $\theta(t)$ in order to express the expected cumulative effect. As long as there is no biological reason to prefer any particular function, we may as well take the exponential one and write

$$1 - \theta(t) = \exp[-\theta(t-t_0)], \quad t > t_0.$$

11. GROWTH REDUCTION IN FOREST TREES

We are then led to a model with a broken linear exponent in which an acidification effect should appear as a difference between the estimates of θ for the two classes of subregions. But of course the possibility that the errors in the assumptions have a significant effect on the difference must then be examined. The model has been applied in the case study mentioned.

ACKNOWLEDGMENT

Thanks are due to Prof. B. Matérn for his kind interest and valuable criticism.

REFERENCES

1. J. Pollanschütz, Die ertragskundlichen Messmethoden zur Erkennung und Beurteilung von forstlichen Rauchschäden, Mitt. Forstl. Versuchsanstalt, Wien, No. 92, 1971.

2. R. Sundberg, Analysis of growth damage in forests in a vicinity of Mosjøen Aluminum Works (in Swedish), unpublished, 1970.

3. B. Jonsson and R. Sundberg, Has the acidification by atmospheric pollution caused a growth reduction in Swedish forests? A comparison of growth between regions with different soil properties, Department of Forest Yield Research, Royal College of Forestry, Stockholm, Research Notes No. 20, 1972.

12

A MATHEMATICAL MODEL FOR PARTICULATE DEPOSITION IN THE RESPIRATORY SYSTEM

H. D. Landahl

University of California
San Francisco, California

0. SUMMARY

A model for the assessment of the hazard of an inhaled airborne material requires (1) an estimate of its distribution of deposition and (2) its subsequent redistribution throughout the body. From a model of the human respiratory system based on anatomical measurements and from the physical laws which govern the impaction, diffusion, and settling of particles, one can make reasonable estimates of the concentration of the particulates in various fractions of the exhaled air. Since these values are in general agreement with experimental data, one can expect the estimates of the amounts of the airborne material deposited in various parts of the respiratory system to be reasonably accurate. These values cannot be directly measured for human subjects.

Some of the factors involved in the redistribution of deposited materials are discussed. A model for these factors is given and illustrated by some examples.

1. INTRODUCTION

Contamination of the atmosphere by gases, vapors, and particulates is one of the more serious aspects of the pollution problem. A major hazard arises because these pollutants are breathed into the respiratory systems of man and animals. Concentration on this aspect does not detract from the importance of the potential hazard of airborne substances on the eyes or skin or their harmful effects on vegetation or their deteriorating effects on property.

In order to assess the human hazard of inspired air pollutants it is necessary to know (1) the amount and location of deposition of the particulates and (2) the subsequent fate of the deposited substances.

2. FACTORS DETERMINING DEPOSITION

We consider first the physical processes involved in determining the deposition. These are impaction, settling due to gravity, and diffusion. These processes in turn depend on properties of the airborne material, and of air, as well as on physiological factors. The important aerosol properties determining deposition are size and shape (quantified in some appropriate manner), density, and hygroscopicity. Thus we may consider a very wide range of particle sizes, a range of densities from UO_2 to materials like lint, and substances ranging from those which do not absorb moisture to those which vigorously pick up water vapor from the ambient air. The most important physical property of air (or any other gas mixture) is its viscosity. Its density also plays a role in determining turbulence, a factor which, in turn, influences deposition.

The important physiological factors determining deposition are (1) the dimensions of the respiratory system, (2) the pattern of respiration (e.g., tidal air, frequency of breathing), and (3) secondary responses in the above factors resulting from the deposition of the particles. Thus an allergenic material can constrict the dimensions of the bronchioles, changing the probabilities of deposition of particles arriving subsequently. Similarly, pathological changes, such as those due to emphysema, bronchiectasis, or sinusitis, can modify the pattern of deposition.

Impaction

Due to its inertia, a particle suspended in the air stream cannot follow the air stream as it bends around in the tortuous passages in the respiratory tract. If a tube with square sides has a right-angle bend, then a particle of mass m and diameter d which enters at the center of the tube will no longer be at the center when it comes out; it will slip a distance s such that the kinetic energy of the particle ($1/2\ mv^2$, v being the air velocity) is all lost because of the viscosity η of the air. Using Stokes' law [1, p. 245] one obtains, for a spherical particle of density ρ,

$$s = \frac{\rho d^2 v}{18 \eta} . \tag{1}$$

12. PARTICULATE DEPOSITION IN RESPIRATION

If $s = R$, the radius of the tube, then the particle initially at the center will just contact the surface and be removed. If there is a uniform velocity distribution across the tube, then 50% of the particles would be removed. The fraction will be different for smooth and for turbulent flow, and this difference must be taken into account. It will also depend on the angle of the bend if it is other than a right angle. In the case of impaction against the nasal hairs, represented by perpendicular cylinders of diameter D_H, we must take into account the fact that not all particles directed toward the cylinder will actually contact it. The probability of a contact can be given approximately by an expression involving the fraction of the cross-sectional area occupied by the nasal hairs, as well as on the quantity ($\rho d^2 v/\eta D_H$), since the slip distance s in Eq. (1) must be of the order of magnitude of D_H.

Settling

Whether the air stream is flowing or not, a particle will fall through the air with a velocity

$$v_s = \rho d^2 g/18\eta \tag{2}$$

for which the gravitational force due to g is balanced by that due to viscosity [1, p. 246]. The proportion of particles which will reach the bottom of a tube will then depend on the orientation of the tube, the time, and the physical properties of the particle and air.

Diffusion

If we consider very small particles, the effect of the random movements of the particle becomes important [2-5]. If D is the diffusion coefficient and t' a duration of time, then the root mean square displacement x is given by

$$x = \sqrt{Dt'} . \tag{3}$$

When this displacement is comparable with the radius R of a circular or a cylindrical region containing the particle, then the probability that a particle near the center of the region will reach the walls becomes close to 1/2. In this way the effect of diffusion, which depends on particle size, can be calculated.

3. CALCULATION OF DEPOSITION IN THE RESPIRATORY TRACT

It is natural to divide the respiratory system into the nasopharyngeal region and the lungs. The lungs can be divided into upper and lower regions or into many divisions: the trachea; primary, secondary, and tertiary bronchi; bronchioles; and alveolar ducts and sacs. For the purpose of calculating deposition it is necessary to use a rather elaborate representation, while for the estimation of the subsequent fate of deposited materials a coarser subdivision can be used.

Retention in the Nasal Passages

The probability of retention of a spherical particle of diameter d and density ρ, for a given air velocity v, can be calculated if we divide the nasal passages somewhat arbitrarily into four regions [6]. The first is the part of the external nares containing nasal hairs. From the area of the opening and from the preassigned total flow rate, the velocity of the air stream is estimated. The product of the total projected cross-sectional area of the hairs and the probability of impaction divided by the area of the opening gives an estimate of the proportion of particles retained by the nasal hairs. Liquid or moist particles have a low probability of being released into the air stream within the period of the experiment. For the second region we consider the first nasal constriction to be represented by a narrow slot with a bend, the values being based on measurements made on the subjects to be used in experiments. In this region only impaction is considered, as is also the case for the third region. In the fourth region, representing the convolutions of the upper nasal passages, the effect of settling becomes appreciable. Diffusion plays a role only for ultrasmall particles, which we shall not consider in this section.

Combining the probabilities of retentions in the various regions yields an estimate of the total retention for a given particle diameter and flow rate. In Fig. 1 the light curve represents the results of such calculations. The circles represent experimental data using corn oil droplets. Samples of air containing the droplets which had passed through the nasal passages at a constant rate of flow were collected on an impacting device, and the amounts were compared with simultaneous control samples.

Because it is difficult to measure the parameters of the nasal passages, the agreement between the experimental and calculated values can be considered satisfactory. A fairly satisfactory agreement was also obtained between theory and experiment for retention on the nasal hairs for particles of several sizes [6].

Since diffusion plays a small role, we might expect from Eqs. (1) and (2) that the important variable can be considered to be ρd^2, or $\sqrt{\rho}d$, which we will refer to as the equivalent impaction diameter. This quantity is directly estimated by an impacting device. In this way it is possible to estimate an equivalent diameter even for aggregated materials, and then

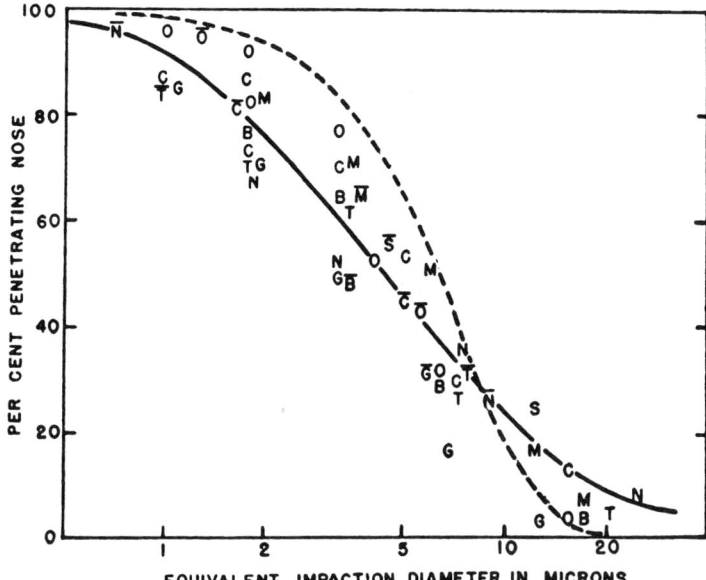

Fig. 1. Penetration of various substances through the human nose as a function of equivalent impaction diameter. Flow rate 17 liters/minute. Symbols O, N, C, M, B, T, S, and G represent data on corn oil, sodium bicarbonate, calcium phosphate, methylene blue, bismuth subnitrate, tyrosine, sodium sulfate, and glycerol, respectively [6].

plot the per cent retention against this diameter. The results should then be independent of density, shape, and degree of aggregation. Experimental results [6] using a number of substances are also shown in Fig. 1. It can be seen that to a first approximation the data from the various substances fall along a single curve. It may be noted that the points labelled G tend to be the lowest. These refer to data on glycerol. Since this is a rather hygroscopic substance one would expect the droplets to increase in diameter in the moist passages and thus to be removed to a greater degree than otherwise.

Lung Retention

In order to estimate the retention in the lungs, it is necessary to have a numerical description of the bronchial tree. Such data are summarized by Findeisen [2], Weibel [7], and Horsfield and Cumming [8].

Modifying these values to obtain the known total volumes for various tidal air volumes and using essentially the expressions for the retention in the various segments of the bronchial tree [9] it was possible to calculate, for various respiratory patterns, the probability of a particle reaching a given region on inspiration and thus the expected fraction retained in that region during inspiration as a result of the various factors. Since the inspiratory pattern always had an approximately constant inspiration rate for a fixed time, a fixed "hold" time, and a fixed expiration time, the combined probability of removal by settling and diffusion could be calculated for each region. For the expiratory phase a similar calculation is necessary. By following a particle in this way the amount retained in each region can be estimated for an entire respiratory cycle.

It is not easy to verify the amounts retained in the various regions. But it is possible to calculate the amounts which are to be expected in various fractions of the expired air [3]. Thus, for example, the expired air can be divided into four equal serial fractions. The first fraction will be expected to have more particles since this fraction will not have penetrated into the smaller regions and since the particles in this fraction will have been the last to enter the mouth and the first to leave. If experimental values [9] for such serial fractions are in agreement with the calculated values, then one is justified in considering the model to be satisfactory and thus the calculated proportion of the particles retained in the various parts of the lung to be approximately correct. In Fig. 2 are shown experimental and calculated values [9] for the relative amounts in equal serial fractions of expired air for three different respiratory frequencies but with the same pattern (3/8 inspiration, 1/8 pause, 3/8 expiration, 1/8 pause) and the same volume flow rate (300 cm^3/sec). The upper, middle, and lower graphs are for 4-, 8-, and 12-sec cycle time, respectively, with corresponding values for the tidal air of 450, 900, and 1350 cm^3. It can be seen that the agreement between theory and experiment is satisfactory. A comparison between theory and experiment has also been made for other respiratory patterns [10]. Here again the agreement was found to be satisfactory to a first approximation.

To the extent that the data agree with experiment we may rely on the calculations of the amounts deposited in various regions of the respiratory tract. These values are not directly observable in human subjects. In Table 1 are listed calculated amounts deposited in various regions under different respiratory patterns, as a function of particle diameter for unit density spheres. The values are for mouth breathing. For nose breathing one can estimate from Fig. 1 the amounts removed before entering the lungs.

The importance of localization is illustrated by the case of the action of adenovirus. Estimates are that as few as three virus particles, if deposited deep in the lungs, corresponds to a threshold quantity, i.e.,

12. PARTICULATE DEPOSITION IN RESPIRATION

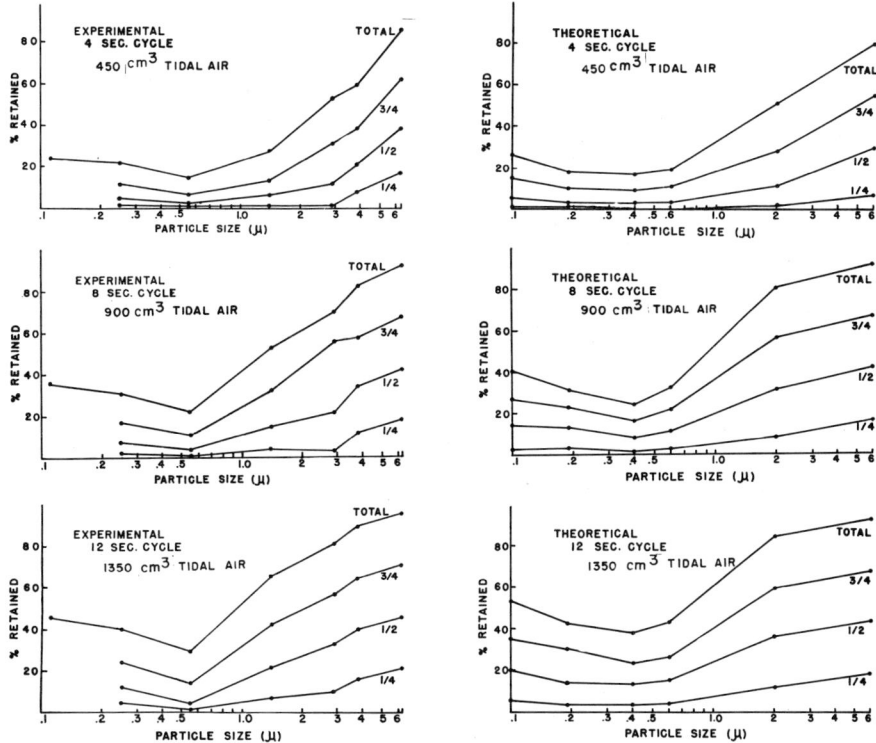

Fig. 2. Experimental and theoretical results on the fate of droplets of various sizes inhaled by human subjects under several conditions. (Reprinted from Ref. 19, Copyright 1951, American Medical Association.)

production of disease symptoms in half the population tested, while it required perhaps seven hundred particles deposited in the nasal passages to produce threshold effect [11, 12]. This is an example of the importance of knowing factors determining deposition [13-16].

While the values calculated are for particles suspended in air and for a subject under the earth's gravitational field, we may calculate what the deposition would be for an individual in space, or for an individual breathing gas mixtures other than air, e.g., a deep sea diver [12, 15, 17].

4. REDISTRIBUTION OF THE DEPOSITED PARTICLES

It is of importance to know the distribution of the deposition of material, since the subsequent fate of a particle depends on the region in which

TABLE 1

Retention in Various Regions of the Respiratory Tract[a]

Air flow	Particle diameter (μ)	Mouth and pharynx	% Retention			
			Upper lung	Middle lung	Lower lung	Total
300 cm³/sec	20	23	41	29	0	93
450 cm³	6	0	7	46	30	83
tidal air	2	0	1	15	25	41
	0.6	0	0	8	8	16
	0.2	0	0	11	11	22
1000 cm³/sec	20	28	60	11	0	99
1500 cm³	6	2	20	42	31	95
tidal air	2	0	3	13	43	59
	0.6	0	0	5	16	21
	0.2	0	0	9	20	29

[a] At a rate of 15 respirations/min.

12. PARTICULATE DEPOSITION IN RESPIRATION

it comes to rest. If it is an insoluble particle deposited in the trachea or bronchi, the ciliary action of the lining of the human lung will rapidly move it upward and it will be swallowed or expectorated. On the other hand, if it is located in the alveoli, it may remain a long time unless it is a material which is engulfed by phagocytes and subsequently brought into the blood or lymph. Thus there are many factors determining the particle's fate. Among the most important properties of the particle are solubility and permeability of membranes to the dissolved component. Thus a very soluble material that can easily cross cell membranes can be found in the blood stream in a very short time. But the physiological reaction it produces, which may enhance its removal is also important. An insoluble substance not consumed by phagocytes may remain indefinitely. If such a material acts locally as a carcinogen, then the stimulation of phagocytosis may be beneficial. But if the material is only toxic to some target organ, increased phagocytosis acts to increase the toxic effect of the airborne material.

Some of the principal variables which determine redistribution of deposited materials are illustrated in Fig. 3 [18]. The broken lines indicate airborne transport, while the solid lines indicate other means of translocation. Thus, for example, (d) represents the rate of transport from the upper lung due to ciliary action, and (j) represents absorption into the blood from the gastrointestinal tract (G. I.). Similarly, (h) and (i) represent passage from the lower lungs into the lymph and out into the blood, including solubility and phagocytic action. Losses from the G. I. tract and blood and metabolic inactivation are not indicated explicitly in the figure.

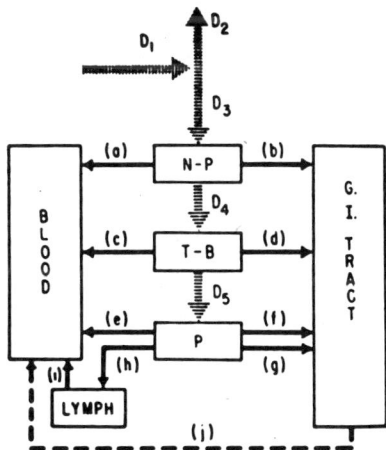

Fig. 3. Schematic representation of the fate of airborne particulates in the human lung. (Reprinted from Ref. 18, Copyright Pergamon Press, 1966).

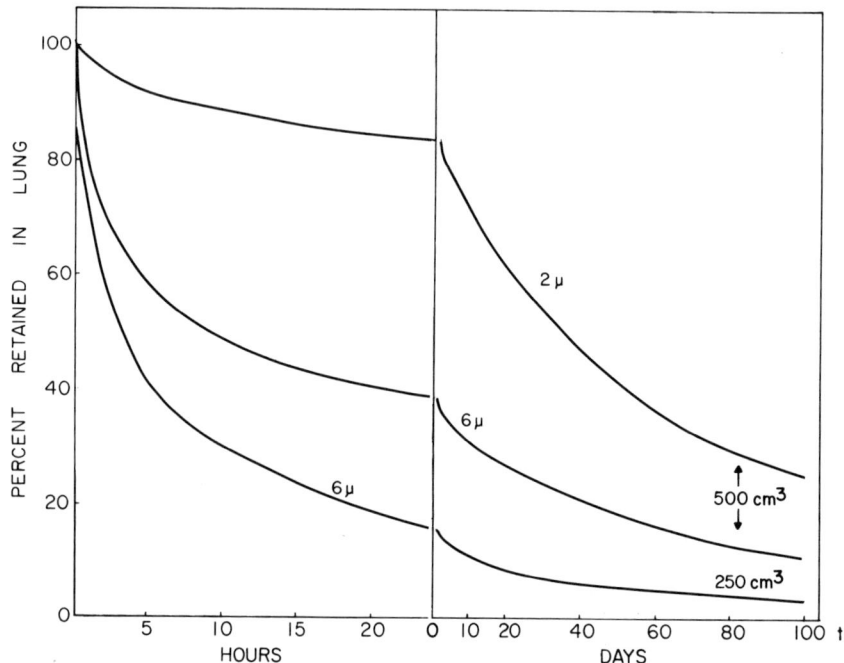

Fig. 4. Percent airborne material deposited in the lung at time 0 which still remains at time t. Values are calculated for 2- and 6-μ-diameter particles when inhaled at 15 respirations/minute with 250 or 500 cm^3 tidal air, using results from Table 1 and a scheme similar to that of Fig. 3.

It may be instructive to give an illustration of the way in which particle size and respiratory pattern can influence the length of time particles will remain in the lungs. We assume that breathing is through the mouth, and we take the values 1 hour, 10 hours, and 100 days for the time constants of particles deposited in the upper, middle, and lower regions of the lungs, respectively. Then, for the 4-sec-cycle 500-ml tidal air pattern, we can calculate the proportion of particles deposited in various regions of the lungs for particles of diameters 2 μ and 6 μ and similarly for a shallow-respiratory-cycle 250-ml tidal air volume. The results of these cases are shown in Fig. 4. It can be seen that the probability of a particle remaining in the lungs depends very much on the particle size and the respiratory pattern, as well as on other factors.

REFERENCES

1. L. Page, Introduction to Theoretical Physics, D. Van Nostrand Co., New York, 1928.

2. W. Findeisen, Über das absetzen kleiner, in der Luft suspendierten Teilchen in der menschlichen Lunge bei der Atmung, Pflügers Arch. Ges. Physiol., 236, 367 (1935).

3. H. D. Landahl, On the removal of airborne droplets by the human respiratory tract: I. The lung. Bull. Math. Biophys., 12, 43 (1950).

4. B. Altshuler, L. Yarmus, and E. D. Palmes, Aerosol deposition in the human respiratory tract, Arch. Ind. Health, 15, 293 (1957).

5. J. M. Beeckmans, The deposition of aerosols in the respiratory tract, Can. J. Physiol. Pharmacol., 43, 157 (1965).

6. H. D. Landahl, On the removal of airborne droplets by the human respiratory tract: II. The nasal passages. Bull. Math. Biophys., 12, 161 (1950).

7. E. R. Weibel, Morphometry of the Human Lung, Academic Press, New York, 1963.

8. K. Horsfield and G. Cumming, Morphology of the bronchial tree in man, J. Appl. Physiol., 24, 373 (1968).

9. H. D. Landahl and T. Tracewell, Penetration of airborne particulates through the human nose. II. J. Ind. Hygiene Toxicol., 31, 55 (1949).

10. H. D. Landahl, Particle removal by the respiratory system. Bull. Math. Biophys., 25, 29 (1963).

11. H. A. Druett, The inhalation and retention of particles in the human respiratory system, in Airborne Microbes (P. H. Gregory and J. L. Monteith, eds.) 17th Symp. of the Soc. for General Microbiol., Cambridge Univ. Press, London, 1967, p. 165.

12. V. Knight, R. B. Couch, and H. D. Landahl, The effect of the lack of gravity on airborne infection during space flight. J. Amer. Med. Assoc., 214, 513 (1970).

13. T. F. Hatch and P. Gross, Pulmonary Deposition and Retention of Inhaled Aerosols, Academic Press, New York, 1964.

14. P. E. Morrow, Particulate matter: General considerations and aerosol deposition in man, in Physiology in the Space Environment, Vol. 2, Respiration, Chap. 16, National Research Council Publication 1485-B, Nat. Acad. Sci., Washington, D. C., 1967, p. 123.

15. T. T. Mercer, P. E. Morrow, and W. Stöber, Eds., Assessment of Airborne Particles. Charles C. Thomas, Springfield, Ill., 1972.

16. N. A. Fuchs, The Mechanics of Aerosols, Pergamon Press, Oxford, 1964.

17. J. M. Beeckmans, Alveolar deposition of aerosols on the moon and in outer space. Nature, 211, 208 (1966).

18. (Task Group on Lung Dynamics), Deposition and retention models for internal dosimetry of the human respiratory tract. Health Physics, 12, 173 (1966).

19. H. D. Landahl, T. N. Tracewell, and W. H. Lassen, On the retention of air-borne particulates in the human lung: II. AMA Arch. Ind. Hygiene Occup. Med., 3, 359 (1951).

DISCUSSION

Stanley V. Dawson

Harvard University
Boston, Massachusetts

The basic approach of this work on the important problem of deposition of airborne particles in lung is very helpful. The agreement between the modelling predictions and the actual measurements seems remarkably good. Nevertheless, some questions are raised in this discussion about details that may affect the reliability of this model of the deposition process. Some of the remarks are closely related to points previously made by Professor Landahl (Refs. 3, 6, and 10 of Paper 12) as well as to points in work as recent as that of Altshuler [1], where additional references are given. The remarks here concern the lung and not the nasopharygeal passages, which have been separately tested.

The geometry of the bronchial tree used in the model has two worrisome aspects. One aspect is that the dimensions are apparently (Ref. 3 of Paper 12) those given by Findeisen (Ref. 2 of Paper 12), which have been heavily criticized because of distortions, for example by Weibel (Ref. 7 of Paper 12). The other worrisome aspect is the assumption of regular branching of the tree, especially in view of extensive measurements of Horsfield et al. [2] showing how highly irregular the branching is. The irregularities of branching affect flow rates, to which inertial deposition is especially sensitive. Also, the use of regular branching would seem to preclude any detailed check of local deposition in the lung, and such a check is important to the validation of the model.

Another geometrical aspect of validation of the work is how the predictions respond to gross changes of normal (nonpathological) lung condition. It is not clear, for example, how lung volume affects the calculations. Also, the model apparently predicts the same sort of deposition for inspiration and expiration and for vertical and horizontal lungs, neither of which may be true.

The method of deriving the effects of the mechanisms of collection indicates implicitly that the class of applicable particles is restricted to those that are nearly spherical and nonhygroscopic and electrically neutral. Apparently the test aerosols satisfied these restrictions, but many naturally occurring particles violate them.

The prediction of the effect of all mechanisms of particle collection depends on the pattern of flow that occurs in the airways. In some recent work, Schroter and Sudlow [3] have obtained details of the flow field during inspiration in a simulated airway, finding a sharp boundary layer on the bifurcation and strong secondary flows downstream of the bifurcation. There have also been direct measurements of inertial deposition in situations that are somewhat analogous to the present modelling, with a finding of a markedly different form of inertial deposition efficiency (Ref. 16 of Paper 12, Fig. 44) than that used by the author (Ref. 3 of Paper 12, Eq. (1)).

A task for future work would be to compare results obtained by the several investigators who have measured and predicted deposition. Some of the gross contradictions already partly indicated, for example, by Fuchs (Ref. 16 of Paper 12, Fig. 62) need explanation.

REFERENCES

1. B. Altshuler, Behavior of airborne particles in the respiratory tract, in Circulatory and Respiratory Mass Transport (G. E. W. Wolstenholme and J. Knight, p. 215, London, 1969.

2. K. Horsfield, G. Dart, D. E. Olson, G. F. Filley, and G. Cumming, Models of a human bronchial tree, J. Appl. Physiol., 31(2), 207 (1971).

3. R. C. Schroter and M. F. Sudlow, Flow patterns in models of the human bronchial airways, Resp. Physiol., 7, 341 (1969).

AUTHOR'S REPLY TO COMMENTS BY STANLEY V. DAWSON

H. D. Landahl

University of California
San Francisco, California

The comments made by Dr. Dawson are very pertinent. A great deal of additional work, of course, needs to be done.

In regard to the comment on lung volume, it should be said that, for the different respiratory patterns, the calculations were based on the values of the dimensions corresponding to the mean lung volume in each instance. The alveolar ducts and sacs were assumed to increase to accommodate the volume change. Increased volume primarily acted through the increased proportion reaching a deeper region, but this is compensated somewhat by greater distances through which diffusion and sedimentation must act. The calculated deposition for inspiration is substantially greater than for expiration, since (1) expiration can only act on what has not been deposited at inspiration and (2) the impaction is not as effective on expiration. The calculations assumed a nearly upright position. Calculated values of deposition for large particles would be increased, especially for a subject lying on his back, assuming no other changes resulting from a different mode of breathing. It would be interesting to compare the retention under a given respiratory pattern for diaphragm vs chest-cage breathing.

13

SOME BASIC PROBLEMS OF ESTUARINE WATER QUALITY CONTROL

J. R. West*

University of Dundee
Dundee, Scotland

and

D. J. A. Williams

University College
Swansea, Wales

1. INTRODUCTION

In the past, the estuarine environment has provided, with little or no interference from man, facilities for navigation, recreation, waste disposal, and food production.

In many areas, in order that estuaries may be used to a maximum economic advantage to man, compatible with current aesthetic standards and acceptable ecological changes, it is becoming of increasing importance to control and manage human activities associated with the estuarine environment. In Scotland, the increasing use and conflicting demands made by man on his limited estuarine resources have led to statutory powers being available to River Purification Boards, so that the use of estuaries for waste disposal may be controlled.

---

*Present address: The University of Birmingham, Birmingham, England.

Such control of estuaries requires

1. a knowledge of existing conditions of the waters,

2. a knowledge of the effects of proposed discharges, and

3. an awareness of the political, social and economic implications of existing and proposed works.

In order to deduce acceptable limits for the volume and concentration of a discharge it is necessary

1. to be able to predict the spatial and temporal distribution of pollutant concentration in the receiving waters, and

2. to be able to assess the ecological, hydrodynamic, aesthetic, and public health consequences of such distributions.

Due to their size, complexity, and rigorous physical conditions, our knowledge and understanding of estuaries is far from being complete. This lack of understanding of many fundamental biological, physical, and chemical processes in estuarine waters often necessitates the use in water quality control of a parametric representation of variables and the use of empirical relationships between these parameters.

Such procedures may be demonstrated by reference to two aspects of water quality control studied recently in the Firth of Tay: first, the use of salt as a naturally occurring tracer to aid pollutant dispersion analysis, and, second, the use of a biological indicator for classifying water for recreational use.

Both of these techniques depend upon the use of statistics because of difficulties in obtaining a detailed understanding of the complex water motion associated with the tidally induced mixing of fresh river water and denser sea water.

2. THE NATURE OF THE PHYSICAL PROBLEMS

Sea water is nearly 3% denser than fresh water [1]. In estuaries the movement of water is dependent on river discharge and the magnitude of the tidal water level oscillations. This dependence has been qualitatively described by Pritchard [2].

13. ESTUARINE WATER QUALITY CONTROL 195

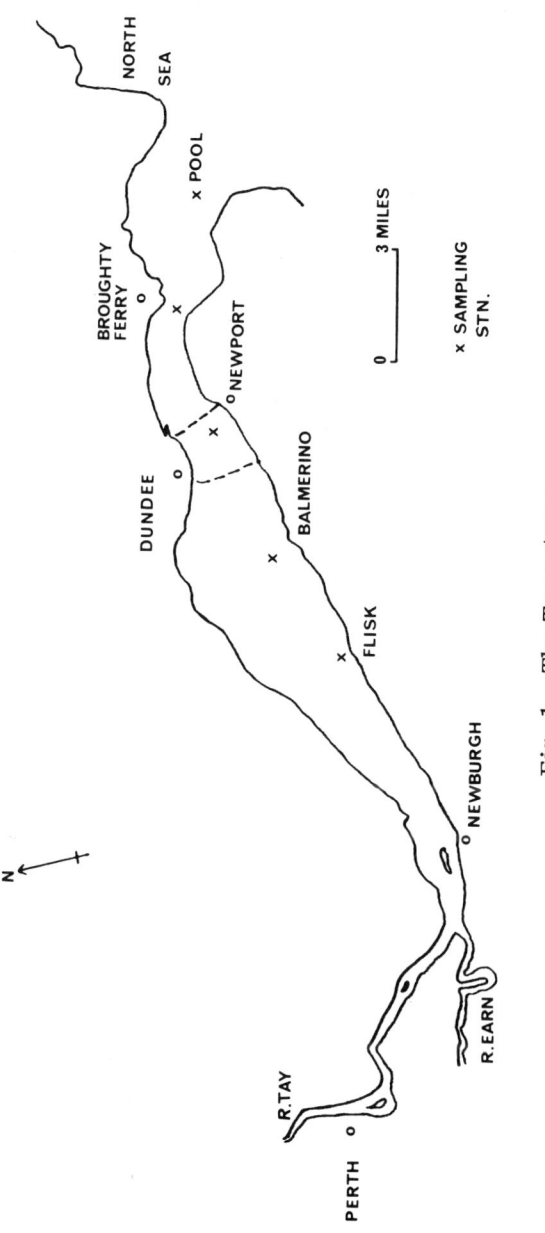

Fig. 1. The Tay estuary.

The Tay Estuary (Fig. 1) links the Rivers Tay and Earn to the North Sea. The limit of tidal motion is near Perth, about 50 km from the mouth of the estuary. The mean tidal range at Dundee is 3.5 m [3]. Salinity intrusion extends for about 30 km during low river flows. The mean daily discharge of the rivers is 180 m^3/sec (Tay 160 m^3/sec, Earn 20 m^3/sec) [4]. The maximum width of the estuary is 5 km and the maximum depth 25 m.

Ideally, for a complete understanding of water movement we must know the three-dimensional spatial and the temporal distribution of salinity and velocity for all the sequences of marine and fluvial influences in which we are interested. A combination of finite economic resources and the geographical size of the Tay led to the necessity of considering the problem of solute movement in only one dimension, and the problem of biological standard definition at a fixed point in space.

3. ONE-DIMENSIONAL REPRESENTATION OF ESTUARINE MIXING

A solute mass balance equation based on fundamental physical principles may be derived for turbulent flow in estuaries. In order to simplify this equation so that it may be compatible with existing field data, it is necessary to reduce it to a semiempirical, temporally and spatially averaged form.

Consider a rectangular element of a continuum with sides of length x_i parallel to a rectilinear coordinate system x_i ($i = 1, 2, 3$). Application of the principle of conservation of mass to the solute gives

$$\frac{\partial}{\partial t} \rho c + \frac{\partial}{\partial x_i} \rho c u_i = 0, \tag{1}$$

where c is the solute concentration of a conservative substance, ρ is the solution density, and u_i is the component of velocity parallel to x_i axis.

The variation of density in estuaries is usually less than 2%; thus it may be assumed constant. Therefore

$$\frac{\partial}{\partial t} c + \frac{\partial}{\partial x_i} c u_i = 0. \tag{2}$$

Following Reynolds' work [12], for turbulent flow, writing each dependent variable as the sum of a mean term and a fluctuating term,

$$u_i = \bar{u}_i + u'_i \quad \text{and} \quad c = \bar{c} + c',$$

13. ESTUARINE WATER QUALITY CONTROL

where

$$\bar{u}_i = \frac{1}{\Delta t} \int_{\Delta t} u_i \, dt \quad \text{and} \quad \bar{c} = \frac{1}{\Delta t} \int_{\Delta t} c \, dt,$$

which implies

$$\int_{\Delta t} u_i' \, dt = 0 \quad \text{and} \quad \int_{\Delta t} c' \, dt = 0,$$

and time-averaging Eq. (2) over a time interval Δt, where Δt is very much greater than the period of molecular oscillations, but much less than the dominant tidal period, gives

$$\int_{\Delta t} \frac{\partial}{\partial t} (\bar{c} + c') \, dt + \int_{\Delta t} \frac{\partial}{\partial x_i} (\bar{c} + c')(\bar{u}_i + u_i') \, dt = 0. \tag{3}$$

Therefore

$$\frac{\partial}{\partial t} \bar{c} + \frac{\partial}{\partial x_i} \bar{c} \, \bar{u}_i + \frac{\partial}{\partial x_i} < u_i' \, c' >_{\Delta t} = 0 \tag{4}$$

where $< \quad >_{\Delta t}$ denotes average over the interval Δt.

Similarly, taking a spatial average over a cross-sectional area A, having used the substitution

$$\bar{c} = c_A + c_A', \quad \bar{u}_i = u_{iA} + u_{iA}',$$

where

$$c_A = \frac{1}{A} \iint_A \bar{c} \, dA, \quad u_{iA} = \frac{1}{A} \iint_A \bar{u}_i \, dA,$$

and assuming that A varies slowly with time, gives

$$\iint_A \frac{\partial}{\partial t} (c_A + c_A') \, dA + \iint_A \frac{\partial}{\partial x_i} (c_A + c_A')(u_{iA} + u_{iA}') \, dA$$
$$+ \iint_A \frac{\partial}{\partial x_i} < u_i' \, c' >_{\Delta t} \, dA = 0. \tag{5}$$

Therefore

$$\frac{\partial}{\partial t}(c_A A) + \frac{\partial}{\partial x_i}(c_A A u_{iA}) + \frac{\partial}{\partial x_i}(A \ll u_i' c' >_{\Delta t} + u_{iA}' c_A' >_A) = 0. \quad (6)$$

Equation (6) may be time-averaged over a tidal period in order to remove the tidal components of velocity and concentration. The algebra is tedious [11], and a similar result may be achieved from Eq. (2) by taking a temporal average over a tidal period T and then a spatial average over a cross-sectional area A_T, giving in turn

$$\frac{\partial}{\partial t} c_T + \frac{\partial}{\partial x_i} u_{iT} c_T + \frac{\partial}{\partial x_i} < u_{iT}' c_T' >_T = 0, \quad (7)$$

where A_T is the maximum cross-sectional area of flow during tidal period T, and

$$c = c_T + c_T', \quad u_i = u_{iT} + u_{iT}'$$

$$c_T = \frac{1}{T} \int_T c \, dt, \quad u_{iT} = \frac{1}{T} \int_T u_i \, dt,$$

and

$$\frac{\partial}{\partial t} c_{TA} A_T + \frac{\partial}{\partial x_i} c_{TA} A_T u_{iTA}$$

$$+ \frac{\partial}{\partial x_i} A_T \ll u_{iT}' c_T' >_T + u_{iTA}' c_{TA}' >_{A_T} = 0, \quad (8)$$

where

$$C_{TA} = \frac{1}{A_T} \iint_{A_T} c_{TA} \, dA \quad \text{and} \quad c_T = c_{TA} + c_{TA}', \text{ etc.}$$

Equations (4), (6), and (8) contain instantaneous velocity and solute concentration terms for which data are not available. Thus it is convenient to relate these fluxes due to fluctuating terms to mean concentration gradients by drawing an analogy with Fick's first law of molecular diffusion. Fick's law may be written

$$F_i = -D_{m_{ij}} \frac{\partial c}{\partial x_j}$$

13. ESTUARINE WATER QUALITY CONTROL

where F_i is the flux due to molecular diffusion in the x_i direction, $\partial c/\partial x_j$ is the concentration gradient in the x_j direction, and $D_{m_{ij}}$ is the coefficient of molecular diffusion (second-order tensor). By analogy, for turbulent flow, for the spatially averaged equation, and for the temporally (T) and spatially (AT) averaged equation, one gets, from Eq. (4):

$$<u_i' c'>_{\Delta t} = -D_{ij}' \frac{\partial \overline{c}}{\partial x_j} ; \qquad (9)$$

and from Eq. (6):

$$<<u_i' c'>_{\Delta t} + u_{iA}' c_A'>_A = -D_{ij}'' \frac{\partial c_A}{\partial x_j} ; \qquad (10)$$

and from Eq. (8):

$$<<u_{iT}' c_T'>_T + u_{iTA}' c_{TA}'>_{A_T} = -D_{ij} \frac{\partial c_{TA}}{\partial x_j} . \qquad (11)$$

Substituting Eq. (11) in Eq. (8) gives

$$\frac{\partial}{\partial t} c_{TA} A_T + \frac{\partial}{\partial x_i} c_{TA} A_T u_{iTA} - \frac{\partial}{\partial x_i} A_T D_{ij} \frac{\partial c_{TA}}{\partial x_j} = 0. \qquad (12)$$

The definition of D_{ij} in terms of fluxes spatially averaged over a cross-sectional area A_T disguises its dependence on the turbulent [Eqs. (9) and (10)] and tidal fluctuations of velocity and solute concentration, though as these variations have not yet been measured for the Tay, in this discussion the point is only of academic consequence. Further, D_{ij} is a function of the scale of measurement through statistical procedures.

Since by definition the lateral and vertical variation of spatial means over a cross section equal zero, under the assumption that $D_{21} = D_{31} = 0$ and that river flow and tidal range are constant (quasisteady state), Eq. (12) may be written

$$D_{11} = \frac{c_{TA} Q_{fT}}{A_T \partial c_{TA} / \partial x_1}, \qquad (13)$$

where $Q_{fT} = A_T u_{1TA} = $ mean (T) river flow.

It is important to realize that this equation does not describe the fundamental physical processes occurring in oscillating estuarine flows, but it does provide an algebraic representation of the effects of these processes which is compatible with existing means of monitoring estuarine water quality parameters. With an adequate knowledge of the salinity distribution in time and space, Eq. (12) can be integrated numerically [14].

4. DATA ACQUISITION AND ANALYSIS

Instantaneous salinity values at a point are likely to be the combined result of the marine and fluvial inputs to the system over a period of time, probably of the order of weeks. In order to establish the magnitude and time scale of these effects, a station continuously monitoring salinity and water level was installed at Newport. This site came closest of the many considered to satisfying the criteria of being accessible in most weather conditions, reasonably free from localized hydrodynamic peculiarities, safe from interference by the public, and not presenting a navigational hazard.

Hourly samples of water were collected by an automatic liquid sampler [7], and stored for analysis of salinity content. Technical difficulties, due mainly to the large suction head experienced at low tide, resulted in a number of gaps in the record. Water levels were recorded at quarter-hourly intervals on computer-compatible punched tape by an automatic tide gauge. River discharge data were supplied by the Department of Agriculture, Fisheries, and Food for Scotland in the form of mean daily discharges. Hydrographic data were taken from the charts prepared by the Dundee Harbour Trust.

The control station data (Fig. 2) were examined for the effects of tidal range and river flow on salinity and an estimate made of the magnitude of these effects. A function of the form

$$y = a_0 + a_1 \sin \omega t + b_1 \cos \omega t, \tag{14}$$

where $\omega = 2\pi/T$, with T = the period, and $A = \sqrt{a_1^2 + b_1^2}$ = amplitude, was fitted by the least squares method to ten months' salinity and water level data. Examination of the amplitudes (Fig. 3) of the approximated equation for values of T from 11.90 to 13.00 hours in steps of 0.01 hours revealed large amplitudes for T = 12.42 hours (semidiurnal component) and for T = 12.00 hours (fortnightly component).

Insight into the salinity-river flow relationship was achieved by filtering out a major part of the tidal effects by filtering Eq. (14) to various intervals of data and taking a_0 as a parametric representation of salinity in

13. ESTUARINE WATER QUALITY CONTROL

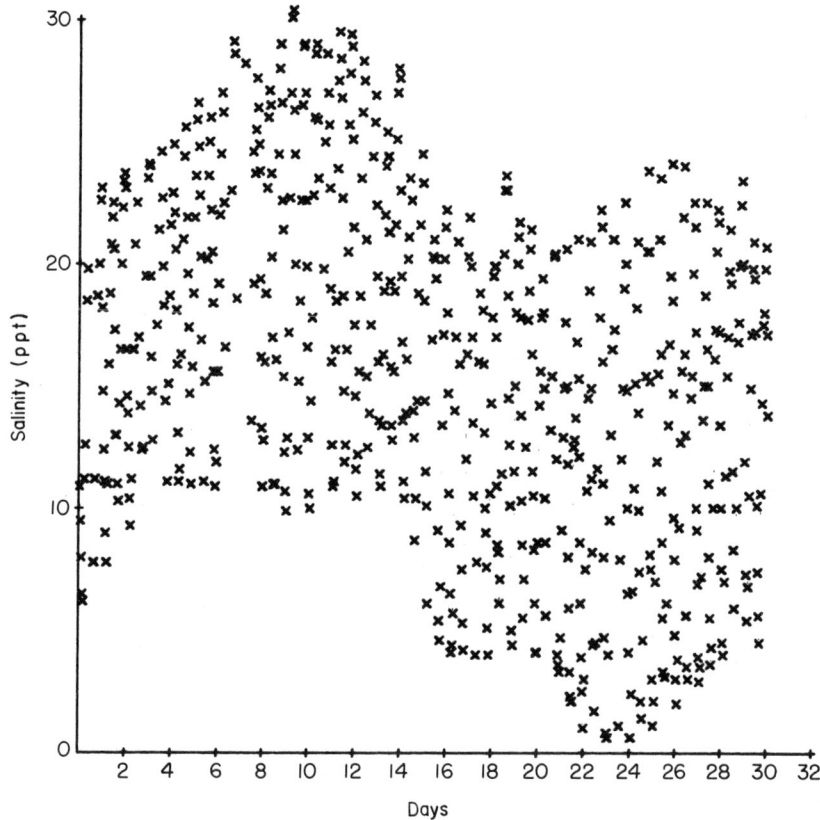

Fig. 2. Salinity variation with time (autosampler data), Newport, January, 1970.

that interval. For intervals of 12 hours (Fig. 4) the distribution of a_0 showed considerable noise, and for intervals of 48 hours (Fig. 5) little noise, but some detail of the record was being lost. A compromise of 24 hours was accepted and values of a_0 were plotted (Fig. 6) against an arithmetic average of the daily mean fluvial flow for the preceding seven days. Study of the records of several floods indicated that seven days was approximately the average time between the peak discharge at the gauging station and the corresponding minimum mean (24 hours) salinity at Newport Pier. An example is shown for the latter part of January, 1970 (Fig. 7).

A linear relationship could be approximated to the salinity-river flow data. Such a relationship must of necessity be of an approximate nature,

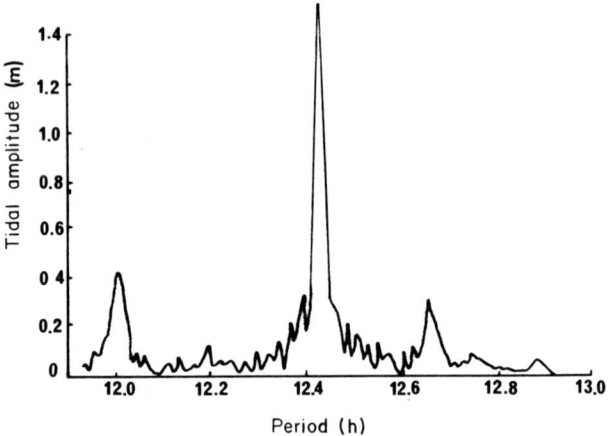

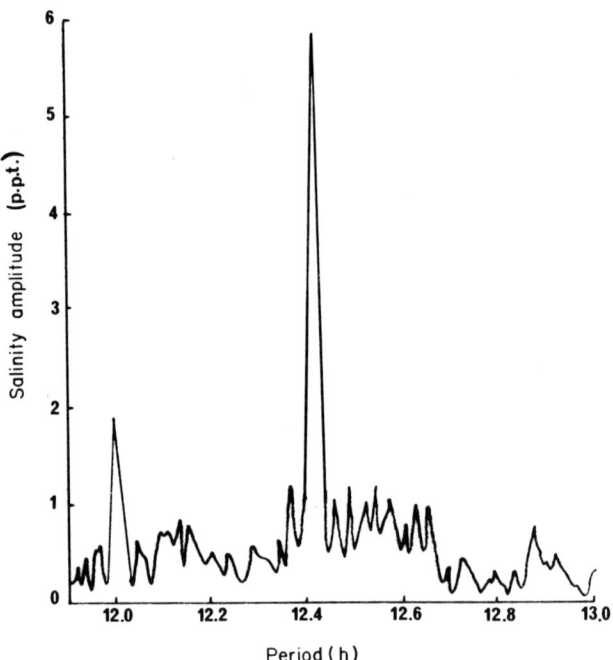

Fig. 3. Variation of amplitude with period for salinity and water level time series.

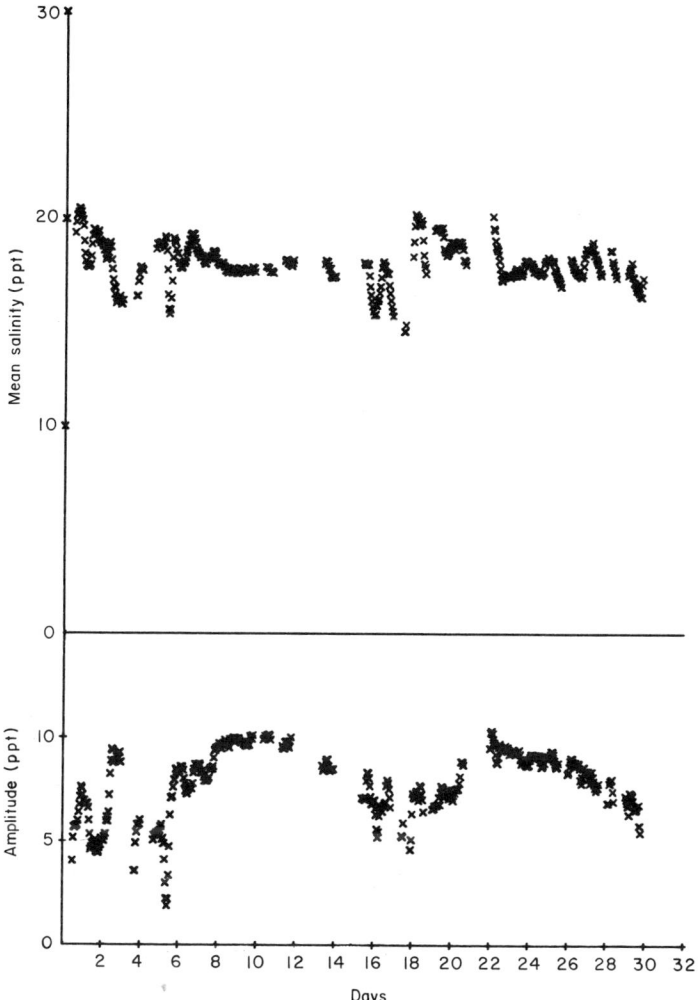

Fig. 4. Mean (12 h) and amplitude of salinity variation with time at Newport, November, 1969. Length of average, 12 h; Step interval of periods, 1 h; Minimum number of readings per period, 10.

because of the dynamic nature of the system and the variability in the maximum flow, volume, and duration of fluvial floods.

Having established that the salinity intrusion might be detected and characterized, salinity data were collected from five stations in the main channel, for several tidal ranges and river flows, and examined for a

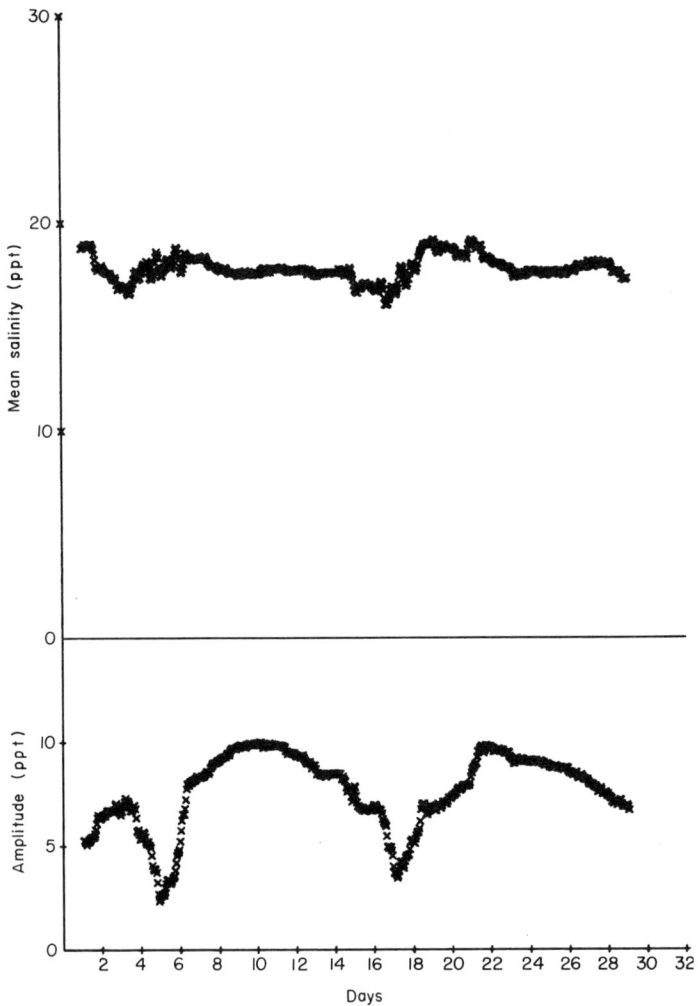

Fig. 5. Mean (48 h) and amplitude of salinity variation with time at Newport, November, 1969. Length of average, 48 h; step interval of periods, 1 h; minimum number of readings per period, 8.

similar relationship. At each station vertical profiles of velocity and salinity were measured at intervals of about 20 min for complete flood and ebb tides. Spatial mean values of salinity were calculated by assuming a linear variation between values and then taking a spatial average. Lateral variation of salinity was assumed to be small.

13. ESTUARINE WATER QUALITY CONTROL

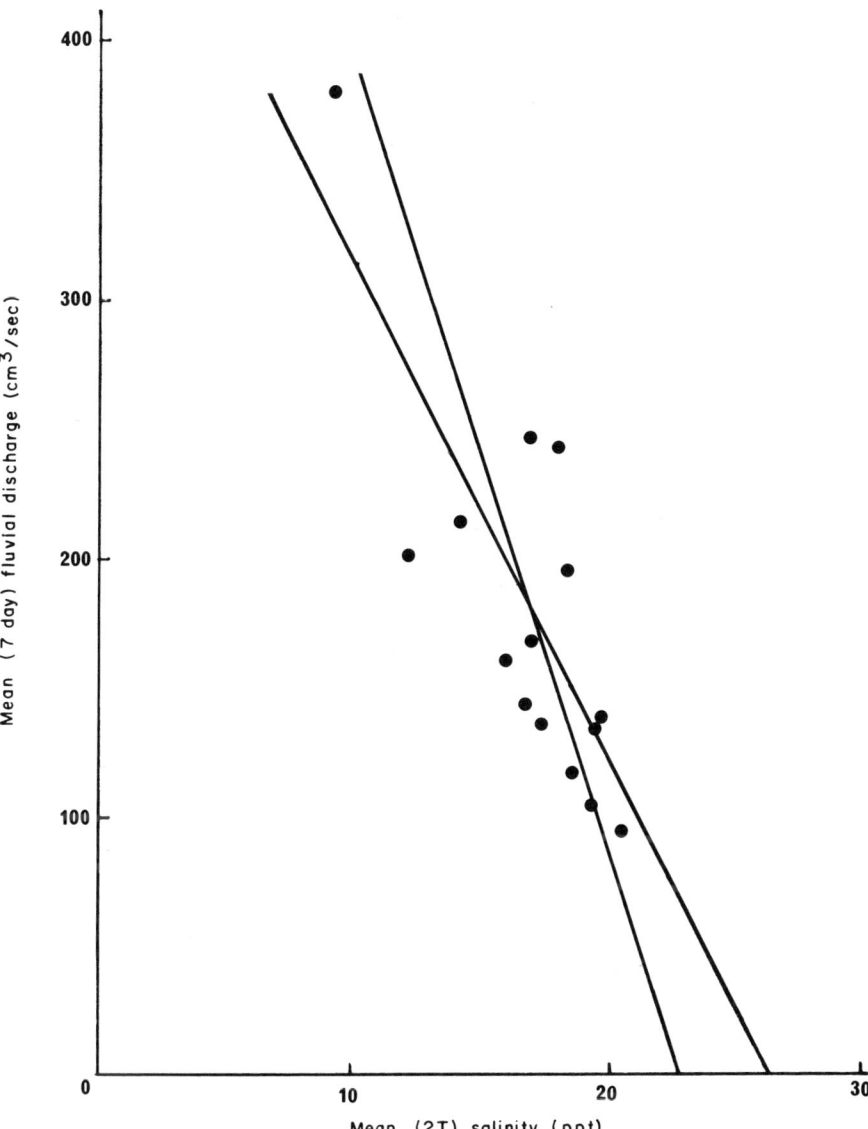

Fig. 6. Variation of mean (2T) salinity with mean (7 day) fluvial discharge.

$$c_A = \frac{1}{2z_N} \sum_{n=0}^{N} (c_n + c_{n-1})(z_n - z_{n-1}) \qquad (15)$$

where z_N is the total depth, N is the number of values in vertical distribution, c_n is the salinity at point n; and $c_0 = c_1$. A sixth-order polynomial was then fitted to the spatial mean values of salinity (Fig. 8) and the mean salinity c_{AT} evaluated from this function by temporal averaging:

$$c_A(t) = \sum_{n=0}^{6} a_n t^n,$$

where the a_n are coefficients of the fitted polynomial and

$$c_{AT} = \frac{2}{T} \int_0^{T/2} c_A(t) \, dt.$$

At four stations the relationship between mean (A, T) salinity and river flow closely approximated to a linear function (Fig. 9). At Balmerino, a nearly constant river flow on all occasions for which salinity was monitored resulted in the slope of the assumed linear function being interpolated with respect to the Newport and Flisk data, and its position fixed by the centroid of the data at that station. Knowledge of these linear relationships makes it possible to evaluate a one-dimensional dispersion coefficient in accordance with Eq. (13).

Salinity may be considered conservative and is readily detectable in situ. Bacteria pathogenic to man and transmitted by ingestion of fecally polluted water are nonconservative and difficult to detect and enumerate. A nonpathogenic organism, Escherichia coli, present in human feces in large numbers and relatively easy to enumerate, is used as an indicator of fecal pollution and the consequent risk of infection.

The survival rate of E. coli in sea water has as yet not been satisfactorily quantified [13]. Attempts by Gameson and co-workers [8, 9] to account statistically for the variability in bacteria counts in terms of the large number of variables involved, e.g., sea roughness, solar radiation, tidal state, and sea temperature, produced inconclusive results. West and Williams [10] have shown, with the aid of an equation similar to Eq. (14) and by choosing a hydrodynamically suitable sampling site, that the variation due to tidal state can be detected and hence the bacterial population may be quantified in the parametric form of a_0 and A. This form can be used to (i) interpret the variation of E. coli over a tidal period and so obtain statistically meaningful data for use in bacteriological and epidemiological studies, and (ii) allow interpretation of other cyclic variables over

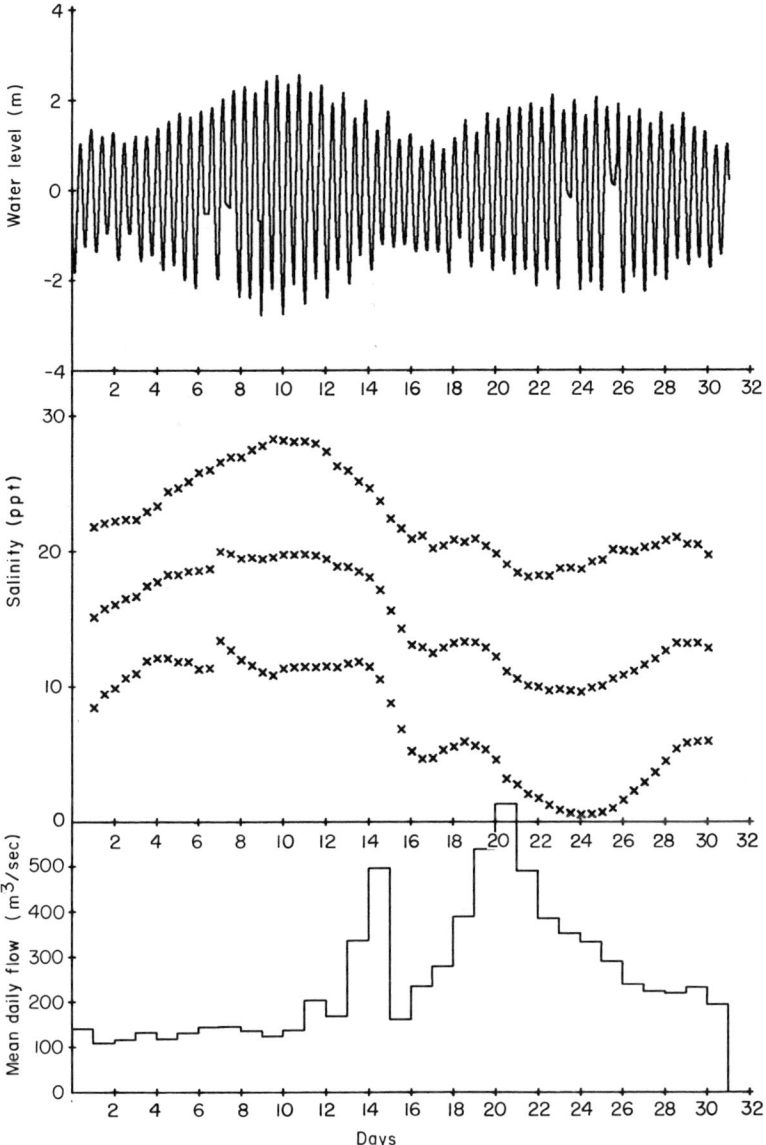

Fig. 7. Water level (Newport), maximum, mean, and minimum salinity (Newport), and river discharge (Tay and Earn) variation with time, January, 1970.

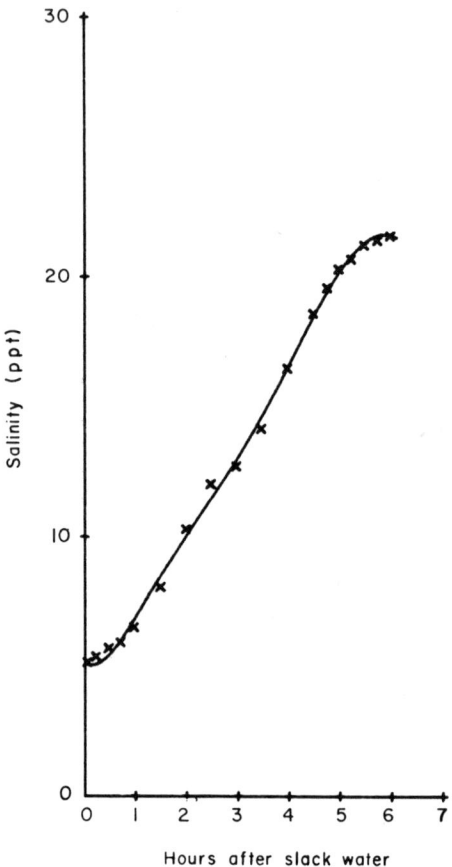

Fig. 8. Mean (z) salinity variation with time, June 3, 1970, Station No. 3; order of fitted polynomial, 6.

a tidal period and so permit aspects of their respective variations with tidal state to be compared. To this end salinity and water temperature data were fitted to the model.

Typical experimental results for E. coli, temperature, and salinity are shown in Figs. 10, 11, and 12, together with the least squares fitted function. The results for salinity and temperature indicate a good fit with the model. Although the observed E. coli values are more scattered, nevertheless the model has allowed determination of such salient features as a mean and maximum in bacterial numbers. The adequacy of the model is well demonstrated by the occurrence of peaks in all three variables near slack water.

13. ESTUARINE WATER QUALITY CONTROL

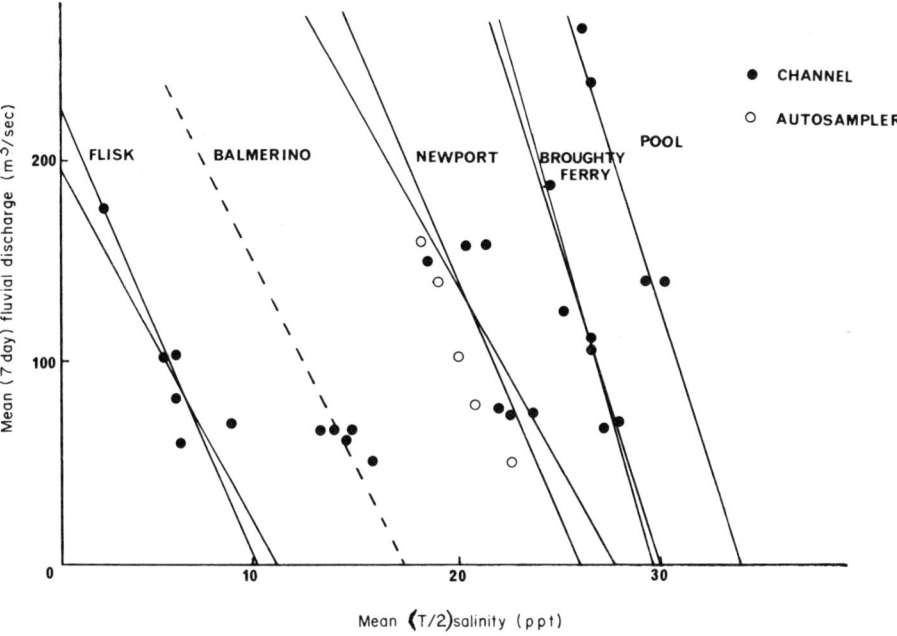

Fig. 9. Variation of mean (T/2) salinity with mean (7 day) fluvial discharge at channel stations.

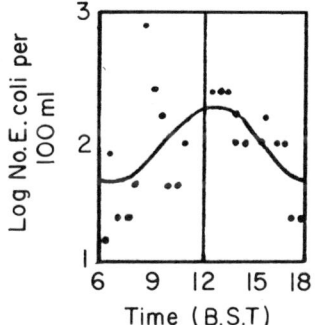

Fig. 10. Coliform (E. coli, Type 1) density variation with time.

5. CONCLUSIONS

The use of statistics has been demonstrated for two water quality control problems. Essentially the first used temporal and spatial averaging techniques in order to quantify, through analysis of field data, the effects of small fluctuations in velocity and salinity, which are at present very difficult to measure.

The second problem illustrates the use of a simple statistical model to make possible the parametric representation of a record which is subject to apparently random fluctuations and appreciable experimental error.

The statistical techniques used may be ordered into three broad categories:

1. the concept of deducing spatial and temporal averages in terms of mean and fluctuating terms,

2. the assumption and approximation of distributions to the field data, and

3. the definition of statistical parameters.

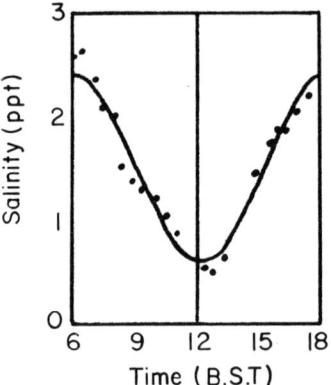

Fig. 11. Salinity variation with time.

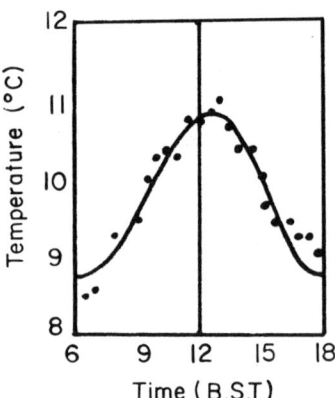

Fig. 12. Water temperature variation with time.

13. ESTUARINE WATER QUALITY CONTROL

The use of these techniques is essential, with existing measurement techniques, if spatially and temporally dependent estuarine phenomena are to be quantified. Such a treatment of instantaneous point values of salinity permits the effects of a number of complex phenomena, generally called turbulent mixing or diffusion, to be evaluated in the form of a coefficient. Studies of less readily measured variables, e.g., bacterial levels, are dependent on such techniques.

It is likely that future work on such problems as the tidal dependence of diffusion coefficients and their spatial dependence in the vertical direction will need a statistical basis for theoretical development, for planning of sampling work, and for the analysis of data.

REFERENCES

1. A. W. Harvey, The Chemistry and Fertility of Sea Waters, Cambridge Univ. Press, London, 1957.

2. D. W. Pritchard, Estuarine circulation patterns, Am. Soc. Civ. Eng., 81, 717 (1955).

3. Admiralty Tide Tables, Vol. 1 (European Waters), Publ. Hydrographer of the Navy (U.K.).

4. Surface Water Year Book of Great Britain, HMSO, London, 1968.

5. A. Fick, Ueber Diffusion, Ann. Phys. (Leipzig), 170, 59 (1855).

6. J. Crank, The Mathematics of Diffusion, Oxford Univ. Press, London, 1956.

7. J. R. West, An investigation into the circulation of the Tay Estuary, paper presented at the Local Meeting of Inst. Civ. Eng., Edinburgh, Feb., 1971.

8. A. L. H. Gameson, A. W. J. Bufton, and D. J. Gould, Studies in the coastal distribution of coliform bacteria in the vicinity of a sea outfall, J. Inst. Water Poll. Control, 66, 501 (1967).

9. A. L. H. Gameson, E. B. Pike, and D. Munro, Studies of sewage dispersion from two sea outfalls, Chem. Ind. (London), 1968, 1582.

10. J. R. West and D. J. A. Williams, The use of a simple mathematical model to account for the variation of coliform density, salinity and temperature with tidal state, in Proc. 5th Water Poll. Control Conf., Hawaii, 1970.

11. A. Okubo, Equations describing the diffusion of an introduced pollutant in a one-dimensional estuary. Stud. Oceanog., 216 (1964).

12. O. Reynolds, On the dynamical theory of incompressible viscous fluids and the determination of the criterion, Roy. Soc. London Phil. Trans., Ser. A, 186, 123 (1895).

13. G. T. Orlob, Viability of sewage bacteria in sea water, Sewage Ind. Wastes, 28, 1147 (1956).

14. W. Boicourt, A numerical model of the salinity distribution in Upper Chesapeake Bay, Tech. Rept. No. 54, Chesapeake Bay Inst., Johns Hopkins Univ., Baltimore, Maryland, 1969.

14

MODELLING CURRENT MOVEMENTS WITHIN BODIES OF WATER

D. Harverson

Edinburgh Corporation
Edinburgh, Scotland

1. INTRODUCTION

The type of model discussed here is one which is designed to reproduce or predict the track of a single, rigid object, such as a drogue. The movement of this object is assumed to be entirely governed by the currents within a particular layer of water in the area with which the model is concerned. The depth of this layer is one of the parameters of the model. Others are the prevailing condition of wind and tide, and the point at which tracking is supposed to start.

A model of this kind can form the basis for predicting a slick of oil or sewage, and can also be used to study the formation of such a slick by a process of continuous discharge [1]. Such a model may be constructed using various techniques. Those which I describe here are appropriate when dealing with broad stretches of water, particularly tidal waters. For rivers and estuaries narrow enough for the flow in any layer to be virtually one-dimensional, a simpler approach is possible [2].

In 1969, a model of this kind was constructed for use in determining the optimum location for the new effluent outfall for the City of Edinburgh [1]. This model was of the layer of water 1.5 m below the surface in an area of approximately 30 km by 16 km in the Firth of Forth. It was based on sets of simultaneous wind and current readings taken during an extensive survey of the area. Final calibration was by means of observed drogue tracks. While developing the model, considerable use was also made of the data from two previous surveys carried out at Tyneside and Teesside in the north of England [3]. In particular the method used for the analysis of wind-induced currents was originally developed with reference to these data.

2. OUTLINE OF TECHNIQUES

Raw data are supplied in the form of a number of current/wind vector pairs $\underline{c}$, $\underline{w}$ taken at various points and under various conditions within the scope of the model. The analysis of these data first attempts to establish the way in which each current vector varies with the independent variables, (i.e., in most cases) with the parameters describing the state of wind and tide. This done, the actual current readings can be replaced by standardized values, and these then extended spatially to all points which the model is to represent.

Although the techniques described are by no means purely statistical, the assumptions made about functional relationships are weak enough to commit us to a considerable reliance on the data. The data must cover the subject adequately with respect to all the independent variables which are significant within the system.

3. CORRELATION OF WIND WITH INDUCED CURRENT

We represent the relationship between wind and induced current by

$$\underline{v} = kA\underline{w},$$

where $\underline{w} = (w_E, w_N)'$ is the wind vector, $\underline{v} = (v_E, v_N)'$ is the induced current vector, k is a scalar, A is the rotation matrix

$$\begin{bmatrix} \cos x & -\sin x \\ \sin x & \cos x \end{bmatrix},$$

and x and k are taken to be constant for the site. That is, we seek an average value for the angle between a wind and the current it induces, and for the ratio of their respective magnitudes, on the assumption that these values may be taken as representative over the whole site.

To obtain a measure of the validity of this assumption, as well as values for x and k, we employ a complex regression technique. This is most readily applicable to data in the form of current/wind vector pairs $\underline{u}$,

14. MODELLING WATER CURRENT MOVEMENTS

$\underline{w}$, where the current vectors $\underline{u}$ contain only a constant basic current element $\underline{b}$ in addition to the wind-induced element $\underline{v}$.

Where the original current readings do contain some other variable element, it is first necessary to eliminate this. Where this element is tidal, it may be eliminated by obtaining the data in sets, each set covering a complete tidal cycle at a particular point, and by forming the vectorial mean of each set. The smaller the time interval between the readings which form the set, the more accurately each residual wind and current vector will represent a wind and corresponding nontidal current at that point. In the studies to which reference has been made, an interval of half an hour was used. Thus each set consisted of 25 or 26 pairs of readings.

Where there are no variable elements other than tidal and wind-induced, these residual current/wind pairs $\underline{r}$, $\underline{w}$ may be related by the equation

$$\underline{r} = kA \underline{w} + \underline{b},$$

where k and A are as defined above, and $\underline{b}$ is the constant basic current vector.

To perform the regression analysis which will give values for x, k, and $\underline{b}$, we express the vectors as complex numbers and write

$$r_E + ir_N = (c_1 + ic_2)(w_E + iw_N) + (b_E + ib_N)$$

or

$$\underline{r} = \underline{c}\,\underline{w} + \underline{b}$$

where

$$\underline{c} = ke^{ix} = k(\cos x + i \sin x).$$

The quantity to be minimized in the regression analysis is then

$$R = \Sigma\,[\underline{r} - (\underline{c}\,\underline{w} + \underline{b})]\,\overline{[\underline{r} - (\underline{c}\,\underline{w} + \underline{b})]}$$

where the bar denotes the complex conjugate. Full details of this technique are given in Ref. 4.

4. VARIATIONS IN THE BASIC DRIFT

In the offshore sites at Tyneside and Teesside, the assumption was made that the basic current was constant over the whole site. This gave correlation coefficients of 0.66 and 0.56, for 37 and 32 sets of readings, respectively. In Edinburgh, where the number of sets of readings was 72, the same assumption led to the unsatisfactory value for the correlation coefficient of 0.32. Grouping the data according to location raised this to 0.55 for a group of 38, with no groupings being found for the rest of the data which raised the significance to a respectable level, although for topographically linked small groups the correlation coefficient was high.

It was assumed that the poor overall correlation was due to variations in the basic current. This was borne out by the topography of the site, which is illustrated in Fig. 1. The curve of the bay, into which a small river flowed, and the considerable range in the depth of water (including the presence of Inchkeith Island), would all contribute to such variations.

In some sites within estuaries, the spread of the readings over time might also cause problems. It might, for example, be necessary to obtain measurements of the flow of river water entering an estuary, and to attempt to correlate the basic current with this.

Returning to the first equation of Section 3, the values obtained at Edinburgh for k and x were 0.01 and -24°, respectively. That is, the current induced at a depth of 1.5 m by any given wind is (on average) rotated clockwise through an angle of 24°, and reduced in magnitude by 1/100.

5. SINUSOIDAL REPRESENTATION OF CURRENT

Having obtained our relationship between wind and wind-induced current, we may now return to our original sets of readings and remove the wind-induced element from each individual wind/current vector. Each of these is then resolved orthogonally, say East and North, so that from each set of readings we obtain two sets of scalar quantities.

We now seek a sinusoidal representation of each of these sets. This procedure summarizes the tidal data at each point, and in doing so, smooths out the scatter caused by the various random elements which no survey procedure can eliminate. It also provides a value for the basic current at that point, which should be compared with any values for basic currents previously calculated by regression analysis.

At Edinburgh, an adequate representation was given by a simple sine curve. In many estuaries and some offshore sites, however, an algorithm to add the appropriate number of harmonics would be required.

14. MODELLING WATER CURRENT MOVEMENTS

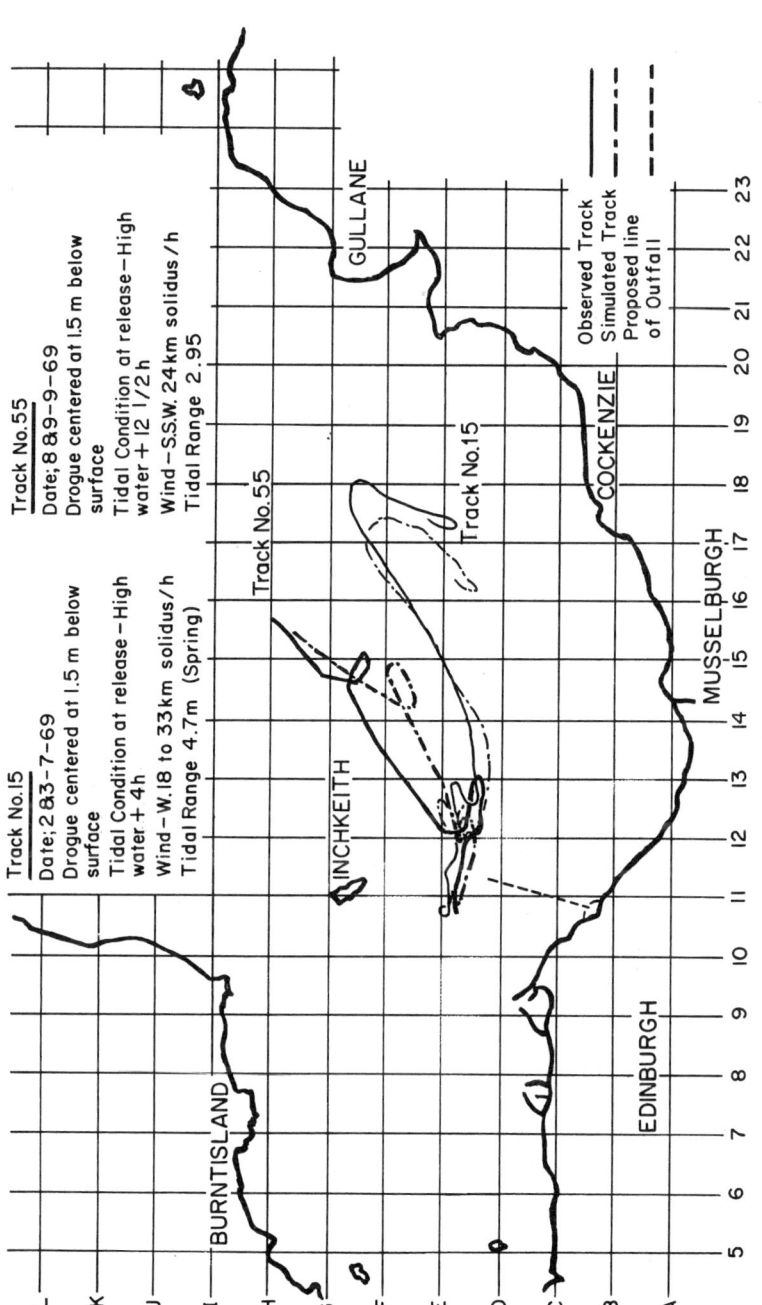

Fig. 1. Comparison of observed and simulated drogue tracks.

Having obtained a sinusoidal representation for the tidal plus basic current at each point at which readings were taken, we may now further summarize the data.

The _period_ of the dominant sine curve is the period of the corresponding tidal cycle. This can be standardized and then ignored.

The _amplitude_ of the curve gives the magnitude at that point of that component of the tidal current at full flow. The period having been standardized, this magnitude will be linearly dependent on the tidal range. We therefore replace it by a constant of proportionality, which enables us to calculate the amplitude corresponding to any given tidal range.

The _displacement_ gives the drift at each point. If this has been found to be dependent on some independent variable, its actual value must also be replaced by a factor or factors defining this relationship.

This leaves the _phase_, which may be defined as the time lag between the sine curve under consideration and that described by the changes in water level at some base point. At many sites, this will be the same for all points and for both directions of resolution, and may therefore be ignored. At the Edinburgh site, there was a marked phase difference between the two directions of resolution at certain points. The major reason for this was thought to be the curve of the bay, which appeared to cause the North-South and East-West tides to turn at different times in the central area of the site.

6. SPATIAL EXTENSION OF THE DATA

For this part of the analysis, a number of techniques are worth considering. When planning the work to be done at Edinburgh, reference to the data from the Teesside and Tyneside studies suggested the practicability of three methods in particular. One involved finding a pair of surfaces, each the best least-squares fit to one component of the data, after it had been prepared as described in Section 5. The other two involved the solution of equations representing the potential nature of the current flow, formulated either as integral equations or as partial differential equations. These two latter are both boundary value problems, and would have required the data to be concentrated on the boundary of the site, or of regions within the site.

However, further acquaintance with the rather complex site at Edinburgh suggested that what was required here were methods allowing

14. MODELLING WATER CURRENT MOVEMENTS

maximum scope for intervention and adjustment. The data were therefore interpolated manually, with topographical information supplementing mathematical best fit requirements.

7. CALIBRATION AND USE OF THE MODEL

After the data have been extended in this way to all points of, say, a half-mile grid covering the area, the model may then be used to simulate recorded drogue tracks. This is done by applying parameters describing the required tidal cycle to the values stored at the grid points, interpolating these, and then adding the current generated by the required wind.

If a large number of tracks have been recorded, they can be used as supplementary data at this stage. That is, the values stored at each grid point can be adjusted to improve the match of the simulated and observed drogue tracks. This process must be used with great care, since obviously it would be possible to create wholly inaccurate data which nevertheless gave a perfect match on selected tracks. However, where a number of drogue tracks taken in good conditions conflict with fixed meter readings of poor quality, or with interpolated values, it may be legitimate to make the adjustment.

At Edinburgh, considerable use was made of drogue tracks in the final calibration of the model. On-line access to the computer enabled us to study the effect of varying parameters on the simulations of a wide range of drogue tracks. Values were eventually reached which were consistent both with these tracks and with all the fixed meter readings which were not in any way suspect.

The model was then used to investigate the suitability of a number of points for the projected sewage outfall. For each of these, tracks were predicted for drogues "released" every half hour over a complete tidal cycle, under a variety of wind conditions. These tracks were analyzed by another computer program, which applied equations for dispersion and die-off. The extent and position of the sewage field which would be produced under these conditions was thereby predicted.

The model used at Edinburgh represented only the layer of water 1.5 m below the surface, which was the estimated center of the sewage field. For a layer at any lower level, exactly the same techniques of recording and analysis could have been used. Fixed meter readings taken less than 1.5 m from the surface, however, usually contain an unacceptably high random element. The study of the movement of surface slicks, therefore, would require some variation in method.

The successful use of drogue tracks in the final calibration of the Edinburgh data suggests one such variation. That is, that the calibration should be done entirely from tracks, which might be made by flexible "sheets" centered only a few inches from the surface. To obtain the large number of tracks which would be required, the tracks could be recorded in batches, using aerial photography. As well as allowing a model to be created for surface layers, this technique might also be appropriate where what is required is a relatively low-cost model of a large area. In these days of widespread oil pollution, some such method of modelling coastal waters in general should certainly be sought.

REFERENCES

1. Forth Hydrographic Survey, J. D. & D. M. Watson Company, Report for the City and Royal Burgh of Edinburgh, February, 1971.

2. R. B. Thorn, Ed., River Engineering and Water Conservation Works, 1966.

3. Teesside sea outfalls, Hydraulics Research Station Report EX393, April, 1966.

4. Physical and Mathematical Studies of Surface Currents, I. W. M. Moir and J. M. Williams, in Proc. Int. Assoc. Water Poll. Reg. meeting, 1970.

5. D. Harverson, in Proc. Int. Assoc. Water Poll. Reg. meeting, 1970.

Part IV

STATISTICAL STUDIES USING REGRESSION AND RELATED METHODS

15

DOES AIR POLLUTION SHORTEN LIVES?*

Lester B. Lave and Eugene P. Seskin

Carnegie-Mellon University
Pittsburgh, Pennsylvania

The Urban Institute
Washington, D. C.

1. INTRODUCTION

The cause of a disease is often difficult to establish. For chronic diseases, establishing a cause and effect relationship is especially difficult. Many studies show that populations exposed to urban air pollution have a shorter life expectancy and higher incidence of lung cancer, emphysema, and other chronic respiratory diseases [1-3]. Yet it is a long step to assert that this observed association between air pollution and ill health is proof that air pollution causes ill health and that steps ought to be taken to abate air pollution for public health reasons. There is disagreement among physicians as to whether air pollution causes lung cancer, bronchitis, or emphysema, or generally reduces life expectancy [1, 4-6]. A related argument is that the effect of air pollution is so slight that it is not a matter of public concern [6].

The situation is similar to the controversy as to whether cigarette smoking causes lung cancer. Both the cigarette smoking and air pollution controversies stem from the fact that many other possible causal factors are present. Urban dwellers live in more crowded conditions, get less exercise, and tend to live more tense lives. Since each of these factors is known to increase the morbidity and mortality rates, care must be taken to control or account for each factor before drawing inferences from the association between air pollution or cigarette smoking and ill health. Some confidence stems from the fact that numerous independent studies with various controls have found an association, thereby endowing it with a level of confidence that no single study warrants [1-3].

---
*Subsequent to the presentation of this paper, the authors undertook further analysis of the data. For details see the note on p. 244.

2. METHOD

Laboratory methods are of limited use in settling these controversies. Instead, the methods required are those of epidemiology. The basic problem is one of controlling all causal variables. Accounting for confounding factors in observed data is one of the purposes of multivariate statistical analysis. Since we have some notion of the model and desire estimates of the partial contribution of each factor, an appropriate technique is multiple regression.

The mortality data for our analysis come from two sources. Mortality data are tabulated in Ref. 7 with more detailed mortality rates reported in Ref. 8. The first source reports total (and total nonwhite) mortality rates, with an additional breakdown for infants under 1 year and under 28 days. The second source disaggregates the rates by race, sex, and age. We have analyzed four age groups: 0-14, 15-44, 45-64, and 65 and older.

Data were collected on 117 Standard Metropolitan Statistical Areas (SMSAs) for 1960.* Air pollution data were gathered by the National Air Sampling Network [9]. Data for 1960 were collected on suspended particulates and total sulfates. Biweekly measurements were taken; we used the smallest and largest readings, as well as the arithmetic mean of the 26 readings. Socioeconomic data were taken from the 1960 Census [10].

Multiple regression attempts to explain the variation in a dependent variable (the mortality rate) in terms of the variation in each of a set of independent or explanatory variables (measured air pollution, and socioeconomic variables). Since we had little a priori knowledge as to which measures of air pollution or which socioeconomic variables would be important, all were initially tried using the 1960 data. Some variables contributed little to the statistical significance of the regression. For example, there were three measures of suspended particulates and three measures of sulfates, some of which were certain to be redundant. Variables whose coefficients were greater than their standard error were retained and the others eliminated, subject to two qualifications. Since interest centered on the air pollution variables, at least one was retained from each set. In reestimating the relation, we often found that the retained air pollution variable was now significant (which is not surprising, given that we eliminated two air pollution measures which were highly correlated with the included one). Sometimes the retained air pollution variable still contributed little to the statistical significance of the regression. Such variables were eliminated, subject to the restriction

---

* These SMSAs were selected because of the availability of pollution data.

15. DOES AIR POLLUTION SHORTEN LIVES?

that at least one air pollution variable was retained in the final equation. We note again that our criterion for retention of a variable was that its estimated coefficient exceed its estimated standard error. Since these estimated relations are ad hoc, they must be viewed with care.

3. RESULTS

3.1. Statement of Results

The first estimated relation is shown in Eq. (1), where the total mortality rate in 1960 is "explained" by: Mean P (the arithmetic mean of the biweekly suspended particulate readings in $\mu g/m^3$), Min S (the smallest of the biweekly sulfate readings in $\mu g/m^3$ x 10), P/M^2 (the population density in the SMSA in people per square mile), % NW (the percentage of the SMSA population who are nonwhite x 10, i.e., the number per thousand), % ≥ 65 (the percentage of the SMSA population 65 and older x 10), and e (an error term):*

$$MR = 19.607 + 0.041 \text{ Mean P} + 0.071 \text{ Min S} + 0.001 \text{ P/M}^2$$
$$(2.53) \qquad\qquad (3.18) \qquad\quad (1.67)$$

$$+ 0.041 \, (\% \text{NW}) + 0.687 \, (\% \geq 65) + e. \qquad (1)$$
$$(5.81) \qquad\quad (18.94)$$

These independent variables explain the total mortality rate across 117 SMSAs extremely well, since 82.7% of the variation is explained (R^2 = 0.827). The effect of each variable on mortality is estimated by the coefficient; each is statistically significant (as shown by the t statistics below the coefficients). Note that both measures of air pollution are significant explanatory factors of the mortality rate across cities. Thus, the data fit the model quite well and the estimated effect of air pollution on mortality is quite substantial. This regression is also reported as the first row in Table 1.

---

*The specification of this relation, and of each of the relations shown in Table 1, comes from Ref. 11. As described above, ad hoc techniques were used to derive a good specification for the aggregate mortality rates in 1960. The specification continued to provide a good fit for 1961 mortality rates [11] and seems to fit the age-, race-, and sex-specific mortality rates quite well.

TABLE 1

Mortality Regressions: Age-, Race-, and Sex-Specific Death Rates[a]

Row no.	Age group	Race	Sex	R^{2}[b]	Const.	Regression coefficients of explanatory variables[c,d,e]						
						Min P	Mean P	Min S	P/M^2	% NW	% ≥ 65	% Poor
1	All ages	W+ NW	M+ F	0.827	19.607		0.041 (2.53)	0.071 (3.18)	0.001 (1.67)	0.041 (5.81)	0.687 (18.94)	
2		NW	M+ F	0.339	9.181		0.186 (3.53)	0.106 (1.49)	−0.003 (−2.16)	0.148 (6.52)	0.547 (4.70)	
3		W	M	0.198	102.405		0.033 (1.81)	0.044 (1.77)	0.001 (1.83)	0.028 (3.51)	0.034 (0.85)	
4		NW	M	0.208	103.813		0.074 (1.55)	0.071 (1.09)	0.001 (0.41)	0.101 (4.87)	0.149 (1.40)	
5		W	F	0.389	61.899		0.035 (2.79)	0.070 (4.02)	0.001 (2.54)	−0.007 (−1.24)	0.023 (0.82)	
6		NW	F	0.340	67.412		0.107 (2.82)	0.191 (3.72)	0.001 (−0.80)	0.088 (5.41)	0.054 (0.64)	
7	<28 days	W+ NW	M+ F	0.271	149.428		0.083 (1.62)	0.120 (1.82)		0.098 (4.04)		0.056 (1.45)

15. DOES AIR POLLUTION SHORTEN LIVES? 227

8		W	M	0.276	175.240				-0.011 (-0.48)	0.097 (2.67)
9		NW	M	0.080	314.580				0.278 (2.85)	-0.368 (-2.39)
10		W	F	0.115	125.836				-0.033 (-1.78)	0.098 (3.31)
11		NW	F	0.130	206.087				0.254 (3.13)	-0.293 (-2.29)
12	< 1 year	W + NW	M + F	0.537	185.802	0.365 (2.82)	0.036 (0.75)	0.026 (0.42)	0.186 (6.52)	0.157 (3.38)
13		W	M	0.070	231.196	0.129 (0.93)	0.147 (0.73)	-0.145 (-0.56)	-0.069 (-2.25)	0.140 (2.82)
14		NW	M	0.126	379.376	0.578 (1.02)	0.053 (1.34)	0.068 (1.34)	0.447 (3.49)	-0.206 (-0.99)
15		W	F	0.174	162.805	0.293 (2.76)	0.270 (1.61)	0.344 (1.61)	-0.086 (-3.67)	0.162 (4.27)
16		NW	F	0.119	251.032	1.125 (2.45)			0.205 (1.98)	0.118 (0.70)
17	0–14 years	W	M	0.185	66.994	0.058 (1.46)			-0.045 (-4.44)	0.063 (4.39)

Additional entry at row 17: -0.107 (-2.67)

TABLE 1 (cont'd)

Mortality Regressions: Age-, Race-, and Sex-Specific Death Rates[a]

Row no.	Age group	Race	Sex	R^2[b]	Const.	Min P	Mean P	Min S	P/M^2	% NW	% ≥ 65	% Poor
							Regression coefficients of explanatory variables[c,d,e]					
18	15–44 years	NW	M	0.096	100.558	0.220 (1.51)				0.070 (1.84)	-0.042 (-0.29)	0.016 (0.29)
19		W	F	0.313	51.000	0.068 (2.59)				-0.041 (-6.06)	-0.111 (-4.16)	0.058 (6.08)
20		NW	F	0.163	53.138	0.293 (2.72)				0.044 (1.58)	0.094 (0.87)	0.078 (1.95)
21		W	M	0.236	52.921		0.036 (2.05)	-0.021 (-0.87)	0.000 (0.59)	-0.023 (-2.35)	-0.062 (-1.52)	0.069 (5.06)
22		NW	M	0.321	49.719		0.150 (1.73)	0.028 (0.23)	0.003 (0.96)	0.140 (2.92)	0.033 (0.17)	0.180 (2.67)
23		W	F	0.164	28.626		0.026 (2.74)	-0.005 (-0.37)	0.001 (1.89)	-0.013 (-2.53)	-0.025 (-1.14)	0.023 (3.13)
24		NW	F	0.345	30.009		0.111 (1.99)	0.068 (0.89)	0.001 (0.89)	0.104 (3.37)	0.054 (0.43)	0.110 (2.52)

15. DOES AIR POLLUTION SHORTEN LIVES?

25	45-64 years	W	M	0.336	245.138	0.127 (1.91)	0.069 (0.75)	0.006 (3.01)	0.107 (2.87)	0.202 (1.31)	0.154 (2.96)
26		NW	M	0.311	237.856	0.252 (1.00)	0.099 (0.29)	0.006 (0.78)	0.326 (2.33)	0.406 (0.71)	0.667 (3.39)
27		W	F	0.395	131.695	0.086 (2.33)	0.185 (3.61)	0.003 (2.84)	-0.028 (-1.36)	0.065 (0.77)	-0.009 (-0.33)
28		NW	F	0.477	141.855	0.373 (1.99)	0.980 (3.85)	-0.004 (-0.68)	0.472 (4.56)	-0.109 (-0.26)	0.333 (2.29)
29	≥ 65 years	W	M	0.385	698.367		0.637 (4.15)		0.103 (2.24)		
30		NW	M	0.051	679.310		0.927 (2.29)		-0.066 (-0.54)		
31		W	F	0.295	470.984		0.806 (6.36)		-0.064 (-1.67)		
32		NW	F	0.108	499.750		0.904 (3.42)		-0.066 (-0.84)		

[a] For means and standard deviations of the relevant variables, see Tables 1a and 1b.

[b] The coefficient of determination. In row 1, the entry 0.827 indicates that 82.7% of the variation in the total mortality rate across 117 cities is explained by the regression.

[c] The entries in each row under the general heading "Regression coefficients of explanatory variables" give the equation for the mortality rate of that segment of the population described by the age, race, and sex entries in that row. Thus row 1 gives the total mortality rate, row 2 gives the rate for all nonwhites, etc.

[d] For example, the entry 0.041 in row 1 under "Mean P" indicates that an increase of 0.1 $\mu g/m^3$ is estimated to increase the total death rate by 0.041 per 10,000 population.

[e] Figures in parentheses are the t statistics for the coefficients (the ratio of the coefficient to its standard error). A value of 1.66 is significant using a one-tailed test.

15. DOES AIR POLLUTION SHORTEN LIVES?

TABLE 1a

Mortality Rates

Race	Sex	All ages	<28 days	<1 year	Mortality[a] ≤14 years	15-44 years	45-64 years	≥65 years	Total
W+NW	M+F								91.3 (15.3)
W	M	115.4 (7.9)	196.9 (20.5)	253.8 (27.5)	66.4 (8.3)	60.9 (7.9)	326.2 (32.2)	741.4 (54.9)	
	F	71.1 (6.3)	148.8 (17.1)	194.7 (22.3)	50.2 (6.0)	32.3 (4.1)	153.2 (18.7)	501.1 (49.8)	
NW	M+F								97.9 (24.8)
	M	141.5 (20.6)	294.5 (84.0)	425.8 (113.5)	118.8 (28.7)	123.6 (40.1)	472.7 (116.6)	714.1 (134.5)	
	F	104.1 (17.9)	233.5 (71.7)	349.7 (91.7)	93.9 (22.0)	85.0 (26.3)	340.1 (98.8)	533.4 (90.6)	

[a] Mean mortality per 10,000 population or per 10,000 live births. Standard deviations are given in parentheses.

TABLE 1b

Means and Standard Deviations of Explanatory Variables in the Regression Analysis of Mortality

Variable	Mean	Standard deviation
Min Particulates ($\mu g/m^3$)	45.5	18.6
Mean Particulates ($\mu g/m^3$)	118.1	40.9
Min Sulfates ($\mu g/m^3$ x 10)	47.2	31.3
Population per square mile	756.2	1370.6
Percentage nonwhite (x 10)	125.1	104.0
Percentage 65 and over (x 10)	83.9	21.2
Percentage poor (x 10)	180.9	65.5

15. DOES AIR POLLUTION SHORTEN LIVES?

Another way of illustrating the regression results is to focus on one of the cities we studied. In 1960, the total mortality rate for Pittsburgh was 103 per 10,000 population, the mean biweekly level of suspended particulates was 166.0 $\mu g/m^3$, the smallest biweekly sulfate reading was 6.0 $\mu g/m^3$, the population density was 788.0 people per square mile, 6.8% of the population was nonwhite, and 9.5% was 65 and older. Fitting these values into the equation leads to a prediction that the mortality rate would be 100 per 10,000 population; this estimate is in error by 3 per 10,000 population.

The percentage of older people is the most important variable. A one point increase (in percentage multiplied by ten) of people 65 and older is associated with a rise in the total death rate of 6.87 per 10,000. Increasing nonwhites in the population (x10) by one percentage point is estimated to raise the total death rate by 0.41 per 10,000. An increase of 1 $\mu g/m^3$ in the minimum biweekly sulfate level or mean biweekly particulate level is associated with a rise in the total death rate of 0.71 or 0.041, respectively. This regression is presented in Table 1 along with regressions explaining the mortality rate for all nonwhites (row 2), for white males (row 3), nonwhite males (row 4), white females (row 5), and nonwhite females (row 6).

Also presented in Table 1 are regressions for the under-28-day death rate (rows 7 to 11), for the under-one-year death rate (rows 12 to 16), for children 14 and younger (rows 17 to 20), for young adults 15-44 (rows 21 to 24), for middle-aged adults 45-64 (rows 25 to 28), and for people 65 and older (rows 29 to 32).

The general model and estimation techniques seem to be appropriate for this analysis. A high proportion of each mortality rate is explained by the independent variables. In almost every case, air pollution is a significant factor in explaining the mortality rate.

3.2. Qualifications

The results of the analysis cannot be accepted at face value. A great deal of care must be taken in interpreting them in view of certain omissions and approximations. For example, the mortality rate is known to depend on characteristics of the individuals at risk, including their habits and exposures, and on the environment [12-14]. Using socioeconomic variables from the Census, some of these characteristics have been controlled. However, no measures are available for occupational exposure, personal habits, and smoking, to name a few important variables. If these effects are not included in measured socioeconomic variables and are related to the level of air pollution, then our pollution estimates will be biased as indicators of causality. Since we have explained much of the variation in the mortality rates, we would argue that the omitted variables would not substantially alter the results. More important, the results seem to be reasonably consistent across the various mortality rates.

It must be noted that measurement errors can be important in these data. For example, generally only a single location in an SMSA was used for sampling air pollution; since there are different pollution sources and varying terrain, a single source is not likely to be representative of the entire area. The mortality data also have a number of problems. For some age-, race-, and sex-specific deaths, the number of reported deaths sometimes exceeded the population at risk; three SMSAs were deleted in analyzing the nonwhite death rates for this reason. Another problem is the mobility of the population. Many of the people injured by pollution migrate to less polluted places while many people who grew up in rural areas, with little previous exposure, migrate to industrial areas. This migration will lead the estimated coefficients of air pollution to be biased downward as indicators of causality. Perhaps the only pure data, from the viewpoint of measured exposure, are for infant death rates where the exposure is known with reasonable accuracy. This and other problems are discussed in Ref. 15.

4. DISCUSSION

Multiple regression appears to be a good tool for explaining variations in mortality rates. Between the air pollution and socioeconomic variables, most of the variation in the mortality rates is explained.* Air pollution is implicated in almost every mortality rate as a significant factor, even after accounting for race, sex, age, and the important socioeconomic variables. This result is evidence that the previously observed "urban factor" is not the only cause of higher morbidity and mortality; air pollution is a significant contributor even after accounting for most of the factors associated with urban living.

---

* In disaggregating the mortality rates by age, sex, and race, there is a large decline in the explanatory power of the regression, R^2. This decline stems from the fact that the principal factors which are known to affect the mortality rate have been used to define more specific mortality classes. This decline in explanatory power is expected and is not important as long as estimated coefficients continue to display magnitudes and signs which are consistent with what is expected from the aggregate mortality rates. It is this consistency that we find most important and reassuring.

15. DOES AIR POLLUTION SHORTEN LIVES?

4.1. A Decision Framework

Is this evidence sufficient to justify government action to abate air pollution as a public health menace? Epidemiologists have been critical of simple cross tabulations or univariate regressions which attempt to show the effect of air pollution on mortality [16]. The questions are: (1) did a particular association occur merely by chance, and (2) given that the observed association is more than a chance occurrence, is this a basis for action? More precisely, the concern is whether the association observed between A and B is a basis for attempting to influence B by manipulating A. Children who get little to eat, especially a diet low in vitamins, have a higher incidence of rickets. To eliminate rickets, we must find a variable (both easy and inexpensive to manipulate) which will prevent it. Classically, we attempt to find what "causes" rickets and then to intercede in the sequence of events. Thus, one "cause" of rickets is malnutrition and the crucial dietary variable to manipulate is the ingestion of Vitamin D. Note that this is not the same as concluding that children must be given a balanced diet, must eat only healthy foods, or must eat regularly. There are many ways of achieving the goal of ingesting the proper amount of Vitamin D.

There has been a rather unfortunate tendency for physicians to fix on the word "proof" and attack various associations as not having been proved to be causal [17]. This is given as a reason why no attempt should be made at manipulation in order to deal with a health problem. However, as soon as the problem is stated in a decision context, the emphasis on the word "proof" can be seen to be out of proportion. While it certainly is correct that interference in a known, deterministic, causal chain is certain to lead to results, it is not generally in the public interest to hold off action until causality is established to everyone's satisfaction. In public health situations where the potential gain is high (as in stopping an epidemic) and the cost is low (as in adding more chlorine to the water), no one would advocate waiting until everyone is sure that the drinking water was really the cause of the epidemic.

The point is that the level of confidence required for general acceptance of proven causality is not required in order to warrant action. For most health situations with relatively high benefits and low costs, a much lower level of confidence is called for. Thus, a prudent man would have thought seriously about giving up cigarettes long before it was generally accepted that cigarettes cause ill health. (We have no confidence that widespread agreement would be obtained today among scientists that it has been proved that cigarette smoking causes lung cancer, emphysema, and heart disease [18-21].*)

4.2. A Comparison across Mortality Rates

The estimated coefficients presented in Table 1 depend on the way dependent and independent variables are scaled. Another way of presenting these results is to show the estimated effect of a 100% increase in the explanatory variables. For example, doubling the mean level of particulates would increase the total mortality rate by 5.29%, and doubling the smallest sulfate reading would increase the total mortality rate by 3.68%. These percentage effects are presented in Table 2. (Coefficients which were quite insignificant are shown in parentheses.)

The first block of percentage effects are for the smallest biweekly sulfate reading. Sulfates have more effect on the very young and the very old, more effect on nonwhites than on whites, and more effect on females than on males. The same pattern of effects holds for the two measures of particulate pollution.†

---

* Another way of viewing the problem stems from observing that A is associated with B. Thus, either A causes B, B causes A, or C causes both A and B and the association between them is spurious because we haven't thought to measure C [22]. However, the human body is so complex that medicine is of very limited help in providing such a theory [4, 23-25]. While we have not heard anyone suggest seriously that lung cancer leads to cigarette smoking, the set of possible causal factors, C, is very large. In general, the only way that evidence can be compiled is by sampling a great many groups with differing characteristics. The convincing evidence is that a factor such as smoking is associated with lung cancer under a wide variety of conditions, in a wide variety of groups. It is the weight of a number of studies which rules out the possibility that the association is a sampling phenomenon; it is the weight of many studies in many settings which dismisses the possibility that some uncontrolled factor is the true cause of lung cancer.

† These effects may be due, in part, to the relationship between air pollution exposure and the socioeconomic distribution within an SMSA. For example, it has been shown that nonwhites reside in areas subject to higher pollution than are areas where whites reside [28]. Thus, the patterns for the different groups may result from doubling different base levels of pollution affecting the groups.

15. DOES AIR POLLUTION SHORTEN LIVES?   237

Interpreting the relative size of the estimated percentage effects in this way is somewhat dubious in that the exact estimate may be more an artifact of the particular variables and data than a true expression of the relation. However, since some of the interpretations tend to shed light on the nature of the relationship, we will present them and caution the reader about putting much confidence in them.

The percentage effects of the variable "percentage of families who are poor" display an interesting pattern. Females tend to be more sensitive than males to poverty in childhood, the reverse later on. Poverty seems to lower the death rate for nonwhite children under 28 days. It is difficult to know whether this result is simply an artifact of the estimation procedure, but it is consistent with the notion that areas with concentrations of the poor have good welfare facilities.

The percentage effects of "proportion of nonwhites in the population" also display an interesting pattern. Concentrations of nonwhites are associated with much higher death rates. Concentrations of nonwhites always lower the female white death rate. For white males, concentrations of nonwhites tend to lower the death rate through age 44. The simplest hypothesis for these effects is one of economic exploitation of nonwhites where there is a sizeable proportion of them, i.e., concentrations of nonwhites lower the white death rate and raise the nonwhite death rate. The results are consistent with a mechanism wherein nonwhite women assume many of the household chores and increase the leisure of white women. For males, nonwhites might assume the jobs requiring particularly difficult labor or involving health risks. There is a striking difference between races for the three categories of death rates among children. White children have a lower death rate when there are many nonwhites (presumably because nonwhites tend them); nonwhite children have a higher death rate (presumably because they are unwatched).

Abating air pollution is one way to reduce the death rate. Other ways are to raise the income of the poor, lessen population density, and deal with some of the ills of urban areas with a high proportion of nonwhites. Some of these other effects can be estimated and are shown in Table 2. For example, doubling air pollution (both sulfates and particulates) is estimated to increase the total death rate by 8.97% (3.68% + 5.29%). Doubling the proportion of poor families in the city, the population density, or the proportion of the population which is nonwhite would be estimated to increase the total death rate by 5.39%, 0.66%, or 5.68%, respectively. Thus, air pollution is the most important of the three factors affecting the total death rate (and by inference morbidity and longevity).

TABLE 2

Elasticities for Each Explanatory Variable

Variable	Sex	Race	All ages	<28 days	Elasticity (×100)[a] <1 year	≤14 years	15-44 years	45-64 years	≥65 years
Minimum sulfates	M+	W+	3.68	3.04					
	F	NW							
	M	W	1.80	(0.63)			(-1.65)	(1.00)	4.06
		NW	2.34	(-2.29)			1.04	(0.98)	6.06
	F	W	4.62	2.16			(0.71)	5.69	7.59
		NW	8.58	6.87			(3.72)	13.44	7.91
	M+	NW	5.06						
	F								
Mean particulates	M+	W+	5.29	5.22					
	F	NW							
	M	W	3.33	(2.16)			7.01	4.61	
		NW	6.15	(5.89)			14.24	6.29	
	F	W	5.80	4.17			9.52	6.63	
		NW	12.09	13.61			15.41	12.92	
	M+	NW	22.32						
	F								
Minimum particulates	M+	W+			6.53				
	F	NW							
	M	W			(2.32)	3.95			
		NW			6.16	8.42			

15. DOES AIR POLLUTION SHORTEN LIVES?

F	W		6.84		6.19		
F	NW		14.60		14.16		
Percentage poor							
M+	W+	5.39					
F	NW			11.17			
M	W		8.90	9.98	17.18	20.60	8.52
M	NW		-22.52	-8.71	(2.37)	26.22	25.47
	W		11.91	15.07	20.96	12.91	(-1.12)
	NW		-22.66	(6.10)	14.91	23.24	17.64
Population per sq. mi.							
M+	W+	0.66					
F	NW						
M	W		0.64			(0.40)	1.44
	NW		(0.30)			(1.53)	(0.95)
F	W		0.99			1.30	1.59
	NW		(-0.64)			(1.32)	(-0.85)
M+	W	-2.53					
F	NW						
Percentage over 65							
M+	W+	63.21					
F	NW	(2.49)					
M	W	8.73			-13.54	-8.47	5.19
	NW	(2.71)			(-2.93)	(2.21)	(7.14)
F	W	(4.32)			-4.16	-6.46	3.56
	NW	46.42			(8.29)	(5.28)	(-2.67)

TABLE 2 (cont'd)

Elasticities for Each Explanatory Variable

Variable	Sex	Race	Elasticity (×100)[a]					
			All ages	<28 days	<1 year	≤14 years	15-44 years	45-64 years	≥65 years
Percentage nonwhite	M+F	W+NW	5.68	6.57	9.15				
	M	W	3.03	(-0.69)	-3.38	-8.50	-4.76	4.10	1.74
		NW	9.11	12.10	13.46	7.51	14.52	8.85	(-1.19)
	F	W	-1.20	-2.80	-5.50	-10.22	-5.20	-2.29	-1.59
		NW	10.88	13.93	7.53	6.01	15.71	17.81	(-1.60)
	M+F	NW	19.36						

[a] Percentage change in the mortality rate that is estimated to occur if the particular explanatory variable were to double in intensity. For example, if the minimum sulfate level doubled, other factors held constant, the total mortality rate is estimated to increase by 3.68% and the mortality rate for nonwhite females, 65 and older, is estimated to increase 7.91%. Where the regression coefficient was less than its standard error, the elasticity is in parentheses. These elasticities are derived from the regression coefficients listed in Table 1.

5. CONCLUSION

The effect of air pollution on mortality is examined in 32 regressions, in which mortality rates are characterized by age (under 28 days, under 1 year, 14 years and younger, 15-44, 45-64, and 65 and older), by race (white and nonwhite), and by sex (male and female). Air pollution has more effect on women than on men, and more effect on nonwhites than on whites. The very young and the very old are most sensitive to air pollution.

In addition to the relations shown here, air pollution continues to be a significant factor when meteorological and heating variables are added [26] or when occupation variables are added [27]. Consistent effects are also obtained in explaining variations in disease-specific mortality rates across the same SMSAs. Finally, a subset of the mortality rates has been gathered for 1961, the results essentially replicating the 1960 ones [11].

Another way to view the effect of air pollution (or of any factor causing ill health) is in terms of its effect on the economy. When people are sick they lose time from work and must spend money on medical care; if illness could be reduced, there would be a savings in these categories. As estimated elsewhere, a 50% abatement in air pollution would reduce the "economic cost" of morbidity and mortality by 4.5% [2]. The importance of this improvement in health might be assessed by noting that the economic cost of all cancers is 5.7% of the total economic cost of ill health in the U.S. Thus, abating air pollution by 50% is estimated to have almost the same effect on economic cost as eradicating all cancer. Such an abatement is probably the most effective way of improving the health of middle-class families. Note that the middle-class family can do something about smoking, but is powerless to lower its exposure to air pollution (except by leaving the city).

These studies show a rather consistent relation between air pollution and mortality across a number of different specifications. All show that mortality rates could be lowered substantially by abating air pollution. For example, lowering the measured levels of minimum sulfate pollution and the mean particulate pollution by 10% would result in a 0.90% decrease in the total death rate.

Even so, there are many reasons to believe that these estimates are gross understatements of the health cost of air pollution. Chronic diseases generally involve long periods of illness. The economic costs, calculated as the sum of lost work and medical expenditures, grossly understate the amount that would be paid to achieve good health for such a chronically ill period. In addition, death may not result from the chronic illness itself, but rather from one or another complication. For example, chronic bronchitis or emphysema is likely to result in death due to heart disease or pneumonia, rather than from the chronic disease.

Perhaps the only good way to estimate the health costs of air pollution would be to analyze morbidity rather than mortality data. It seems certain that such an investigation would give a higher health cost, since no one can die of emphysema or other chronic illnesses who has not suffered them, while some of the people with chronic illnesses die from other causes. In addition, such an investigation would detect increases in morbidity rates, such as those of simple respiratory diseases, which may occur long before death is a consideration. Other much less severe illnesses known to be associated with air pollution are unrelated to mortality. For example, eye irritation is a common reaction to acute pollution; costs from this malady will never be reflected in mortality statistics (except possibly for accidents).

We have concentrated on the health effects of air pollution, without alluding to costs associated with cleaning and deterioration of inert materials, with vegetation and animal damage, and with the aesthetic effects of living in a dirty, uncomfortable, overcast world. These costs can be substantial [15].

In view of the resources devoted to attaining better health, it seems clear that social welfare would rise substantially with the spending of resources to abate air pollution. It is time that we pressed forward with a program of abatement.

ACKNOWLEDGMENTS

This research was supported by a grant from Resources for the Future, Inc. We thank Myrick Freeman, Martin Geisel, Edwin Mills, and Robert Strotz for helpful comments. Any errors and opinions are those of the authors.

REFERENCES

1. D. Anderson, Can. Med. Assoc. J., 97, 528, 585, 802 (1967).

2. L. Lave and E. Seskin, Science, 169, 723 (1970).

3. U. S. Public Health Service, Air Quality Criteria for Sulfur Oxides, Pub. No. AP-50, Nat. Air Poll. Control Admin., U. S. Gov. Printing Office, Washington, D. C., 1970.

15. DOES AIR POLLUTION SHORTEN LIVES? 243

4. M. Battigelli, J. Occup. Med., 10, 500 (1968).

5. I. Greenwald, Arch. Ind. Hygiene Occup. Med., 10, 455 (1954).

6. W. H. O., Seminar report, Bull. W.H.O., 23, 264 (1969).

7. U. S. Dept. of Health, Education, and Welfare, Vital Statistics of the United States, 1960, U. S. Gov. Printing Office, 1961, Washington, D. C.

8. E. Duffy and R. Carroll, United States Metropolitan Mortality, 1959-1961. Pub. No. 999-AP-39, U. S. Public Health Service, Nat. Center for Air Poll. Control, 1967, Cincinnati, Ohio, 1967.

9. U. S. Public Health Service, Analysis of Suspended Particulates, 1957-1961, U. S. Gov. Printing Office, Washington, D. C., 1962, Pub. No. 978.

10. U. S. Dept. of Commerce, County and City Data Book, U. S. Gov. Printing Office, Washington, D. C., 1962.

11. L. Lave and E. Seskin, J. Amer. Stat. Assoc., 68, 284 (1973).

12. B. Ferris, Jr., Arch. Environ. Health, 16, 511 (1968).

13. J. Kosa, A. Antonovsky and I. Zola, Eds., Poverty and Health, Harvard Univ. Press, Cambridge, Mass., 1969.

14. J. Lave and L. Lave, Law Contemp. Prob., 35, 252 (1970).

15. L. Lave in Environmental Quality Analysis (A. Kneese and B. Bower, Eds.), Johns Hopkins Press, Baltimore, Maryland, 1972, p. 213.

16. P. Lawther, J. Inst. Fuel, 36, 341 (1963).

17. B. MacMahon, T. Pugh, and J. Ipsen, Epidemiological Methods, Little, Brown and Co., Boston, Mass., 1960.

18. K. Brownlee, J. Amer. Stat. Assoc., 60, 722 (1965).

19. S. Cutler, J. Amer. Stat. Assoc., 50, 267 (1955).

20. R. Fisher, Brit. Med. J., 2, 297 (1957).

21. U. S. Public Health Service, Smoking and Health, Pub. No. 1103, U. S. Gov. Printing Office, Washington, D. C., 1964.

22. H. Simon, J. Amer. Stat. Assoc., 49, 467 (1954).

23. M. Amdur and D. Underhill, Arch. Environ. Health, 16, 460 (1968).

24. A. Goetz, Int. J. Air Water Poll., 4, 168 (1961).

25. A. Goetz, in Inhaled Particles and Vapours, (C. Davies, Ed.), Pergamon Press, New York, 1961, p. 295.

26. L. Lave and E. Seskin, Amer. J. Pub. Health, 62, 909 (1972).

27. L. Lave and E. Seskin, Swed. J. Econ., 73, 76 (1971).

28. A. M. Freeman, III, in Environmental Quality Analysis (A. Kneese and B. Bower, Eds.), Johns Hopkins Press, Baltimore, Maryland, 1972, p. 243.

Note added in proof: The mortality rate analyzed in Eq. (1) was defined as the total number of deaths divided by the total population of an SMSA. It was not adjusted for age, sex, or race. Similarly, the mortality rates analyzed in Table 1 for all ages, 0-14, 15-44, and 45-64 were not adjusted for age. These rates were computed without taking account of the differences in the population at risk of the individual age categories which were aggregated. These adjustments in the mortality rates have now been made and are reported in Chapter 4 of L. Lave and E. Seskin, Air Pollution and Human Health, in press for Resources for the Future, Inc., Johns Hopkins University Press. The results indicate that while the estimated parameters change, the qualitative conclusions described here remain unchanged. For example, lowering the levels of minimum sulfate pollution and mean particulate pollution by 10 percent would be associated with a 0.63 percent decrease in the adjusted total mortality rate instead of the 0.90 percent decrease in the unadjusted total mortality rate documented in this paper.

DISCUSSION

William Fairley

Harvard University
Cambridge, Massachusetts

The authors have in my view wisely separated the scientific question of the existence of a link between air pollution and ill health from the decision problem of whether or not to abate air pollution. A decision to abate pollution might rest on a likelihood or even only a suspicion of the important health effects suggested by their analysis and by those of others. And of course a decision to abate air pollution may be based on other undesirable effects, such as reduced visibility, eye irritation, effects on materials, local or worldwide climatological effects, etc.

In this paper Lester Lave and Eugene Seskin discuss the implications of their research on the effects of air pollution on age-, sex-, and race-specific mortality. I will direct my discussion to the strategy adopted for establishing a causal link between pollution and mortality. The investigations reported here use published data on air pollutant levels and on mortality in 1960 in 117 urban areas of the United States (specifically, in 117 Standard Metropolitan Statistical Areas or SMSAs). The statistical method employs multiple regression to study the relation of mortality to air pollutant levels in urban areas while controlling for a variety of other variables which might affect the relationship. It should be noted that this is a study of mortality only, not morbidity or ill health generally, and that it is a study of the effect on mortality of two selected air pollutants - particulates and sulfates - not of air pollution generally.

This regression study supports the existence of a causal link between levels of particulates and sulfates on the one hand and mortality rates on the other. I will organize my comments around two questions. First, are the regression coefficients for levels of particulates and sulfates statistically significant? Second, are there other plausible factors which might account for the observed association between mortality and the pollutant levels? The two principal issues, then, are significance and control.

Taking up the question of significance first, I find the statistical analysis presented convincing as far as it goes, but I would suggest three extensions. I would note that further analysis is not just gilding the lily in this case, because in the regression of total mortality on pollutant levels and other variables the quoted t values of 2.53 for mean particulates (Mean P) and 3.18 for minimum sulfates (Min S), while large, are not so large as to defy reversal after reanalysis. Direct evidence for this position comes from the range of numbers quoted in the paper for these t values as different mortality rates and different sets of independent variables are examined. For the particulates coefficient this range is 0.73 to 3.53 and for the sulfates coefficient it is -0.87 to 4.02.

First, the interpretation of the quoted t statistics rests on assumptions about the statistical model underlying the least squares fitting procedure. The authors rely on the robust nature of least squares analysis, but it is possible to make some checks on the adequacy of the assumptions that make a least squares fit of a multiple regression equation reasonable. There are no references in this paper to examinations of the data and/or formal statistical tests for heteroscedasticity, correlation of error terms (for example, geographic), nonlinearities, or gross departures from normality.

Second, alternative techniques for assessing the variation in the coefficients could be tried. One of these is the jackknife, in which in essence successive blocks of data are left out, the coefficients computed on the remaining data, and the resulting coefficients compared. In the same spirit would be a comparison of the coefficients for different subgroups of the 117 SMSAs, e.g., between regional groupings of the SMSAs.

A third comment on significance levels is that they are in this case computed from variations in essentially one replicate of the data. The data is almost entirely for a single year and is for a single country. Analysis for additional years and within different countries is called for.

Turning now to the issue of control for other variables that might either mask or spuriously suggest a causal link, I find the paper to be very clear on the requirements for good control. Control is sought in this and other papers through a combination of (1) entering additional explanatory variables such as climate and occupation into the models, and (2) studying the models within narrower population subgroups, that is through age-, sex-, race-, and disease-specific equations. One device for control that was not used is blocking of the SMSAs (e.g., regionally), as suggested earlier in the discussion on significance.

The questions that remain with control can be grouped in two parts: (1) questions about variables left out and (2) questions about variables

DISCUSSION

included. As to variables which might help to explain mortality differences between SMSAs it is not hard to think of additional candidates for study, such as absolute city size, extent of migration, indices of health care, to name three, but, of course, trying other variables is a never ending game. At the same time, the subject of explaining mortality differences between SMSAs is apparently sufficiently young that there will no doubt be surprises ahead. And, again, the significance levels of the pollutant levels are not so large that one can suppose they will be unaffected by better models of SMSA mortality differences.

To be more confident that the relation of mortality and pollutant levels is not spurious it would be very helpful to better understand the relation of air pollution itself to other variables. That is, it would be helpful to understand why SMSAs differ in levels of particulates and sulfates, and then to check whether the causes of pollutant levels were or were not independent causes of mortality.

Turning now to a discussion of the variables included in the equations for mortality, I will make one or two comments.

The age variable "percentage over 65" is the most important variable for explaining differences in total mortality. If this variable is significantly associated with pollutant levels, inclusion of the entire age distribution by appropriate variables instead of the single variable "percentage over 65" might both improve the mortality model and make substantial changes in the coefficients for the pollutant levels.

The same point can be made about the partial measurement of income distribution in the SMSAs by a "percentage poor" variable, although this variable did not appear nearly as important as age.

THE USE OF MULTIPLE REGRESSION EQUATIONS FOR THE ASSESSMENT AND SHORT-TERM FORECASTING OF URBAN AREA BACKGROUND POLLUTION

M. Benarie, D. Badellon, T. Menard, and A. Nonat

Institut National de Recherche Chimique Appliquée
Vert-le Petit, France

0. SUMMARY

Multiple linear regression equations were computed with pollutant concentrations as dependent variables and meteorological factors as independent variables. The former are mean readings of the concentration of strong acidity and the smoke-shade index, taken every twenty-four hours. The data are from a survey we have been doing since 1967 in the Rouen district. The following meteorological factors were taken into consideration: season, direction and velocity of wind, temperature, duration of fog, duration of sunshine, a turbulence coefficient expressing the vertical temperature gradient, and wind shear. All these variables were mean values measured over 24-hour or shorter periods of time by the Meteorological Office.

In contrast with similar previous work, the regressions did not include

1. the calm periods, i.e., when the arithmetic mean of the eight three-hourly wind velocity measurements was below 0.125 m/sec;

2. the days showing too much variability in wind direction, which total less than 3% of the observed period.

When only four meteorological variables were used (season, wind direction, wind velocity, and temperature), the mean error of the computed dependent variable was about the same as when all aforementioned independent variables were put into the equations. No further improvement was obtained by using logarithmic regression.

The use of the regression equations is shown in the following cases.

1. The problem of authorizing new sources. The regression equations allow the computation of background pollution for various wind sectors, wind velocities, heating conditions, etc., on the basis of relatively short surveys. On the other hand, classical atmospheric diffusion theory gives estimates of ground concentrations for any conditions of stability and wind velocity. If the two computations are implemented for identical meteorological conditions, the assessment of the relative importance of a new source may be improved.

2. The possibility of computing pollutant concentrations for days or periods when data are lacking because of instrument failure or for any other reason ("forecast a posteriori").

3. "True forecasting." As far as meteorological forecasting is reliable, its numerical estimates, introduced as independent variables, give fair short-term estimates of pollutant concentrations. If the topographic distribution and output of the sources change, the efficiency of the forecast is obviously diminished.

For a test period, the correlation coefficient of the forecasts with the actual pollutant concentrations was significant at the 1% level.

1. INTRODUCTION

A recent paper by Benarie [1] gave a bibliographical review of methods of forecasting urban air pollution, so there is no need to give one here. That paper expressed the opinion, in full agreement with Marsh and Withers [2] and others that the estimates arrived at by regression formulas are rather poor. It seems now, as will be explained below, that better results are obtained from regression formulas by restricting their generality and thus defining their specific applicability. This is the argument presented here.

2. THE OBSERVED DATA

Between March, 1967, and December, 1969, measurements of the 24-hour means of acidity (by the non-specific SO_2 method) and smoke (by

16. URBAN AREA BACKGROUND POLLUTION 251

the "soufre-fumées" or SF, French standard AFNOR NF X 43 005) were made at seven stations in the Rouen region of France. The sampling stations were:

1. Social Security building, 8th floor, 22 m from the ground,

2. same building, 15th floor, 45 m from the ground,

3. Sotteville, hospital roof, bordering on a park,

4. County Hall (Prefecture), 15 m from the ground,

5. Mt. St. Aignan, 2nd floor of the Faculty of Science,

6. Maromme, Town Hall (Mairie) first floor, 80 m from the nearest traffic, and

7. Petit-Couronne, on a watertower, 18 m from the ground.

Figure 1 shows the positions of these stations, with a schematic indication of the topography and the main pollution sources.

3. VARIABLES USED IN THE MULTIPLE REGRESSION EQUATIONS

3.1. Principles

We used only data routinely observed by the Meteorological Office: wind direction and velocity, temperature, a figure for precipitation, duration of sunshine, and wind shear (as measured at Trappes). Further data, such as vertical temperature gradient, height of the inversion layer, etc., were not measured in Rouen as a daily routine, and thus no local information was available for this paper. These and other data, which are no doubt very important for the study of air pollution, could be obtained only by making special meteorological observations, which were financially out of the question. Consequently such data were not taken into consideration. Our purpose was not to look for the regression that would fit best theoretically or to explain pollutant concentrations meteorologically, but to discover how to obtain useful estimates from available data.

It is certainly possible to choose a better set of independent variables than we did, and under some circumstances the form in which the meteorological variables are introduced could be improved. But to do this, long-term observations of various local meteorological data would be needed,

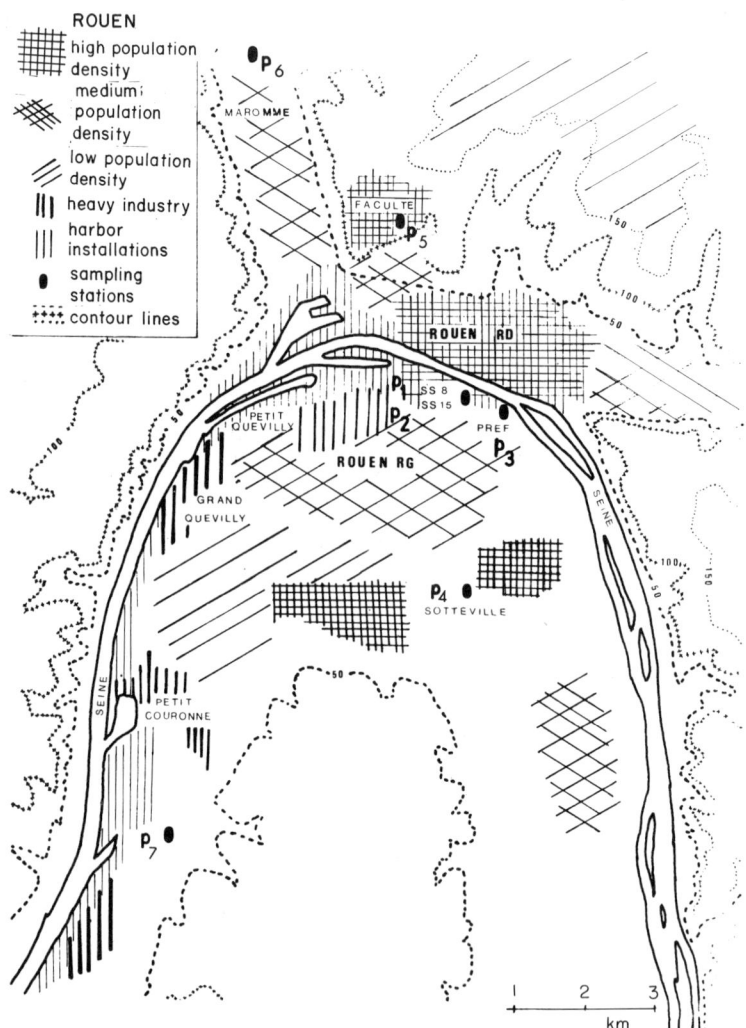

Fig. 1. Map of the sampling area.

16. URBAN AREA BACKGROUND POLLUTION

which, although available in Rouen, we fear do not exist in most places with air pollution problems. Thus scientific depth would be gained, but general practical usefulness lost. On the other hand, computer programs are available in which the order of inclusion of the independent variables is defined by their partial correlation coefficient with the dependent variable. We chose, instead, as will be seen, a program with a fixed order of independent variables. It happens that some of the meteorological variables which were available to us have a very slight influence on pollutant concentrations. This may be the case either because there is no causal relationship between that variable and pollution, or else because the data are, like the gradients measured at Trappes (about 200 km away), from a position too distant to be of real importance. Small partial regression coefficients, in connection with insignificant partial correlations, will be reliable signs of this kind of situation. Another way to judge possible improvements of the estimate is to observe the progressive reduction of the mean absolute error as further independent variables are introduced into the equations.

3.2. Season

The computation of the regression equations was done only for two well characterized periods of the year:

1. November, December, January, February, and March, when space heating contributes heavily to pollution levels. Atmospheric dispersion is also of winter type.

2. The months without space heating: June, July, August, and September: emissions are of exclusively industrial character and the dispersion is of summer type.

For April, May, and October no regression equations were computed for the time being, mainly because not enough homogeneous data could be pooled.

3.3. Wind

The modulus of the mean wind vector V_m, expressed in m/sec, is introduced into the computations in form of $1/V_m$, since pollutant concentrations are very approximately inversely proportional to wind velocity. This is a formal expression and should not be taken as a physical affirmation of a kind that regression equations cannot provide. Its computation is straightforward from the three-hourly wind observations of the Meteorological Office.

Calm Periods

Total absence of measurable wind is comparatively rare. (Absence of measurable wind is defined as a stagnation period of at least 24 hours, when the arithmetic mean V_a of eight three-hourly observations is exactly zero, and is characterized by steeply rising pollutant concentrations.) Periods without measurable wind are not taken into account for the computation of the regression equations. This is one of the ways specified in the Introduction of restricting generality in order to obtain a better fit for the main body of the data. During the reference period only a single interval of 24 hours was dead calm. All the other 24-hour intervals were taken into account, even when there was only a single three-hourly wind registration of 1 m/sec and seven that were equal to zero.

Wind Direction

The direction of the wind is defined as that of the mean wind vector for 24 hours. Accordingly, each 24-hour period belongs to one of the following wind sectors (see Fig. 2):

Sector 1, from 305° to 80° (counting clockwise from North = 0°), zone of medium wind frequency;

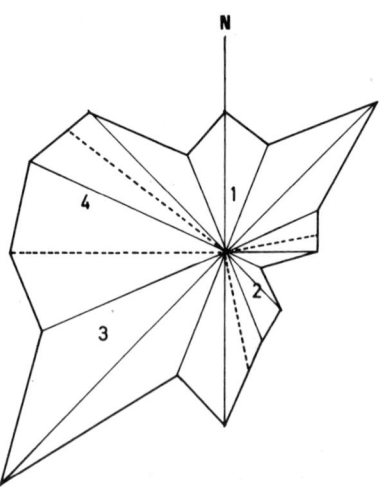

Fig. 2. Wind rose (frequency of observations of wind velocities above 2 m/sec at Rouen-Rouvray) and a representation of the four wind sectors used.

16. URBAN AREA BACKGROUND POLLUTION

Sector 2, from 80° to 170°, zone of low wind frequency;

Sector 3, from 170° to 270°, zone of highest wind frequency; and

Sector 4, from 270° to 305°.

This division into sectors, arbitrary at first look, is a compromise between the local wind rose and the geographical distribution of the main pollution sources. It was dictated by the technical needs of a specific survey in the area, and was maintained unchanged only to economize in a new program. As the purpose of this paper is to present the principles of a method and not to discover specific facts, any sector division will do. It would be quite different if meteorological inferences or a very precise location of source(s) were being pursued. We are doing such work concomitantly, with subdivision into 16 equal sectors.

Changing Wind Direction

Define a parameter for wind variability, following Grisollet [3, p. 8]:

$$c = \frac{\text{Modulus of mean wind vector}}{\text{Arithmetic mean of wind velocity}} = \frac{V_m}{V_a}.$$

If $c = 1$, clearly the wind blew from a constant direction during the 24-hour period; the farther the actual value is from this maximum, the greater the wind variability. During the approximately one thousand days of the survey, the observed distribution was:

$c > 0.6$ 91%

$c > 0.8$ 75%

$c < 0.4$ 2.7%.

As a second restriction of generality, the days with $c < 0.3$ were not taken into account. For these rare days of highly variable wind direction, 24-hour means are too long and regression equations computed from shorter than 24-hour time averages should be used.

3.4. Other Meteorological Data

Following the principles explained in Section 3.1, we further considered the following data.

Mean temperature T_m, introduced into the computation in tenths of degrees. It should be remembered that T_m is both a meteorological variable and an emission factor. As the temperature declines, pollution from space heating becomes more important. Thus, a more complex behavior of this independent variable should be noted. From the formal point of view of regression analysis, this physical dependence is of no consequence.

Duration of fog, D_B, during the 24-hour period, expressed in hundredths of hours. This is the best numerical indication of inversions we can get from routine data.

Duration of precipitation, D_P, in tenths of an hour.

Duration of sunshine, D_S, also in tenths of an hour, as an indication of the vertical exchange. This variable is far from being statistically independent of the duration of precipitation.

A variable DTM/DVM, which is the temperature gradient between the ground and 500 m divided by the wind gradient (shear) for the same levels. These gradients are measured twice a day, at noon and at midnight at Trappes, 200 km southeast of Rouen, and were used tentatively because of the lack of nearer observations.

4. MULTIPLE LINEAR REGRESSION EQUATIONS

A regression equation was computed for each season (factor of two levels: winter and summer), each wind sector (factor of four levels), each of the seven sampling stations, and for the two pollutants sampled by each. This computation was done in steps, by successively introducing the aforementioned independent variables:

$$J_0 = a_{00} + \frac{a_{10}}{V_m} \tag{1}$$

$$J_1 = a_{01} + \frac{a_{11}}{V_m} + a_{21} T_m \tag{2}$$

$$J_2 = a_{02} + \frac{a_{12}}{V_m} + a_{22} T_m + a_{32} D_B \tag{3}$$

$$J_3 = a_{03} + \frac{a_{13}}{V_m} + a_{23} T_m + a_{33} D_B + a_{43} D_P \tag{4}$$

16. URBAN AREA BACKGROUND POLLUTION

$$J_4 = a_{04} + \frac{a_{14}}{V_m} + a_{24} T_m + a_{34} D_B + a_{44} D_P + a_{54} D_S \tag{5}$$

$$J_5 = a_{05} + \frac{a_{15}}{V_m} + a_{25} T_m + a_{35} D_B + a_{45} D_P + a_{55} D_S + a_{65} \frac{DTM}{DVM} \tag{6}$$

all with $V_a \geq 0.125$ m/sec and $c > 0.3$.

Table 1 gives some of the 112 full equations computed, and the standard errors σ_i of the computed coefficients.

4.1. The Significance of the Meteorological Variables

Regression Coefficients and Range of the Independent Variables

The mean wind velocity ranges from 0.125 m/sec (conventional minimum) to about 10 m/sec. As can be seen from Table 1, the contribution of the corresponding term of the regression equation is important if wind velocity is not too high, even when a_1 is exceptionally small (Eqs. 14, 67, 71, 84 of Table 1).

It is not too difficult to explain the few negative values of a_1: if the wind blows slowly from a sector which contains only a few sources, it diminishes the pollutant concentrations. Strong winds, with a correspondingly stronger lateral turbulence, diminish the concentration somewhat less. The rare value $a_1 = 0$ (Eq. 112 of Table 1) is for a sector with absolutely no source, so that wind, strong or weak, is without influence on concentration.

The range of the mean temperature during one season is of the order of 300 tenths of a degree. Thus the term $a_2 T_m$ represents a meaningful amount if $a_2 > 0.1$. This is generally the case in the winter and seldom in the summer.

Theoretically, the duration of fog can vary between 0 and 2400 hundredths of an hour. Thus a_3, with the general range from 0.01 to 0.07 (very rarely about 0.001, in Eqs. 23, 67, 100 of Table 1), can contribute to the result if there is a fog of a few hours duration.

TABLE 1

Coefficients of 24 of the Regression Equations Computed for This Study and Their Standard Errors

Equation number[a]	Wind sector	Station number[b]	Pollutant[c]	a_0	σ_0	a_1	σ_1	a_2	σ_2	a_3	σ_3	a_4	σ_4	a_5	σ_5	a_6	σ_6
1	1	1	S	110	13	35	6	-0.47	0.19	0.025	0.012	-0.25	0.17	0.48	0.21	0.0077	0.016
2	1	1	F	65	10	16	4	-0.28	0.13	0.029	0.008	-0.11	0.11	0.26	0.14	-0.0022	0.01
13	1	7	S	93	13	24	6	-0.13	0.19	0.015	0.011	-0.14	0.16	0.16	0.22	0.0077	0.02
14	1	7	F	40	6	7	3	-0.11	0.08	0.012	0.005	-0.03	0.07	0.32	0.09	-0.0022	0.008
19	2	3	S	58	10	13	3	-0.31	0.13	-0.011	0.006	-0.05	0.12	0.37	0.20	-0.0064	0.015
23	2	5	S	116	19	13	6	-0.68	0.26	0.001	0.013	-0.35	0.25	-0.04	0.39	-0.0246	0.03
27	2	7	S	73	11	20	4	-0.06	0.15	-0.012	0.007	-0.14	0.14	0.18	0.12	-0.012	0.017
29	3	1	S	91	22	78	28	-0.06	0.22	0.052	0.02	-0.20	0.15	0.43	0.26	-2.406	1.3
35	3	4	S	107	24	59	30	-0.14	0.24	0.044	0.02	-0.033	0.15	-0.23	0.30	-2.289	1.4
36	3	4	F	27	7	39	8	-0.11	0.07	0.070	0.01	-0.05	0.05	0.11	0.09	-0.430	0.4
37	3	5	S	70	15	75	19	-0.21	0.15	0.058	0.015	-0.11	0.1	0.54	0.19	-1.722	0.85
47	4	3	S	105	19	30	15	0.04	0.25	0.033	0.016	-0.60	0.18	0.02	0.25	-0.031	0.025
48	4	3	F	54	15	36	11	0.07	0.19	0.025	0.012	-0.27	0.14	0.01	0.18	-0.017	0.018
67	1	6	S	38	9	5	5	-0.08	0.06	0.001	0.006	-0.10	0.068	0.03	0.04	-0.027	0.04
68	1	6	F	27	10	13	6	-0.03	0.06	0.003	0.006	-0.08	0.07	-0.02	0.04	0.044	0.04

16. URBAN AREA BACKGROUND POLLUTION

69	1	7	S	68	33	45	17	-0.11	0.20	0.040	0.019	-0.32	0.23	0.06	0.13	-0.126	0.13
71	2	1	S	73	34	4	2	-0.08	0.17	-0.038	0.04	-0.56	0.45	-0.17	0.13	0.697	0.53
84	2	7	F	29	13	2	2	-0.05	0.07	0.015	0.014	0.08	0.17	0.01	0.05	-0.010	0.23
90	3	3	F	36	8	27	4	-0.13	0.05	0.012	0.008	-0.05	0.04	-0.04	0.04	-0.018	0.24
91	3	4	S	30	26	35	15	0.16	0.16	0.014	0.02	-0.04	0.13	0.12	0.13	-0.346	0.9
99	4	1	S	110	31	-16	8	-0.22	0.21	0.012	0.017	-0.017	0.10	-0.05	0.15	0.145	0.10
100	4	1	F	13	13	9	3	0.08	0.08	0.001	0.007	-0.02	0.04	-0.04	0.06	-0.038	0.04
111	4	7	S	93	100	30	25	-0.02	0.7	0.037	0.05	0.33	0.32	1.21	0.53	0.334	0.29
112	4	7	F	-2	23	0	6	0.25	0.16	0.009	0.01	-0.07	0.07	-0.12	0.12	0.014	0.07

[a] Season for Eqs. (1)–(48), winter; Eqs. (67)–(112), summer.

[b] For station numbers, see Fig. 1.

[c] S = strong acidity, F = smoke.

The range of precipitation is 240 tenths of an hour; that of sunshine 160 tenths of an hour. It can thus be seen that the contribution of the corresponding terms is practically unimportant except in very rare cases.

The values of DTM/DVM during the survey were between -120 and +30. The coefficient a_6 is very occasionally less than -1; mostly it is between -0.1 and 0. Thus, the quotient of the thermal gradient and the wind gradient (as measured at Trappes) can only rarely, or rather by chance, have an influence on the result of a regression equation.

The Significance Level of the Regression Coefficients

A few strictly pragmatic remarks resulting from the comparison of the regression coefficients with their standard errors:

1. The partial correlation coefficient belonging to the wind vector is generally high. It is (except in a few cases, Eqs. 111 and 112) at least twice, often three or four times its standard error.

2a. During the heating season, the independent variable "mean temperature" rather often has a significant partial regression coefficient. (See also the remark about T_m in Section 3.4.)

2b. In summer, on the other hand, the mean temperature is without significant influence on the pollution levels. Then, for the most part, the partial correlation coefficient has a very low negative value.

3. In winter, the partial regression coefficient belonging to the variable "duration of fog" is generally significant, being about twice its standard error. During the summer, due mainly to the small number of observations, this partial regression coefficient is only rarely significant. The partial correlation is often significant.

4. It seems that in winter, as in summer, rain diminishes the pollution levels. The corresponding partial regression coefficient is rather small; for this reason, any stronger affirmation concerning rain would be premature. In any case, the rain (or precipitation) must last several hours to have an appreciable effect on pollution levels. Generally the partial correlation coefficient is small.

5. As for the duration of sunshine, one can be even less positive than for rain. During the winter, the partial regression coefficient is generally positive with a rather low level of significance. The high degree of association, in the sense of a negative correlation, of this variable with precipitation and fog should also be taken into account.

16. URBAN AREA BACKGROUND POLLUTION

During the summer, the partial regression coefficients for sunshine are mostly negative. The correlation is so insignificant that no valid conclusion can be drawn.

6. On the other hand, the partial correlation of DTM/DVM with the dependent variable is, with very few exceptions, very nearly zero. It should be investigated in the future whether this is due to the distance (as the data are obtained at Trappes), or to the fact that observations are made only twice a day.

Successive Estimates Obtained by Introducing the Meteorological Variables One after Another

As mentioned at the end of Section 3.1, one may also examine possible improvements of the estimate as one meteorological variable is introduced after another.

The mean absolute error, $|\text{obs} - \text{calc}|$, was computed for the estimates $J_0, J_1, \ldots, J_5$ (see above). As an example, Table 2 gives the relative error*

$$R_i = \frac{100 \, \Sigma \, |\text{Obs} - \text{Calc}|}{\Sigma \, \text{Obs}}, \quad i = 0, \ldots, 5,$$

corresponding to the successive approximations J_0 to J_5 for the winter season and wind sector 1, only. The same table also gives the relative errors obtained with logarithmic regression equations (see Section 4.2 below).

Table 2 gives the same picture as the first two parts of this subsection. The introduction of precipitation, sunshine, and the quotient of the temperature gradient by the wind gradient (at Trappes) contributes but little to the improvement of the successive estimates. As a result, the order of introduction of these variables becomes also a topic of only slight interest.

If one takes into account only the season, the direction and mean velocity of the wind, and perhaps the mean temperature, one obtains by regression an accuracy of estimation which compares favorably with that obtained by some mathematical models or empirical equations [1,2]. This improvement is mainly due to the elimination of the calm periods and of the days with highly variable winds, and thus involves some loss of generality.

---

*On the topic of numerical forecast verification, see Erikson [6] and Panofsky and Brier [5, pp. 200-201].

TABLE 2

Progression of Relative Errors with Successive Introduction of Meteorological Variables [a]

Station number	Pollutant [b]	Form of regression	R_0	R_1	R_2	R_3	R_4	R_5
1	SO_2 (131)	lin	34.3	32.1	31.3	30.6	28.4	28.3
		log	35.1	32.3	31.0	30.0	27.9	28.0
	Smoke (80)	lin	38.3	37.5	33.6	33.2	32.5	32.6
		log	38.5	38.3	34.1	33.5	21.5	32.5
2	SO_2 (145)	lin	34.1	30.6	30.1	29.3	29.2	29.4
		log	33.7	30.3	28.3	28.2	27.9	28.0
	Smoke (87)	lin	38.0	36.8	32.8	31.6	31.3	31.3
		log	38.9	37.3	32.6	31.7	31.25	31.2
3	SO_2 (86)	lin	47.6	46.2	44.6	44.3	43.8	44.0
		log	52.0	49.2	45.7	45.4	44.7	44.7
	Smoke (77)	lin	34.8	34.9	31.9	31.7	30.4	30.4
		log	35.7	35.7	31.4	31.4	30.5	30.6

16. URBAN AREA BACKGROUND POLLUTION

4	SO$_2$(154)	lin	38.2	37.5	35.4	35	34.9	34.8
		log	37.4	37.3	34.8	34.0	33.9	34.0
	Smoke(73)	lin	45.5	46.1	41.4	40.8	40.4	40.3
		log	43.0	43.3	39.6	39.5	38.6	38.8
5	SO$_2$ (73)	lin	39.7	36.4	33.7	33.4	32.7	32.7
		log	43.75	38.9	35.6	35.1	33.7	33.7
	Smoke(44)	lin	36.1	36.3	35.5	34.9	32.9	33.1
		log	37.6	37.7	35.9	33.2	31.8	31.8
6	SO$_2$ (95)	lin	33.2	31.0	31.1	30.0	28.9	28.7
		log	33.9	32.0	31.0	29.2	28.4	28.1
	Smoke(75)	lin	34.0	34.0	33.3	32.4	29.6	29.2
		log	34.4	34.7	33.3	32.1	30.0	29.4
7	SO$_2$ (109)	lin	37.2	37.3	37.3	36.3	36.0	36.0
		log	38.2	38.1	37.0	35.9	35.0	35.0
	Smoke(53)	lin	36.5	36.3	34.8	34.7	31.9	31.9
		log	37.0	36.5	35.0	34.3	32.6	32.6

[a] Example covers winter for wind sector 1.
[b] Number of observations.

4.2. Practical Form of the Multiple Regression Equations

It follows from Section 4.1 that a regression equation having the form of Eq. (2) for each wind direction (sector) and season is enough to give a fair estimate of urban pollution levels. This conclusion is close to that of Kolar [4], who worked with logarithmic regression equations, but with fewer data. Our trial with logarithmic regression showed no noticeable improvement.

The conclusions of Kolar [4] are based on few observations, and these were specially selected for specific meteorological conditions. On the other hand, the negative appraisal of Marsh [2], also concerning logarithmic regression, is based on a considerable data bank taken fully, without any selection or exclusion.

We think that the more optimistic findings of the present paper result from having a fairly great number of observations, from which, as already mentioned, a small percentage were excluded, corresponding to specific conditions for which the regression method should expressly not be used. These latter cases should be treated separately by different methods.

5. USES OF REPRESENTATION BY MULTIPLE REGRESSION EQUATIONS

5.1. For Authorizing New Sources

Often, for the authorization of a new source, the computed concentration (as obtained by known methods of atmospheric dispersion estimation) is compared with a "background pollution," which is usually some long-term mean value.

The mathematical models for atmospheric dispersion give reasonable estimates of short-term concentrations on the ground, at least for level terrain. These values correspond to well defined meteorological conditions (stability classes). The concentrations thus computed should logically not be compared with a long-term mean as background, but to the typical concentration belonging to the specific meteorological conditions. So it seems evident that a background concentration valid for some short period but nevertheless of a general character should be used instead.

The multiple regression equations allow the computation of a good estimate of a typical (or mean) background concentration for a given wind direction and a given season, etc. It seems obvious that without proceeding by multiple regression, it would be quite impossible to obtain similar

16. URBAN AREA BACKGROUND POLLUTION

results except by taking measurements for a sizable number of days, each of them characterized by almost exactly the same meteorological conditions.

5.2. Use in Interpolation and Extrapolation

A Posteriori Forecast

We term "a posteriori forecast" the estimation of pollution concentrations from observed meteorological data any time these data are already available, i.e., certainly after the episode. This case, besides being a convenient and usual forecast verification, is also an interpolation method for cases when air pollution observations are lacking, whether because of instrument failure or for any other reason. It is almost superfluous to stress again that this interpolation assumes constancy of emissions.

Table 2 is an example of forecasting a posteriori, specifically of the evolution of its mean absolute error.

A slightly different case, in principle, of forecasting a posteriori is when regression equations computed for a given period, here from data of 1967-1969, are used retrospectively, but for another period, e.g., 1970, which was not used for their establishment. Table 3 shows a few examples of this.

Table 4, on the other hand, gives the mean absolute error of the first two equations (in the same way as Table 2) for the winter 1970-71, which was not used for the computation of the regression equations. Wind sector 3 was selected for this example because, as will be shown later, a very important emission source is located in this sector. It can be seen from Table 4, that with the exception of SO_2 for Station 7 (and also, to a smaller extent, for Station 6), the mean absolute error is very nearly the same as in Table 2. (SO_2 for Stations 7 and 6 is analyzed in more detail directly below.)

The Detection of Exceptional Emission Sources

The multiple regression equations are a shorthand expression for the mean pollution for different situations during the reference period, here from March, 1967, to December, 1969. If there is a strong new emission, its effect on the ambient pollution concentration in one or more stations and for a given wind direction will be a systematic concentration excess. This is just a quantitative expression for the qualitative assessment frequently made when, knowing the emission inventory and the elements of the local micrometeorology, we say, e.g., that such a pollution concentration, in this season and for that wind direction, is far too high and must have a definite cause, such as increased emission from some factory.

TABLE 3

Comparison of Forecasts with Observations

Season	Wind sector[a]	Station and pollutant[b]	Obs.	J_0	Diff. 1	J_1	Diff. 2
Winter	1	1 S	113.	123.	10.	145.	32.
	(March 9, 1970)	1 F	89.	81.	13.	88.	1.
		2 S	97.	76.	21.	166.	69.
		2 F	56.	134.	78.	94.	38.
		3 S	103.	76.	27.	95.	8.
		3 F	60.	71.	11.	80.	20.
		4 S	176.	171.	5.	170.	6.
		4 F	81.	72.	9.	89.	8.
		5 S	72.	63.	9.	83.	11.
		5 F	24.	40.	16.	41.	17.
		6 S	92.	86.	6.	106.	14.
		6 F	45.	70.	25.	73.	28.
		7 S	97.	102.	5.	109.	12.
		7 F	15.	51.	36.	58.	43.
	3	1 S	64.	90.	26.	88.	24.
	(Nov. 11, 1970)	1 F	12.	32.	20.	31.	19.
		2 S	54.	75.	21.	69.	15.
		2 F	11.	31.	20.	31.	20.
		3 S	50.	60.	10.	52.	2.
		3 F	24.	34.	10.	33.	9.
		4 S	93.	93.	0.	85.	8.
		4 F	20.	26.	6.	22.	2.
		5 S	95.	66.	29.	56.	39.
		5 F	17.	24.	7.	19.	2.
		6 S	48.	61.	13.	44.	4.
		6 F	23.	40.	17.	28.	5.
		7 S	198.	182.	16.	190.	8.
		7 F	9.	17.	8.	18.	9.

16. URBAN AREA BACKGROUND POLLUTION

TABLE 3 (cont'd)

Season	Wind sector[a]	Station and pollutant[b]	Obs.	J_0	Diff. 1	J_1	Diff. 2
Summer	1	1 S	30.	35.	5.	34.	4.
	(June 5, 1970)	1 F	29.	29.	0.	28.	1.
		2 S	32.	39.	7.	40.	8.
		2 F	28.	31.	3.	30.	2.
		3 S	13.	25.	12.	23.	10.
		3 F	28.	31.	3.	30.	2.
		4 S	40.	88.	48.	93.	53.
		4 F	20.	23.	3.	22.	2.
		5 S	8.	27.	19.	28.	20.
		5 F	13.	13.	0.	13.	0.
		6 S	33.	28.	5.	27.	6.
		6 F	24.	22.	2.	22.	2.
		7 S	14.	63.	49.	63.	49.
		7 F	17.	21.	4.	22.	3.
	2	1 S	31.	37.	6.	39.	8.
	(June 3, 1970)	1 F	22.	36.	14.	44.	22.
		2 S	30.	46.	16.	41.	11.
		2 F	23.	40.	17.	43.	20.
		3 S	26.	25.	1.	26.	0.
		3 F	25.	34.	9.	38.	13.
		4 S	36.	96.	60.	75.	39.
		4 F	20.	26.	6.	29.	9.
		5 S	18.	36.	18.	34.	16.
		5 F	14.	20.	6.	21.	7.
		6 S	30.	37.	7.	38.	8.
		6 F	20.	35.	15.	37.	17.
		7 S	21.	52.	31.	52.	31.
		7 F	13.	21.	8.	24.	11.

[a] Date of observation given in parentheses.

[b] S = strong acid, F = smoke.

TABLE 4

Relative Errors R_0 and R_1 for a Period Which Was Not Used for the Establishment of the Multiple Regression Equation [a]

Station number	Pollutant	Number of observations	R_0	R_1
1	SO_2	114.0	33.6	33.9
	Smoke	35.0	43.3	43.1
2	SO_2	115.0	36.1	36.6
	Smoke	35.0	44.4	44.4
3	SO_2	71.0	44.8	46.9
	Smoke	35.0	40.5	40.5
4	SO_2	102.0	29.9	31.7
	Smoke	31.0	33.6	32.8
5[b]	SO_2	84.0	38.1	39.2
	Smoke	-	-	-
6[b]	SO_2	72.0	44.3	45.1
	Smoke	-	-	-
7	SO_2	312.0	55.8	55.4
	Smoke	20.0	33.3	34

[a] Winter 1970-71, wind sector 3.

[b] Observations of smoke were not made at this station during winter 1970-71.

Exactly this case is exemplified by Table 5. For the three days chosen as an example, all of them for the same wind sector, we see that the observed concentrations are strongly above the computed ones at Station 7 only. It is difficult to think that the error is in the estimation, since for the other stations the computed values are not particularly bad. Even for the other pollutant, smoke, Station 7 fails to show any anomaly. It is difficult to avoid the conclusion that, on these days, some factory southeast of Station 7, i.e., situated in wind sector 3, emitted more SO_2 than usual.

16. URBAN AREA BACKGROUND POLLUTION

TABLE 5

Comparison of Forecasts with Observations [a]

Date	Station and pollutant[b]	Obs.	J_0	Diff. 1	J_1	Diff. 2
March 30, 1970	1 S	115.	88.	27.	88.	27.
	1 F	10.	31.	21.	31.	21.
	2 S	112.	73.	39.	73.	39.
	2 F	10.	30.	20.	30.	20.
	3 S	52.	58.	6.	58.	6.
	3 F	12.	33.	21.	33.	21.
	4 S	108.	91.	17.	91.	17.
	4 F	15.	25.	10.	25.	10.
	5 S	33.	64.	31.	64.	31.
	5 F	9.	23.	14.	22.	13.
	6 S	30.	59.	29.	59.	29.
	6 F	12.	38.	26.	37.	25.
	7 S	675.	182.	493.	182.	493.
	7 F	18.	17.	1.	17.	1.
Dec. 1, 1970	1 S	120.	100.	20.	100.	20.
	1 F	33.	38.	5.	39.	6.
	2 S	125.	86.	39.	87.	38.
	2 F	36.	37.	1.	37.	1.
	3 S	89.	67.	22.	68.	21.
	3 F	37.	41.	4.	41.	4.
	4 S	120.	101.	19.	103.	17.
	4 F	38.	31.	7.	31.	7.
	5 S	54.	76.	22.	77.	23.
	5 F	19.	28.	9.	29.	10.
	6 S	33.	69.	36.	72.	39.
	6 F	30.	46.	16.	47.	17.
	7 S	615.	184.	431.	183.	432.
	7 F	33.	19.	14.	19.	14.
Dec. 2, 1970	1 S	119.	91.	28.	91.	28.
	1 F	29.	33.	4.	33.	4.
	2 S	110.	76.	34.	74.	36.
	2 F	27.	32.	5.	32.	5.
	3 S	146.	60.	86.	58.	88.
	3 F	38.	35.	3.	35.	3.
	4 S	114.	94.	20.	91.	23.
	4 F	19.	27.	8.	25.	6.
	5 S	70.	67.	3.	64.	6.
	5 F	15.	24.	9.	23.	8.
	6 S	17.	62.	45.	57.	40.
	6 F	19.	40.	21.	36.	17.
	7 S	496.	182.	314.	185.	311.
	7 F	21.	17.	4.	18.	3.

[a] Winter, wind sector 3.   [b] S=strong acid, F=smoke.

This fact was indeed confirmed by a factory which had stepped up its SO_2 emission by 20 tons per 24 hours during the period.

It would be of interest to have a quantitative assessment of the significance of given differences between computed and observed values. For the time being, we cannot give a definite answer to this rather complicated question. With a very dense network of stations, each monitoring more than one pollutant, if only one pollutant shows an unusual rise, and it is indicated by more than one station, then a very small difference between observation and computation might already be significant. With only a few stations and pollutants, the difference must be quite considerable before allowing a definite judgment. The mathematical analysis not yet having been made, we estimate from past experience that for the Rouen survey, at least a doubling of the computed concentration for a single station, direction and pollutant is needed before making a definite affirmation. Less difference would be considered enough, if it persists for several days in the same wind sector.

When the exceptional emission occurs not only during a few days, but persists over most of the year, as was the case in Rouen, its effect is also felt in an increased mean absolute error: see the line for "Station 7, SO_2" in Table 4.

A Priori Forecast

One might try to estimate pollution levels with the aid of the multiple regression equations from meteorological forecasts. Table 6 is self-explanatory and gives a few examples of the results.

No forecast from the regression equations should be attempted when the meteorological forecast indicates calm or very slight winds. This case is outside of the scope of the present method; the treatment of calms has previously been outlined [1].

For an a priori forecast one chooses the equation corresponding to the season, and to the meteorologically forecast wind sector. Then numerical values for wind velocity and mean temperature are put into the equation. On a sample of 24 such forecasts, the correlation coefficient between observed and forecast values was 0.64, corresponding to the 1% significance level for a test of independence. We are continuing this forecast to arrive at a greater body of data for a more valid assessment.

16. URBAN AREA BACKGROUND POLLUTION 271

TABLE 6

A Priori Forecast for Station 1, Pollutant SO_2

Daily forecast	Est. num. values		Obs. values		SO_2 ($\mu g/m^3$)	
	V(m/sec)	T(°C)	V(m/sec)	T(°C)	Forecast	Obs.
Medium strong SE winds Rain Jan. 9, 1970	8	3	5	6	110	126
Medium strong NE winds Same weather as yesterday March 15, 1970	8	1	8	4	132	63
Weak to modest N winds Temp. rising March 16, 1970	4.5	6	2.5	5.5	104	103
Modest winds, temporarily quite strong from S Temp. decreasing Nov. 12, 1970	6	8	3.5	6	80	94
Modest winds changing from S to SW Temp. slightly decreasing Nov. 14, 1970	5	5	5	5	99	62

TABLE 6 (cont'd)

A Priori Forecast for Station 1, Pollutant SO_2

Daily forecast	Est. num. values		Obs. values		SO_2 ($\mu g/m^3$)	
	V(m/sec)	T(°C)	V(m/sec)	T(°C)	Forecast	Obs.
Winds from S and W, temporarily strong Temp. decreasing Nov. 16, 1970	7	3	5	4.5	94	90
Winds from S and W, temporarily strong Temp. rising Nov. 17, 1970	7	8	6.5	9.5	92	93
Weak or moderate NE winds Cooler air, sunshine June 13, 1970	4	15	4.5	17	37	33
Moderate NE winds Sunshine, temp. rising June 15, 1970	5	18	2.5	18	33	31
Moderately strong SW winds Changing and cool weather Sept. 10, 1970	10	11	5	12	63	76
Winds from SW to W, moderately strong Temp. slightly decreasing Sept. 14, 1970	10	13	5.5	16	54	46

REFERENCES

1. M. Benarie, Essai de prevision synoptique de la pollution par l'acidité forte dans la région rouennaise, Atm. Environ., 5, 313 (1971).

2. K. J. Marsh and V. R. Withers, An experimental study of the dispersion of the emissions from chimneys in Reading, Atm. Environ., 3, 281 (1969).

3. H. Grisollet, Etude du vent et de la "brise diurne" à PARIS au sommet de la Tour Saint-Jacques, Monographies de la Météorologie Nationale No. 46, Météorologie Nationale, Paris, 1965.

4. J. Kolar, Der Einfluss der meteorologischen Grössen auf die Schwefeldioxid-Immission in Städten, Wärme, 73, 72 (1967).

5. H. A. Panofsky and G. W. Brier, Forecast verification, in Some Applications of Statistics to Meteorology, Pennsylvania State Univ. Press, University Park, Pa., 1968.

6. B. Erikson, Forecast verification, in Statistical Analysis and Prognosis in Meteorology, Technical Note No. 71, WMO-No. 178 TP. 88, World Meteorological Organization, Geneva, 1962, p. 107.

17

ANALYSIS OF HEALTH EFFECTS DATA:
SOME RESULTS AND PROBLEMS

V. Hasselblad, W. C. Nelson, and G. R. Lowrimore

National Environmental Research Center
Environmental Protection Agency
Research Triangle Park, North Carolina

1. INTRODUCTION

The Community Research Branch in the Environmental Protection Agency has been designing health effects studies for several years. The series of studies to be discussed here has been given the acronym CHESS, for Community Health and Environmental Surveillance System. The purpose of these studies is to relate air pollution levels to both acute and chronic human health effects.

Communities selected for study must have an exposure gradient ranging from low to high for the pollutant of interest. Other factors, including age and sex composition, smoking habits, and socioeconomic levels of residents, must be similar or easily adjusted for when the data are analyzed. Pollution is measured daily as a 24-hour integrated sample at several sites, although some continuous monitors are now being used.

Both one-time retrospective surveys and prospective studies of panels, which report every week or so, are used as health indicators. Like the British Research Council questionnaire, the one-time surveys include questions on chronic illness, cough, and phlegm production. Studies on chronic effects also use pulmonary function measurements of elementary school children. The panels include asthmatics, as well as elderly persons with and without chronic heart or lung disease. The panel studies give daily information for a period of about 9 months and are repeated each year. Some approximate study sizes for each indicator are given in Table 1.

TABLE 1

CHESS Study Population Sizes

Indicator and frequency	FY 71 population	FY 72 population
Chronic respiratory disease (every 2 years)	30,000	120,000
Lower respiratory disease (every 2 years)	30,000	120,000
Acute respiratory disease (every 2 weeks)	15,000	36,000
Lung function (twice a year)	7,000	75,000
Asthmatics (once a week)	750	1,350
Elderly panels (once a week)	450	525
Irritation symptoms (3 times a year)	10,200	5,000
Pollutant burdens (every 2 years)	3,900	40,000

The biometric phase of these studies includes problems of data collection and processing, and statistical analyses. In the following paragraphs some examples of these problems are outlined briefly, along with the solutions which we have used. A brief summary of the results of a few studies are shown. We also list a few of our unsolved problems in the hope that someone will come up with the solutions we need.

2. DATA PROCESSING

Collection and processing of health information are major activities in the studies. To avoid an enormous amount of coding and keypunching, some of our questionnaires are printed on optical-scan forms. Even this method requires reading, rereading, editing and correcting. Other questionnaires filled out by the panelists are keypunched. Most of the questionnaires are being converted to optical scanning. Naturally the sheer volume of data presents a storage and retrieval problem even after the data have been corrected and stored in the computer.

3. ANALYSIS OF REPEATED MEASUREMENTS

Although data processing represents a major portion of our efforts, much time is spent in determining appropriate statistical analyses for the data. The data are not always amenable to classical analyses because of repeated measurements on the same persons, serial correlation, autocorrelation, or discreteness of the variables. An example of the repeated measurements problem occurred in a study we ran on second-grade school children in Cincinnati using pulmonary function measurements [1]. Three-quarter-second vital capacity (FEV 0.75) was measured for each child, and some marginal means from the study are given in Table 2. The measurements were taken at three different times; in the fall, winter and spring. Pollution monitoring stations were placed in the vicinities of the schools. Although simple correlations were computed, it was felt that a multivariate analysis of variance would allow for the repeated measurements and adjust for the covariates of socioeconomic status, height and sex. These results are given in Tables 3, 4, 5, and 6.

4. CATEGORICAL DEPENDENT VARIABLES

In many studies the dependent variables are categorical in nature. For example, our chronic respiratory disease questionnaire contains questions about cough and phlegm production. The responses are graded on a severity scale ranging from 1 to 7. Since almost one-half the responses were in one category it was impossible to use standard regression analysis. Instead, a categorical data analysis using a linear model as described by Grizzle, Starmer, and Koch [2] was used. This was applied to the data from five cities in Montana and Idaho that had varying levels of trace elements from mining and smelting (Table 7). No meaningful differences were found by city, although smoking was found to have a large effect on cough and phlegm production. This particular analysis does not consider all the relevant factors: that would have exceeded the capacity of the programs that use this method. As an alternative to total consideration, the severity scores were transformed to the cumulative frequency distribution of the study population. A standard regression analysis was performed on the transformed data, and the results of this analysis are also given in Table 7. The tests for significance are very similar for the two methods. The latter method has the advantage that it can be easily expanded to include a larger number of factors.

5. THRESHOLD ESTIMATION

Since our studies are designed for use in the setting of standards, the lowest level at which effects can be measured should be determined. Most statistical analyses assume that the relationship between two or more variables is a monotonic function. For example, a correlation assumes a straight line. Thus a significant relationship between a health indicator

TABLE 2

Marginal FEV 0.75 Means for 198 Second-Grade
Children in the Cincinnati School Study

Males	1.194
Females	1.122
Low pollution area	1.176
High pollution area	1.139
White	1.196
Black	1.081
November	1.147
February	1.141
May	1.184

TABLE 3

Correlation Matrix for Pulmonary Function Readings for the
Cincinnati School Study

	Nov. FEV	Feb. FEV	May FEV	Econ.	Sex	SO_x	Ht. (in.)
Nov. FEV	1.000	0.885	0.857	0.072	0.229	−0.222	0.685
Feb. FEV		1.000	0.900	0.109	0.327	−0.366	0.688
May FEV			1.000	0.192	0.285	−0.338	0.660
Econ.				1.000	−0.058	−0.168	0.177
Sex					1.000	−0.043	0.219
SO_x						1.000	−0.282
Ht. (in.)							1.000

and a pollutant would imply that the effect occurs at all levels of exposure.
As a simple alternative to the straight line, we hypothesized a segmented
line with zero slope up to a point X_0 and a straight line beyond X_0 (Fig. 1).
This relationship has been called a "hockey stick" function. The technique
for obtaining the least-squares solution for a more general problem has

TABLE 4

Regression Coefficients for the Cincinnati School Study

Factor	Coefficient		
	Nov.	Feb.	May
Econ.	-0.0089	-0.0039	0.0172
Sex	0.0275	0.0406	0.0359
SO_x	-0.0119	-0.0605	-0.0520
Ht.	0.0564	0.0536	0.0535

TABLE 5

Covariance and Correlation Matrices for the Cincinnati School Study
(Correlations Given Below the Diagonal)

	Nov.	Feb.	May
Nov.	0.02223	0.01736	0.01818
Feb.	0.7912	0.02165	0.01932
May	0.7488	0.8064	0.02651

TABLE 6

Results from Linear Model Analysis of the Cincinnati School Study

Factor	Degrees of freedom	Wilks' U	P Value
Econ.	3,1,193	0.9433	0.0091
Sex	3,1,193	0.9305	0.0025
SO_x	3,1,193	0.8712	< 0.0001
Ht.	3,1,193	0.5648	< 0.0001

TABLE 7

Analysis of Severity of Respiratory Symptoms for Males in the Montana-Idaho Study

Factor	Degrees of freedom	Analysis of categorical data		Transformed analysis of variance	
		x^2	P Value	F	P Value
Cities	3[a]	14.52	0.0023	3.83	0.0094
(gradient)	1	1.02	0.3125	0.63	0.4274
Ages	2	8.27	0.0160		
(linear)	1	7.53	0.0061	7.16	0.0075
Smoking	3	157.22	< 0.0001		
(linear)	1	128.38	< 0.0001	182.87	< 0.0001
Fit of model	18	14.20	0.7160		

[a] Only four cities were included in this analysis because of small cell sizes.

been worked out by Quandt [3]. For a particular problem relating decreased performance of runners with photochemical oxidant pollution, we concluded that there probably was a threshold at about 12 pphm oxidant, with a 95% confidence interval of 6.8 - 16.3 pphm.

6. OTHER PROBLEMS

Many of our studies deal with acute upper and lower respiratory illness rates. Comparisons of these rates are usually made by standard methods. Upper respiratory disease rates were found to be higher in Chattanooga for both children and adults living in high concentrations of NO_2 as compared with those who did not [4,5]. The use of school absence rates, however, has not given any reliable estimates of respiratory illness, so these studies have been dropped.

Techniques for analyzing panel studies have not been satisfactorily developed. These observations on groups of individuals over an extended period of time present the usual problems of longitudinal studies. Seasonal variation of both the health indicators and the pollution measurements

17. ANALYSIS OF HEALTH EFFECTS DATA 281

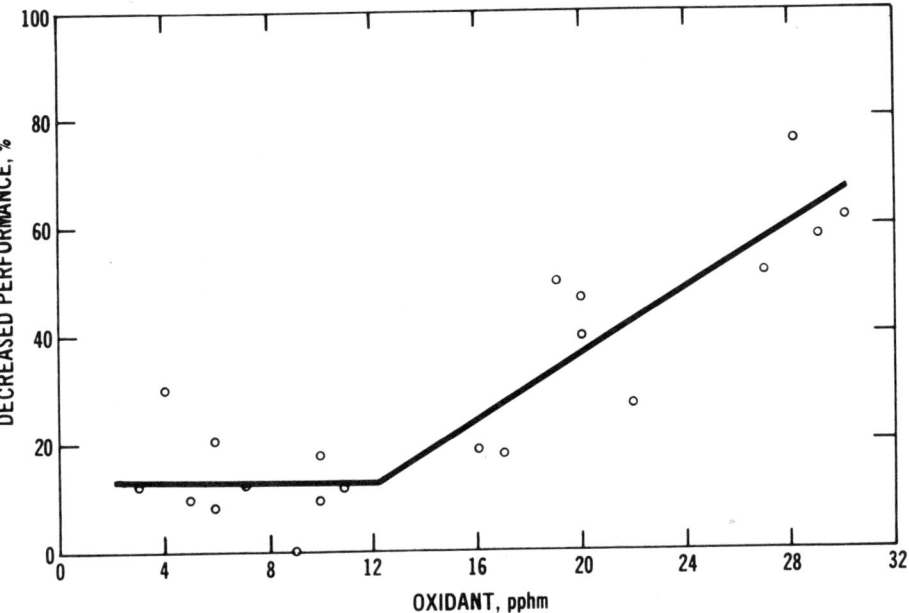

Fig. 1. Percent decreased performance vs oxidant level, with fitted "hockey stick" function.

further confound the studies. Some types of probit analysis are being investigated.

Estimating the pollutant exposure of a human being is extremely difficult. In the case of elementary school children, we can measure pollution at their school and assume that all the children going to that school are exposed to that pollution level. In some cases we have to estimate the pollutant exposure of an adult from his residential address, using a few pollution values from several miles away from his residence. We may not know how long he has lived at that address, where he works, or a number of other important factors. For most pollutants, there are no emission inventories to give additional information on exposure. Statistical models that will give exposure estimates based on the available information are needed.

These are just a few of our more serious statistical problems. We feel that these are important problems, and their solution could benefit anyone doing this type of epidemiologic research. We are always open to suggestions or constructive criticism.

REFERENCES

1. C. M. Shy, V. Hasselblad, R. M. Burton, C. J. Nelson, and A. A. Cohen, Air pollution effects on ventilatory function of US schoolchildren, Arch. Environ. Health, 27(3), 124 (1973).

2. J. E. Grizzle, C. Stamer, and G. G. Koch, Analysis of categorical data by linear models, Biometrics, 25(3), 489 (1969).

3. R. E. Quandt, The estimation of the parameters of a linear regression system obeying two separate regimes, J. Amer. Stat. Assoc., 53(4), 873 (1958).

4. C. M. Shy, J. P. Creason, M. E. Pearlman, K. E. McClain, and F. B. Benson, The Chattanooga school children study: effects of community exposure to nitrogen dioxide. I. Methods, description of pollutant exposure, and results of ventilatory function testing, Air Poll. Control Assoc. J., 20(9), 539 (1970).

5. C. M. Shy, J. P. Creason, M. E. Pearlman, K. E. McClain, and F. B. Benson, The Chattanooga school children study: effects of community exposure to nitrogen dioxide. II. Incidence of acute respiratory illness, Air Poll. Control Assoc. J., 20(9), 582 (1970).

18

A STATISTICAL MODEL OF A POLLUTED ESTUARY

D. W. Mackay and J. Gilligan

Clyde River Purification Board
Glasgow, Scotland

1. INTRODUCTION

The highly polluted estuary of the River Clyde in Western Scotland has been the subject of study by the authors and their colleagues for several years. The object has been to provide a rational plan for the economic rehabilitation of the estuary and, in particular, to gauge what reduction would be required in polluting load from specified sewage treatment works to maintain dissolved oxygen levels in the estuary at desired levels during critical periods. The situation is a highly complex one and our solution was to develop a model based on the analysis of recorded field data, rather than to attempt the formulation of a mathematical model from first principles. This paper deals mainly with the variabilities inherent in an estuarial system, rather than with the practical applications of our findings, which have been described elsewhere [2].

The variables dealt with in terms of their ability to influence levels of dissolved oxygen in the estuarial system are freshwater flow, temperature, and tidal range. Due to the large quantity of data available and the statistical techniques which can be applied, it has been found possible to separate and define quantitatively the importance of each of the variables concerned. Each of these can be influenced by man's activities, and any attempt to improve the condition of the estuary by reducing the polluting load must recognize the importance of these other factors in influencing fluctuations in the levels of dissolved oxygen in the estuary.

2. GENERAL NATURE OF THE SITUATION

Figure 1 illustrates the complex nature of the system with the estuary of the River Clyde discharging to a deeply indented coastline and fringed by fiordic sea lochs of varying depth and alignment. This results in circulation patterns in the lower reaches of the estuary which are not easily predictable by the techniques which might be applied to an estuary discharging on an open coastline. The River Clyde drains the industrial heart of Scotland

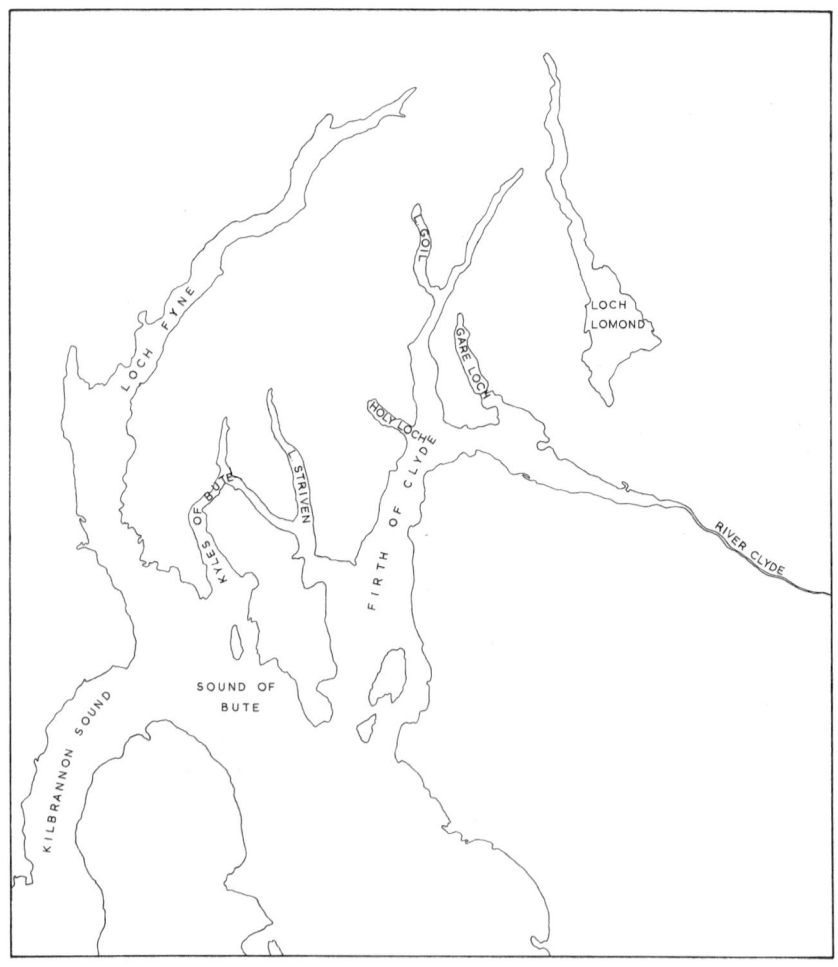

Fig. 1. The Clyde sea area.

18. STATISTICAL MODEL OF A POLLUTED ESTUARY

with a population of over two million in the catchment area, and a dense concentration of industry, mainly centered in steel production and heavy engineering. Figure 2 shows in diagrammatic form the freshwater flow (in m^3/sec) and the polluting load (in terms of five-day biochemical oxygen demand = B.O.D.) entering the estuary from the upper tidal limit at Glasgow to the open firth some 38 km seawards. The channel has a mean depth of 10 m, and for the purposes of this survey 13 sampling stations were established at 3-km intervals.

In terms of salinity distribution the estuary varies between being highly stratified, with fresh water at the surface and sea water at greater depths, and being quite well mixed, with little salinity gradient between surface and bed. For this reason each level of dissolved oxygen quoted in this paper represents the mean of five samples taken, within a very short time interval, in the vertical plane at individual stations on each sampling occasion. Dissolved oxygen levels in the polluted stretch of estuary vary considerably, as illustrated in Fig. 3.

One of the first obvious features to emerge from our investigations was that dissolved oxygen levels were linked with freshwater input. However, since four major rivers discharge to the estuary, we were faced with the difficulty of deciding whether to correlate against the flow values at the time of sampling or for some period before that. Additionally, it seemed unlikely that dissolved oxygen levels at a sampling station some distance from the entry point of a river would react as quickly to a change in river flow as those at a station immediately adjacent to the river mouth. To reduce this potential source of error, the regression of dissolved oxygen against freshwater input was calculated for each station, with the flow measured on the day of sampling, and then averaged up to a period of six days before sampling by steps of one-day units. The calculations were laborious, involving the summed flows of several rivers averaged over seven time periods and applied to thirteen stations on each of over 100 sampling dates, but by this iterative process, the length of flow period most appropriate to each station and yielding the best relationship between levels of dissolved oxygen and freshwater flow was established. In practice the time period varied between three and seven days.

This search could produce some "artificial" improvement in the later multiple regression analysis, not reflected in the standard error of estimate, but this should not be serious, since only dissolved oxygen and flow were used, and the analyses produced a very compact set of results, with significant, but small, differences between alternative flow periods. While periods ending one or more days before sampling could also have been tried, our knowledge of flow patterns and retention times within the system dictated against such an approach.

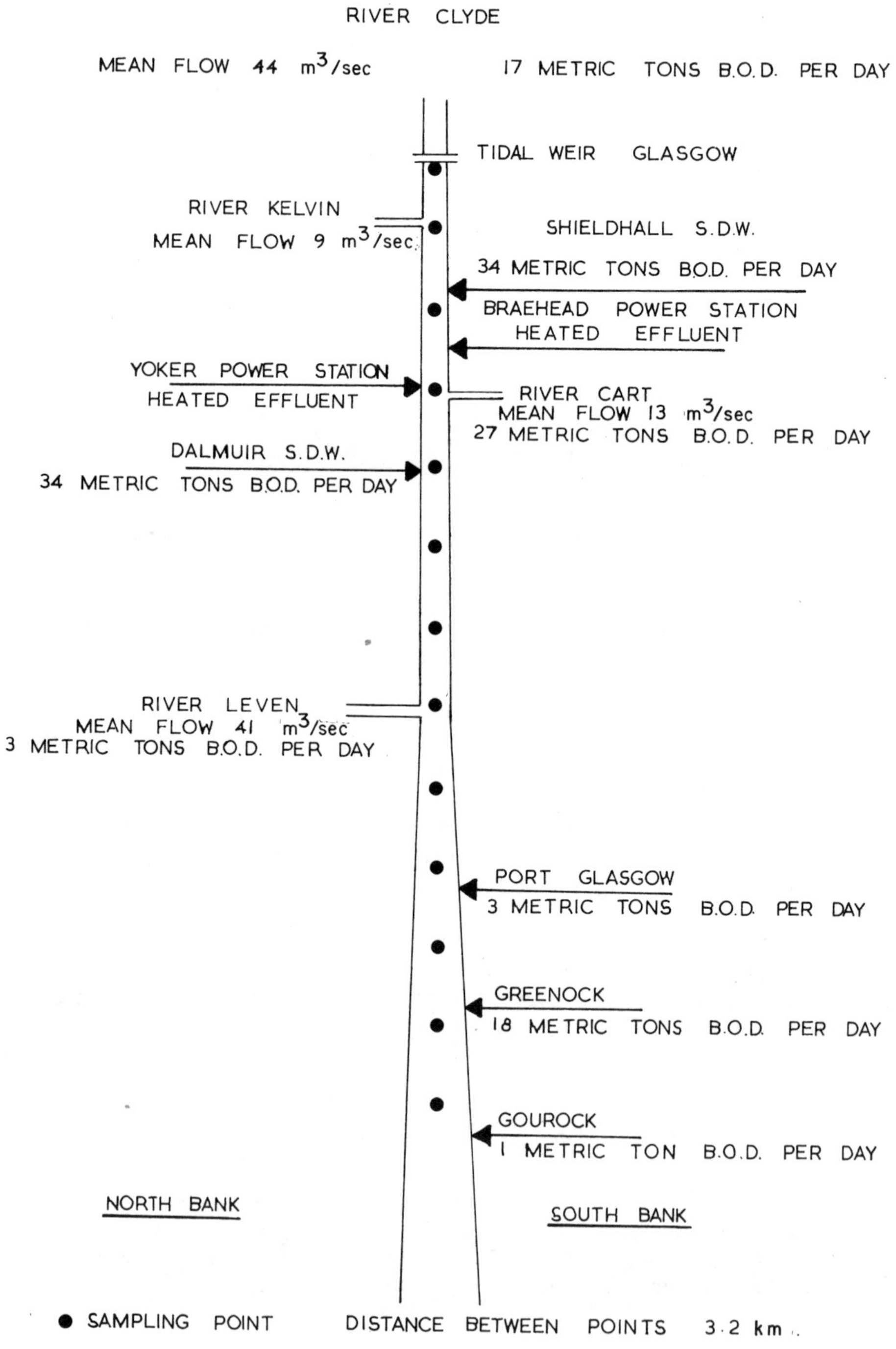

Fig. 2. The polluting load on the Clyde estuary.

18. STATISTICAL MODEL OF A POLLUTED ESTUARY

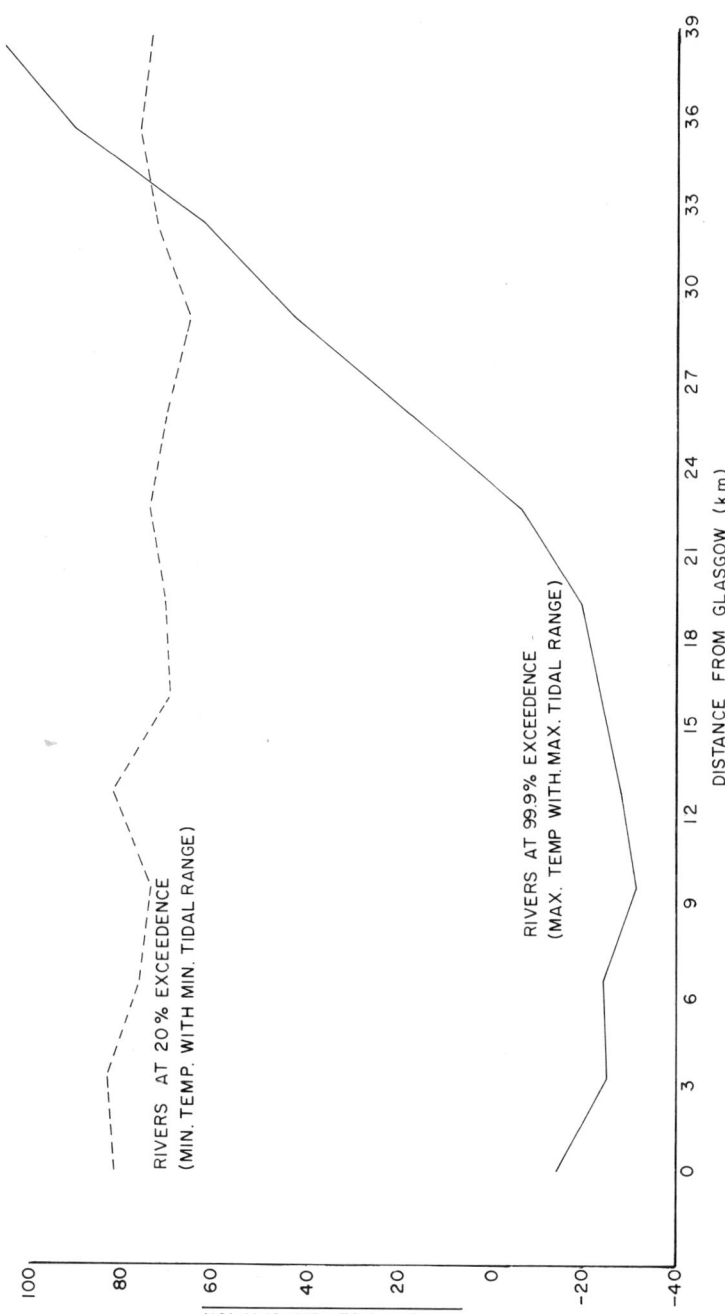

Fig. 3. Levels of dissolved oxygen under varying conditions of freshwater flow.

3. THE EFFECT OF TIDAL RANGE

Having established the flow period most appropriate to each station, it was now possible to divide the survey data into groups. The first division was into groups where the flow regime had been similar for the appropriate number of days preceding the surveys. These groups were then subdivided according to tidal range, and the dissolved oxygen levels were compared. The results are expressed in Table 1.

To state that a river is at 99% exceedence flow (see Table 1) means that in an average year that flow would be exceeded for 99% of the time. It is noteworthy that at times of high tidal range (spring tides) dissolved oxygen levels are depressed. It might be expected that a high tidal range would result in more rapid flushing of a basin as polluted water was removed and replaced by clean water. For the Clyde estuary we know that retention time can be high, up to 25 days, and that the estuary tends to the well-mixed condition during spring tides and to the stratified condition during neap tides. It would appear, therefore, that the stratified condition may allow more rapid escape of the polluted fresh water in the surface layers, and that during times of spring tides polluted material is trapped in the upper reaches, where deoxygenation occurs.

4. MULTIPLE REGRESSION ANALYSES

The relationship between dissolved oxygen levels in the estuary and the three variables (freshwater flow, temperature, and tidal range) was now established by multiple regression analysis. The data for each of the 13 sampling stations were presented to the computer as a matrix having 85 lines, corresponding to 85 approximately weekly observations, and 4 columns, one for each variable, and the program was selected from the large suite made available by the computer proprietors.

The program gives the mean, variance, and standard deviation of each variable. It then correlates each variable with every other variable individually and a correlation matrix is printed out.

In the example shown in Table 2 a good correlation is indicated between y (dissolved oxygen) and x_1 (log flow) and x_2 (temperature), and a much less significant correlation with x_3 (tidal range). The matrix can also be used to identify relationships between the other variables. For example, in this case there is an inverse relationship between x_1 (flow) and x_2 (temperature).

The final result is a series of equations (Table 3) describing dissolved oxygen levels in terms of the other three variables at each station and including the standard errors of estimate.

TABLE 1

The Effect of Tidal Range on Dissolved Oxygen Levels

Tidal range (m)	Dissolved oxygen (% saturation)		
	Rivers at 99% exceedence flow	Rivers at 90% exceedence flow	Rivers at 50% exceedence flow
<1.9	32	38	48
1.9 - 2.3	27	33	47
2.4 - 2.7	18	27	45
>2.7	18	26	42

TABLE 2
Example of a Correlation Matrix

	y (dissolved oxygen)	x_1 (log flow)	x_2 (temp.)	x_3 (tidal range)
y	1.000	0.799	-0.769	-0.251
x_1	0.799	1.000	-0.589	-0.091
x_2	-0.769	-0.589	1.000	0.081
x_3	-0.251	-0.091	0.081	1.000

The way in which tidal range varies is quite predictable, and tables of predicted tide levels are published by the Admiralty and used for navigation purposes. Temperature varies in two ways. The first is seasonal, the average temperature for each month changing little from year to year. Superimposed on this are smaller changes closely linked with freshwater input. Therefore, knowing the flow pattern of the influent rivers, so long as the polluting load remains within the regime which applied during the period of surveys we can predict with reasonable accuracy the levels of dissolved oxygen which are likely to occur at any station in the estuary during a year of average rainfall. It is also possible to predict the effects on levels of dissolved oxygen at each station, of drought, of abstraction of water from the rivers, and of adding heated effluents. Obviously, great caution must be used when extrapolating beyond the limits of recorded data.

TABLE 3

Predictive Equations for Dissolved Oxygen in Clyde Estuary[a]

Station[b]	Const.	a_1	a_2	a_3	s[c]
0	-18.67	57.89	-2.46	1.25	11.8
3	-19.20	59.83	-2.26	-3.08	11.1
6	-6.24	51.40	-2.47	-4.07	11.5
10	26.64	35.77	-2.72	-10.11	11.3
13	18.75	38.59	-2.61	-9.09	10.5
16	16.66	36.86	-2.30	-8.66	10.9
19	22.55	31.31	-2.45	-8.79	11.3
22	42.11	25.31	-1.72	-12.53	10.5
26	68.89	9.45	-1.65	-8.60	11.2
29	59.45	6.36	-0.72	-3.12	10.6
32	106.30	-10.22	-0.90	-4.66	10.8
35	141.05	-26.50	-0.26	-3.54	13.6
38	119.99	-22.45	1.17	-1.05	17.5

[a] The form of the equations is

$$y = \text{const.} + a_1 x_1 + a_2 x_2 + a_3 x_3$$

where

y = dissolved oxygen expressed as percentage saturation

x_1 = $\log_{10}$ of upstream freshwater input to estuary (m^3/sec)

x_2 = temperature (average of readings at five depths, °C)

x_3 = tidal range (m).

[b] Distance from tidal weir (km).

[c] Standard error of estimate.

Also, there is undoubtedly the risk that one variable may be acting as proxy for several, e.g., temperature for several seasonal factors, so that caution must be exercised in predicting the effects that a deliberate change in one variable might produce. However, in practice this has usually not appeared to be the case for the model described here.

A previous publication [1] dealt with the effects of changes in polluting load on dissolved oxygen levels in the estuary under average conditions. The calculations were based on the increases or decreases in the biochemical oxygen demand of the effluents to the system, coupled with the retention time in the estuary of discharges at various distances downstream of the tidal weir. The calculated changes have been found to be correct in practice when there was a major change in one of the principal polluting discharges. However, these changes could only be calculated for long-term average conditions, and Fig. 4 illustrates clearly the very considerable variations which can take place in a relatively short time, mainly related to the other variables included in the model.

An example of predicted values plotted against actual field data for one particular day is presented in Fig. 5. However, the predicted values correspond more nearly to actual values when the independent variables fall in the area illustrated, with considerable deviation occurring in the upper range of values of dissolved oxygen, probably due to the smaller quantity of data available in this range and its outlying nature.

5. IMPORTANCE OF VARIABLES

If we now consider each variable in turn we may determine their relative importance in relation to the dissolved oxygen regime.

Freshwater Input

The influence of freshwater input is most marked in the upper reaches where, at mean tidal range and mean temperature, levels of dissolved oxygen vary between 0 and 60% saturation. Further downstream the effect becomes gradually less marked, and seaward of the 30 km distance from Glasgow Bridge the effect is reversed and dissolved oxygen levels decrease with increasing freshwater input. This may be explained by the fact that at times of high freshwater flow, retention time of polluted material in the upper estuary is reduced from over 20 days to less than three days, so that biochemical oxygen demand which would normally be exerted in the upper estuary has its effect much further seaward.

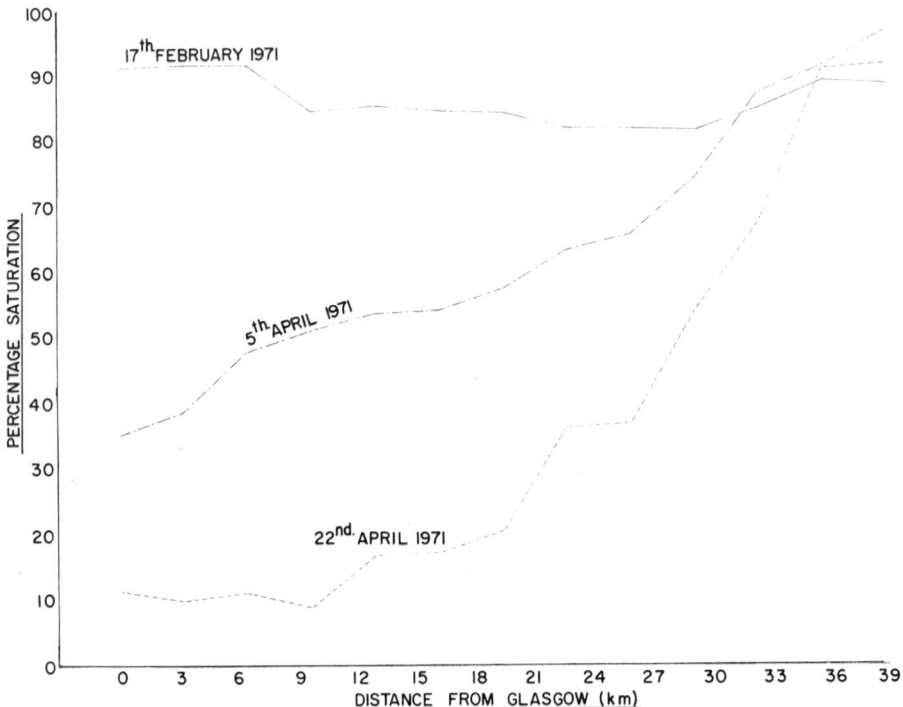

Fig. 4. Variation of dissolved oxygen levels with time in the Clyde estuary.

Temperature

The maximum range of temperature is recorded in the upper reaches of the estuary with top levels around 20 °C and minimum values around 3 °C. The range narrows somewhat as we proceed seaward, with a maximum of 17 °C and minimum of 4 °C at the outermost station. The effects of temperature on dissolved oxygen levels are not so pronounced as those of flow, but are nevertheless important. In the upper reaches, at 50% exceedence flow and mean tidal range, the maximum fluctuation in temperature can produce changes in the dissolved oxygen levels of up to 45% in terms of percentage saturation. This effect is maintained to the 22 km station. Thereafter the influence of temperature falls off rapidly until at 38 km the effect is reversed and high temperatures appear to result in increased levels of dissolved oxygen. The theory that this reflects a proxy effect and is in fact due to photosynthesis, which might be expected to coincide with high temperatures, rather than to temperature alone, is reinforced by the recording of levels well in excess of 100% saturation in

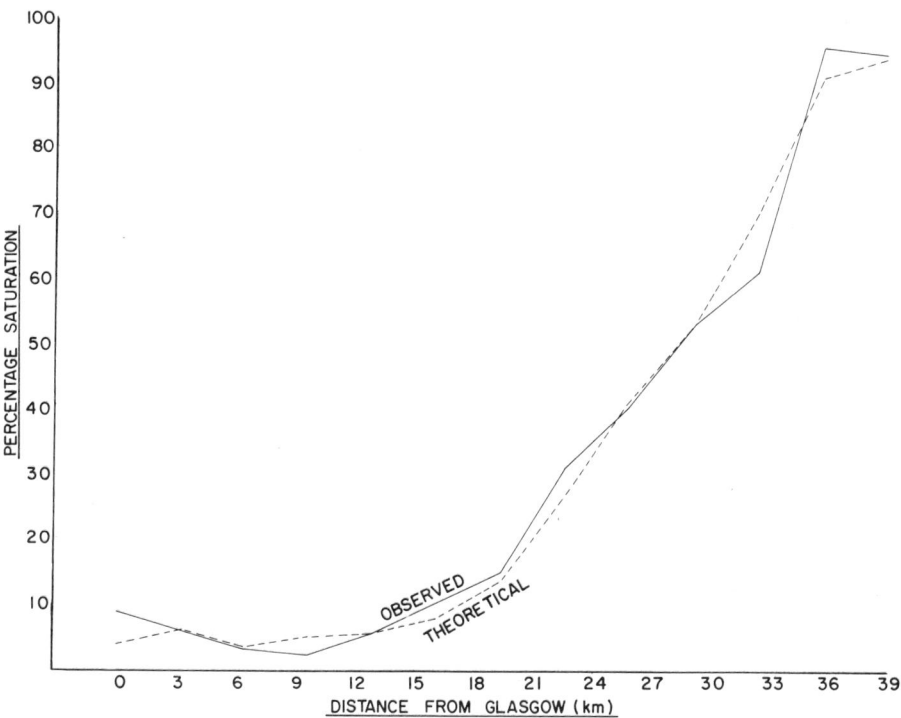

Fig. 5. Predicted values of dissolved oxygen plotted against field data.

this area. Plankton capable of photosynthesis do not occur in the upper reaches of the estuary.

Tidal Range

The effect of tidal range (2.0 - 3.5 m) on dissolved oxygen levels in the estuary as a whole unit has already been illustrated in Table 1. A more detailed examination of the data shows that at the station farthest upstream, i.e., the dock area in the center of the city of Glasgow, increased tidal range has a beneficial effect on levels of dissolved oxygen. At times of high spring tides the levels of dissolved oxygen will be about 3% higher in terms of percentage saturation than at times of neap tides, other factors being equal. Further downstream the occurrence of spring tides depresses dissolved oxygen levels by between 7% and 20% saturation, the deficit increasing generally between stations 3 and 26. Thereafter the effect decreases until at station 38 the deficit is reduced to 3% saturation.

6. CONCLUSIONS

The three variables considered above are important in influencing levels of dissolved oxygen in the estuary. The one vital variable not considered in this paper is the polluting load. This may be considered constant in the short term, since it results from a large population and the discharges from many industrial sources, without significant seasonality. In the long term, however, improvements are being effected as major sewage treatment works are constructed or are made more efficient by the installation of new equipment. The improvement may be measured, as illustrated in Fig. 6, by comparing a series of current surveys with predicted values derived from those carried out some time previously.

When several major improvements have been carried out, and their results measured in this way, the model can then be used to predict the

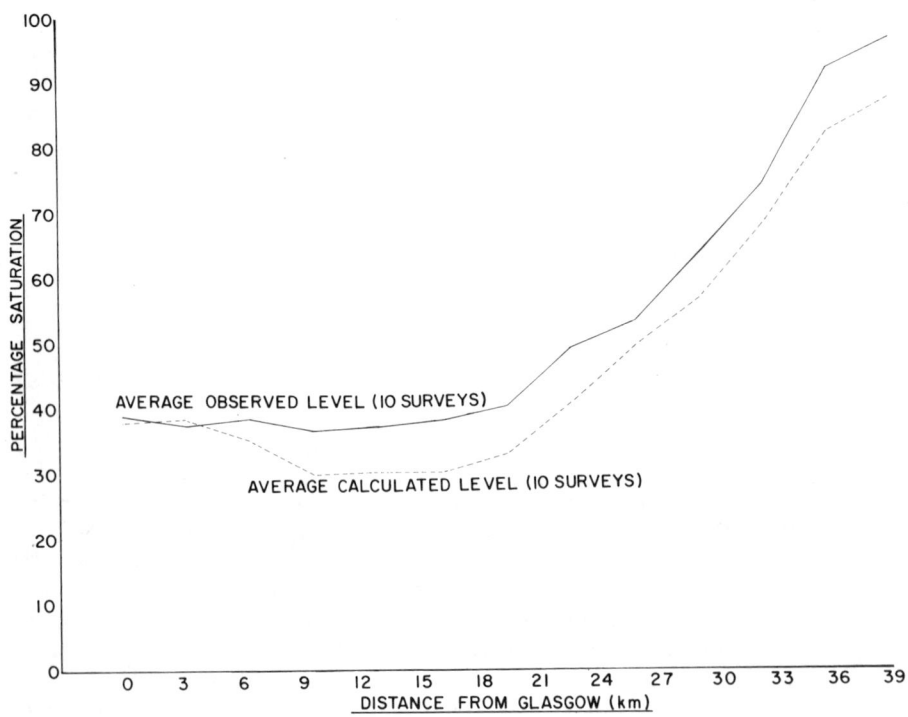

Fig. 6. The improvement in estuarine water quality related to reduced biochemical oxygen demand.

effects of proposed improvements for the critical stages when the estuary is approaching a suitable condition for the passage of anadromous fish or other functions.

REFERENCES

1. D. W. Mackay and G. Fleming, Correlation of dissolved oxygen levels, freshwater flows and temperatures in a polluted estuary, Water Res., 3, 121 (1969).

2. D. W. Mackay and J. I. Waddington, Quality predictions in a polluted estuary, Advances in Water Pollution Research, Vol. 2, Proceedings of the Fifth International Conference, Pergamon Press, New York, 1971, p. III-7.

SOME STATISTICAL ANALYSES OF WATER QUALITY IN THE DELAWARE RIVER

Robert V. Thomann

Manhattan College
Bronx, New York

1. INTRODUCTION

The "science" of water quality management is still very much in its infancy. At the present time interest centers primarily around long-term capital improvement projects to restore bodies of water with poor water quality. In addition, a considerable degree of planning and legislative activity is directed toward the problem of maintaining those water bodies that are presently at a good or high level of quality. All of these efforts involve the specification of water quality criteria in varying degrees. Concurrently with long-range planning and environmental control activity, increased attention is being paid to the short-term water quality forecasting problem. "Short-term" is taken to mean time-varying problem contexts on the order of weeks to three- or four-month seasonal scales.

The purpose of this paper is to present some results of a statistical analysis of variations in water quality as measured by the dissolved oxygen and temperature of rivers and estuaries, and in waste input, as measured by the oxygen demand. Further, the paper discusses the implications of such statistical variations for the achievement of standards and for the short-term forecasting problem.

2. AVAILABLE DATA

The primary data base used in this paper draws on the data obtained from a water quality monitoring network on the Delaware River operated by the U. S. Geological Survey. A review of the types of data collected at these stations and the output format is given in Refs. 1 and 2. The location of the stations is shown in Fig. 1 and Table 1 lists the mile-points of each of the stations. Mean daily data were available for 1968 and 1969 and in some cases hourly data were available for the months of July, August, and September of 1968.

Data were also available on waste discharge variability from other work [3] and from analysis of new data from an industrial discharger. These discharges, however, are not located in the Delaware Basin.

The water quality monitors, as seen in Fig. 1 and Table 1, reflect a wide range of conditions, from an upland station in a clean, free-flowing river of depth about 3-6 ft (E. Stroudsburg) to a highly polluted estuary station with depth about 30 ft (Philadelphia). This affords an opportunity to examine the "natural" variability in DO (dissolved oxygen) and temperature. Such variability is of obvious importance in the setting of stream quality standards and further in determining whether such standards are contravened by a waste discharge.

Data gaps occur in the records and are especially serious when analyses are made of the hourly data. The gaps tend to be less significant when the mean daily data are analyzed. In the time series analysis of the mean daily data, no smoothing or interpolation of missing data was attempted. Rather, all analyses of the mean daily data were carried with gaps in the data.

Some typical data records are shown in Figs. 2 and 3. The seasonal regularity in the mean daily data can be noted, as well as some important day-to-day transient phenomena. The hourly time series records showed variations representative of river and estuary situations. For example, at E. Stroudsburg, the upriver station, diurnal effects of photosynthesis on DO are present and hourly changes in water temperature are evident. At an estuary station (e.g., Bristol) such effects tend to be dampened by the approximately semidiurnal tidal effects. For the estuary stations then, periodicities of about 12.42 hours and 24 hours are present. In all cases, however, the day-to-day transients, as shown in Fig. 3, dominated the variability. One of the tasks of the statistical analysis, therefore, is to provide insight into the relative degree of day-to-day variability in quality as opposed to the "high-frequency," "hour-to-hour oscillations. This has obvious implications in any forecasting and management scheme.

3. STATISTICAL ANALYSES

Data available for analyses in water quality management, as indicated above, often have "gaps" in the record, although for some situations,

19. WATER QUALITY IN THE DELAWARE RIVER

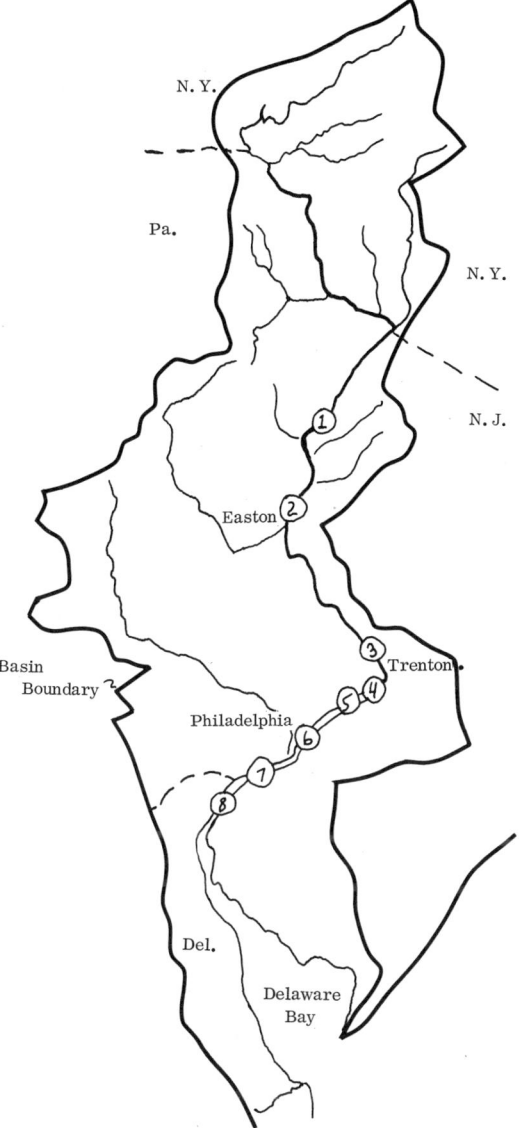

Fig. 1. Delaware River basin, showing location of water quality monitoring stations.

e.g., waste plant effluent, data without gaps may be available. The analyses performed on the data generally consisted of the following sequence:

TABLE 1

Location of USGS Water Quality Monitoring Stations

Station No.	Name	Mile point [a]	Remarks
1	E. Stroudsburg, Pa.	220.0	Northernmost station - river section
2	Easton, Pa.	184.7	Below entrance of Lehigh River
3	Trenton, N. J.	134.5	Above estuary
4	Bristol, Pa.	119.2	Estuary station
5	Torresdale, Pa.	110.0	Estuary station
6	Philadelphia, Pa.	100.1	Estuary station - poorest quality
7	Chester, Pa.	82.3	Estuary station
8	Delaware Memorial Bridge	68.7	Estuary station - in recovery zone

[a] Mileage above mouth of Delaware Bay.

1. examination of record to determine a priori any deterministic components,

2. least squares regression to remove deterministic (periodic) components,

3. construction of residual record,

4. autocorrelation and power spectrum analyses during several stages in the analysis, and

5. for selected records, a first-order autoregressive fit to data residuals.

19. WATER QUALITY IN THE DELAWARE RIVER

In all cases reported on in this paper, a single sinusoidal function was used where inspection of the record indicated a periodicity of this type was dominant. The sinusoidal model is therefore

$$y = A_0 + S_1 \sin \omega t + C_1 \cos \omega t + e(t), \tag{1}$$

where y is the water quality observation; ω is a given input frequency; A_0, S_1, and C_1 are regression coefficients; and e(t) is the residual error. For equally spaced data the coefficients of Eq. (1) are obtained directly as a simple single-frequency Fourier analysis [3,4]. For unequally spaced data, three normal equations must be solved simultaneously.

The estimate of the autocorrelation function was obtained by

$$\hat{R}(\tau) = \frac{\Sigma y(t) y(t + \tau)}{N - 1 - \tau} \tag{2}$$

where the product term, for unequally spaced data, is computed only for real data pairs, thereby excluding products of a data point with a "gap".

After estimation of the autocorrelation function, a standard Fourier transformation was performed to obtain an estimate of the power spectrum, following Ref. 5. In addition to analysis of cyclical trends, autocorrelation, and spectral computations, a first order autoregressive analysis was carried out on residual water temperature records. Again, gaps in the data were observed and the autoregression carried out only for complete pairs of data. The scheme for the water temperature records was, therefore,

$$T(t) = A_0 + S_1 \sin \omega t + C_1 \cos \omega t + R_1 (t-1) + e(t), \tag{3}$$

where R_1 is the first-order autoregression coefficient. An excellent review of diagnostic procedures suggested for analysis of quality time series is given by McMichael and Hunter [3]. It should be stressed, however, that many of the analyses reported on herein included records with gaps. Correlations at some lags, then, cannot be well estimated, and the results of the diagnostic procedures have to be taken with caution.

4. RESULTS OF ANALYSIS

Since water quality responses are closely associated with nearby waste discharges, it is well to review the nature of the variability of such discharges. The author [4] has previously indicated the wide variation in

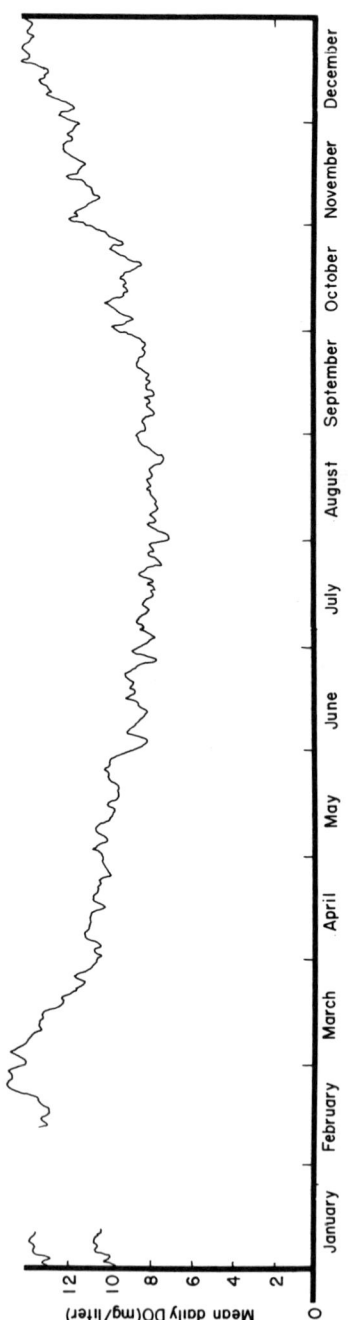

Fig. 2. Mean daily DO (mg/liter) at E. Stroudsburg, 1968.

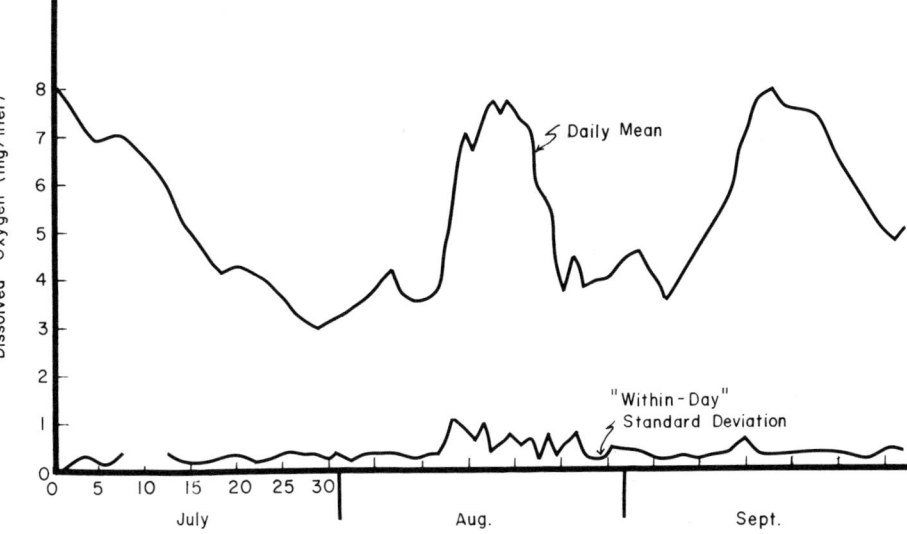

Fig. 3. Daily mean DO (mg/liter) at Bristol, July-Sept., 1968.

effluent load that can occur from municipal waste treatment plants. Coefficients of variation of effluent biochemical oxygen demand (BOD) greater than unity are not uncommon. In addition, spectral estimates of discharges indicate some variance at high frequencies, apparently associated with industrial wastes. As a further illustration, Fig. 4 shows the variance spectrum of the influent, primary effluent, and final effluent from a waste treatment plant of a large paper mill. The distribution of variance is similar to that of municipal plants, showing considerable variance at low frequency and a substantial peak at a frequency corresponding to a period of about 7 days. Note that higher variance occurs at the low-frequency end of the spectrum even after complete treatment.

Typical results of the analyses of the hourly DO data during the summer months are presented in Fig. 5, which shows the variance spectra for the DO at E. Stroudsburg, the most northerly station in the system and the station least subject to waste discharge influences. Figure 5 also shows the DO at Bristol, a station located in the tidal portion of the Delaware and subject to surrounding discharges. The difference between the two stations is highlighted by the spectra. For E. Stroudsburg, with mean of 8.1 mg/liter and standard deviation of 0.52 mg/liter, an analysis of the 24 hour component indicated an average daily amplitude of 0.44 mg/liter (35% of variance) with a maximum occurring at 1600 hours. This strongly indicated the presence of photosynthesis activity by aquatic plants. The variance breakdown of the record is shown in Table 2. The dominance of the low-frequency, i.e., day-to-day, variation is indicated by these results, although there is significant within-day variability resulting from photosynthesis.

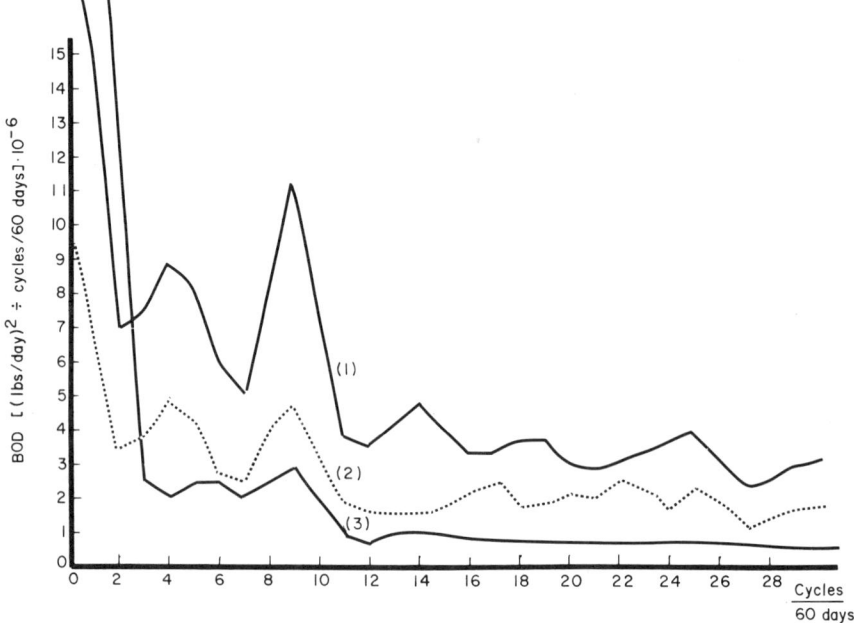

Fig. 4. Influent and primary and secondary effluent variance spectra, Paper Mill: (1) influent spectrum; (2) primary effluent spectrum; (3) secondary effluent spectrum.

The dotted-line plot of Fig. 5 (mean 5.27 mg/liter, standard deviation 1.52 mg/liter) indicates the tidal effect and diurnal effect on the DO at Bristol. Both are present and indicate a combination of diurnal photosynthesis and tidal effects together with semidiurnal variability and a hint of some higher harmonics. As before, there is a dominance at the low-frequency end of the spectrum. More than 50% of the total variance during the three-month period is concentrated in the frequency band from 0 to 4 cycles/6 days. The total variance is greater than at the upstream, nontidal, nonpolluted station by a factor of about 10. These results, from the examination of the hourly data, indicate the importance of the longer-term day-to-day variability in DO. Subsequent analysis therefore concentrated on the daily average data, with some analysis of the within-day standard deviation.

The results of the analyses of daily average DO and temperature are summarized in Figs. 6 and 7. Specific attention is directed toward the upper portions of both figures. As indicated, the residual standard deviation after removal of the annual harmonic represents a significantly greater variation than the average within-day standard deviation. This indicates

19. WATER QUALITY IN THE DELAWARE RIVER

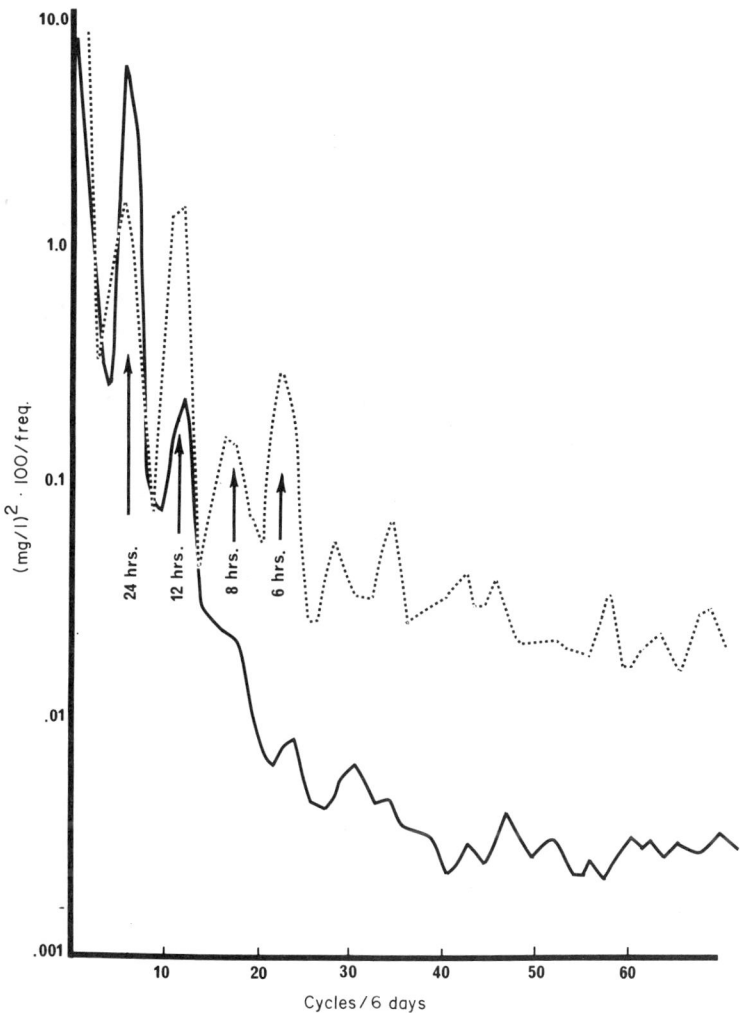

Fig. 5. Variance spectra. DO at E. Stroudsburg, July-Aug., 1968 (solid line). DO at Bristol, July-Aug., 1968 (dotted line).

the importance of the day-to-day changes in water quality in contrast to the within-day or hour-to-hour changes. It is also interesting to note that the residual DO standard deviation at E. Stroudsburg, the unpolluted river station, is about 0.6 mg/liter. This indicates that water quality standards which call for absolute minima in dissolved oxygen must recognize a background "noise" level of at least 0.5 - 0.8 mg/liter standard deviation.

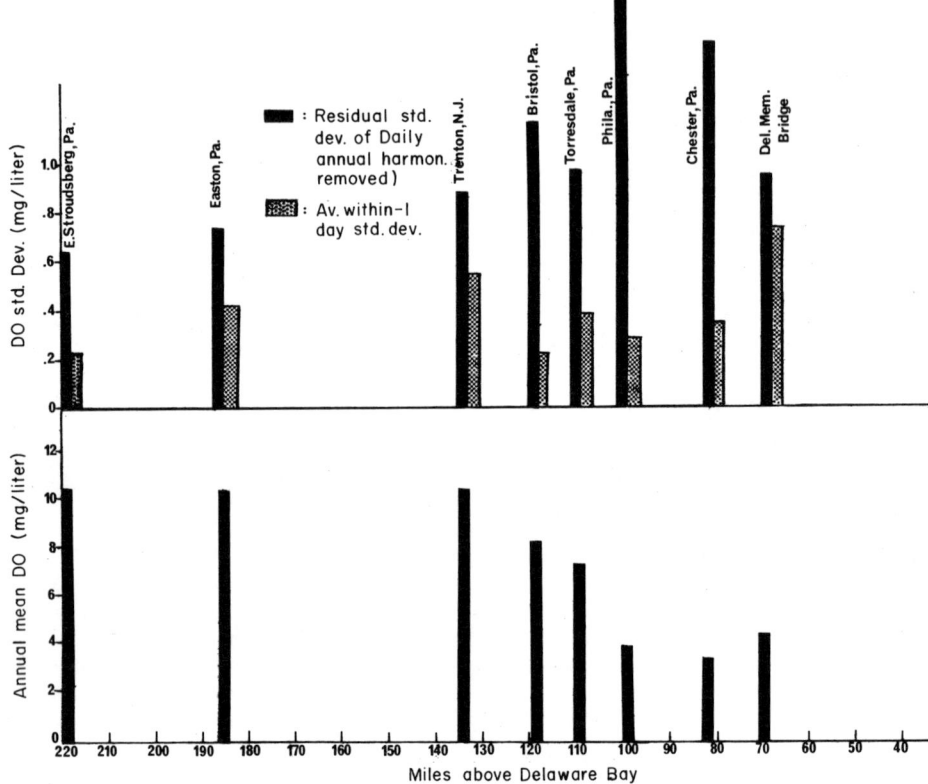

Fig. 6. Summary of analysis of average daily DO, 1968.

Figure 6 also indicates an apparent relationship between high residual standard deviation and low annual means, although, since only eight stations are available for this single series, the determination of a definitive relationship must await more data.

Figure 7, representing analyses of average daily water temperature, also illustrates several points. There is an apparent longitudinal trend in average temperatures from about 11.5° C at E. Stroudsburg to 16° C at Philadelphia. This may reflect increased utilization of the estuary for waste heat disposal. Further, the significantly greater day-to-day variability in water temperature over the hour-to-hour variability is evident in the upper part of Fig. 7. The residual day-to-day standard deviations are only about 0.2° C.

The residual average daily temperature data were also subjected to autocorrelation and spectral analysis. A typical autocorrelation function

19. WATER QUALITY IN THE DELAWARE RIVER 307

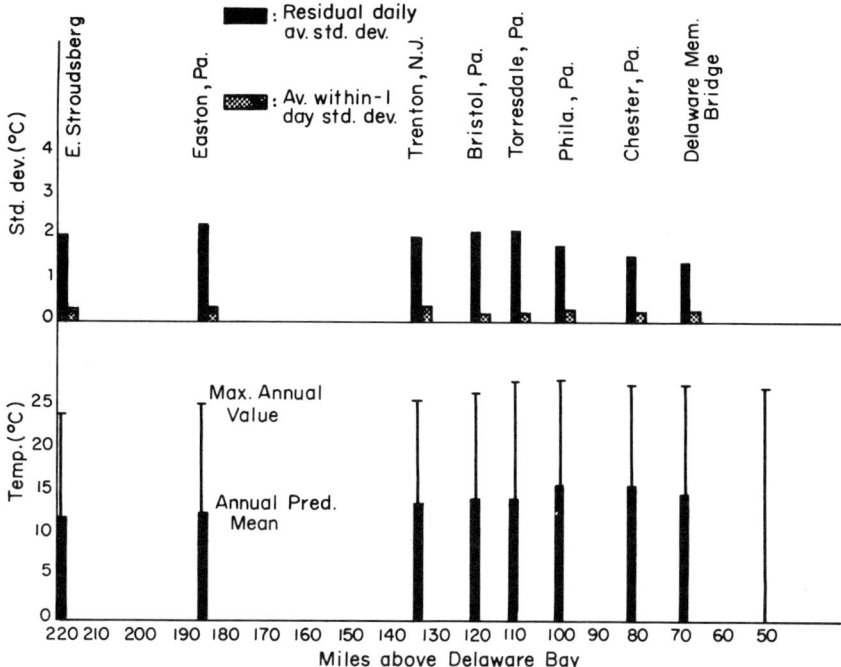

Fig. 7. Summary of analysis of average daily temperature, 1968.

before and after removal of the annual harmonic is shown in Fig. 8(a) and (b). The latter function is approximately exponential, suggesting a first-order autoregressive model [6]. The lag-one correlation coefficients are shown in Table 3. These coefficients were used to generate a new residual record according to a first-order autoregressive scheme. A typical autocorrelation function after first-order analysis is shown in Fig. 8(c), and indicates that a first-order model coupled with an annual harmonic is a suitable model. This is similar, therefore, to the temperature model given by McMichael and Hunter [3]. Table 3 summarizes the results of the analysis of the temperature data. As shown, for temperature analyses, a model composed of a single sinusoid and a first-order autoregression results in better than 99.5% removal of the initial variance for almost all stations. It is interesting to note that the addition of the first-order autoregression term reduced the residual standard deviation from about 1.5-2°C after harmonic removal to 0.5-0.8°C after autoregression, a further reduction in variance of about 85%. Some analysis and inspection of the results indicate that the residuals are approximately normally distributed. One could expect, therefore, that the range due to the completely random variability in water temperature is about ±1.5°C at the 95% probability level. This range has important implications for assessing the degree to

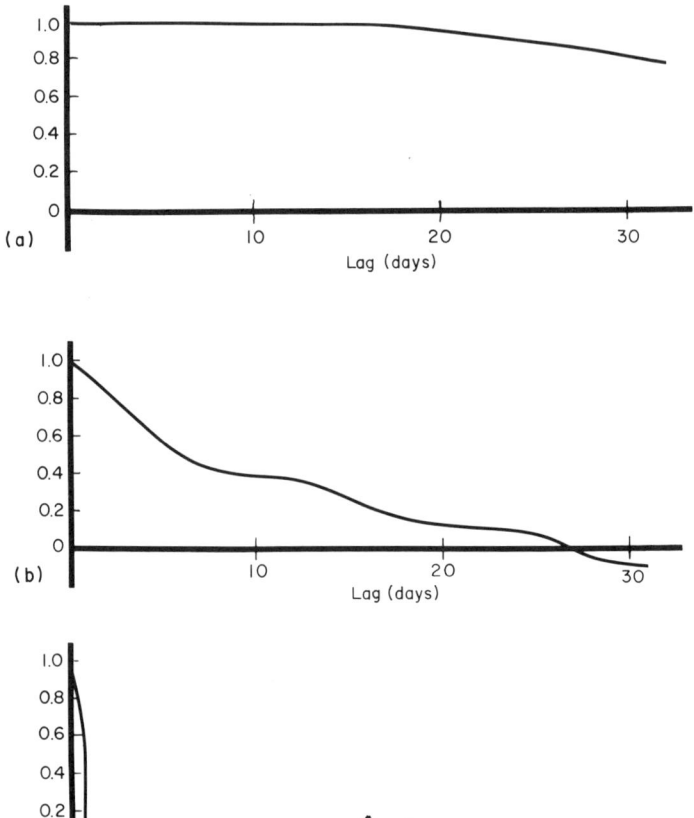

Fig. 8. (a) Autocorrelation function, E. Stroudsburg, temp. (°C) before harmonic removal. (b) Autocorrelation function, E. Stroudsburg, temp. (°C) after harmonic removal. (c) Autocorrelation function, E. Stroudsburg, temp. (°C) after first-order autoregression.

TABLE 2

Approximate Distribution of DO Variance[a]

Freq. range (cycles/6 days)	Period range (h)	% of total variance
0 - 4	∞ - 36	62
4 - 8	36 - 18	35
> 8	< 18	3

[a] Hourly values - July-Sept., 1968, E. Stroudsburg, Pa.

TABLE 3

Summary of Temperature Analyses

Station	Initial variance (°C)	Residual variance[a] (°C)	Residual standard dev. (°C)	Lag-one coeff.	Standard dev. after auto-regression	Variance removal (%)
E. Stroudsburg	55.0	4.00	2.00	0.932	0.76	99.1
Easton	71.5	5.29	2.30	0.950	0.83	99.2
Trenton	81.9	3.96	1.99	0.927	0.81	99.4
Bristol	81.9	4.45	2.11	0.958	0.63	99.6
Torresdale	77.7	4.71	2.17	0.963	0.55	99.6
Pier 11	87.4	3.39	1.84	0.910	0.60	99.7
Chester	71.6	2.69	1.64	0.936	0.55	99.5
Del. Mem. Bridge	93.8	1.79	1.39	0.902	0.48	99.9

[a] After removal of annual harmonic.

which a projected heated water discharge will exceed required temperature maxima.

5. SUMMARY

These results indicate that a large degree of the variability in dissolved oxygen and temperature of the Delaware River and estuary is associated with day-to-day, low-frequency oscillations. The results from an analysis of an unpolluted river station at E. Stroudsburg indicated average daily residual dissolved oxygen concentrations (after removal of low-frequency oscillations) of about 0.6 mg/liter standard deviation and a within-day standard deviation of about 0.2 mg/liter. Spectral estimates of hourly DO data at this station showed a strong diurnal peak with some associated higher harmonics. A diurnal amplitude of 0.4 mg/liter is estimated during the summer months. Analysis of temperature data indicates that a model composed of an annual periodicity with a first-order autoregression accounts for better than 99.5% of initial variance. The standard deviation of the residual temperature variations after application of this model ranged from about 0.5°C to 0.8°C. In general, the results indicate the need to recognize background variability when establishing water quality criteria and attempting short-term forecasting and management schemes.

REFERENCES

1. R. W. Paulson, in Experience with Computer Use in Managing Water Quality Data in the Delaware River Basin, Proc. Nat. Symp. on Data and Instrumentation for Water Quality Management, Univ. of Wisconsin Press, Madison, Wis., 1970, p. 454.

2. C. F. Merk, "A Graphical Summary of Dissolved Oxygen Data for the Delaware River Estuary for Water Years 1965-69," U. S. Geological Survey, Water Res. Div. Penn. Dist., Open File Report, Sept., 1970, 15 pp.

3. F. C. McMichael and J. S. Hunter, Stochastic Modeling of Transport Processes in Rivers, Int. Symp. on Stochastic Hydraulics, Pittsburgh, Pa., 1971, 57 pp.

4. R. V. Thomann, Variability of waste treatment plant performance. Proc. Amer. Soc. Civil Eng., Sanit. Eng. Div. J., 96 (SA 3), 819 (1970).

5. R. V. Thomann, Time series analyses of water quality data, Proc. Amer. Soc. Civil Eng., Sanit. Eng. Div. J., 93 (SA 1), (1967).

6. J. S. Bendat and A. G. Preisal, Measurement and Analysis of Random Data, John Wiley & Sons, New York, 1966.

Part V

OTHER STATISTICAL METHODS

ASPECTS OF SAMPLING ATTITUDES TOWARDS SOLID WASTE PROBLEMS

Albert J. Klee

Solid and Hazardous Waste Research Laboratory
Environmental Protection Agency
National Environmental Research Center
Cincinnati, Ohio

1. GOALS AND OBJECTIVES

This study was part of a larger project [1] involving the development of communication programs that would be effective in helping the public understand and appreciate the sanitary landfill as a method of disposing of solid waste in those areas in which it is appropriate. Consequently, the primary objective of this study was to obtain information about the attitudes of the public and the factors affecting these attitudes, e.g., cognitions and perceptions of a sanitary landfill.

2. INFORMATION SOUGHT

Information was sought about characteristics of the public that could lead to better understanding and improved communication between solid waste managers and citizens. It included the following:

Attitude. How do people feel about sanitary landfills, either positively or negatively? Are the attitudes expressed by vocal persons representative of the attitudes of the majority?

Perception. Do people perceive the solid waste problem as severe or relatively mild? What reasons do they give for their beliefs?

Awareness. Are people aware of solid waste disposal sites in their neighborhood and are they aware of the disposal methods used in their own cities?

Knowledge. Do people know the difference between a sanitary landfill and a dump?

Strength of feeling. Do people have very intense feelings, i.e., are their attitudes deeply rooted, or do they hold little commitment to their views of solid waste disposal problems and specific disposal methods?

Objections. What objections do people have to the use of the sanitary landfill as a means of disposing of waste?

Characteristics of public officials. How do city officials rank these objections?

Demographics. What are the social and economic characteristics of people living in the area near a sanitary landfill as compared with those who live farther away, or those who represent the general population in cities with dumps?

It was also desired to know the relationships among the various items named above. For instance, does distance between place of residence and disposal facility have any relationship to attitude?

Differences between householders and civic officials are important in that, if they differ in goals, perceptions, or knowledge, etc., communication problems may be serious. Therefore, essentially the same information was obtained for householders and civic officials as separate groups in order to identify differences that might exist.

Other information desired included determining how those who oppose sanitary landfills compare with others. It was also desired to understand the characteristics of organizations which have opposed specific disposal operations and to know the nature of their objections. As part of this process of gaining understanding, case studies of three cities were made.

3. SOME CONCEPTS USED IN THE STUDY

In this study, frequent reference is made to the following terms. Terms are used with the definitions given here.

Sanitary Landfill. For the purposes of this study a sanitary landfill was defined as a disposal site at which solid waste is compacted and covered daily and, when finished, covered with at least two feet of compacted earth

20. ATTITUDES TOWARDS SOLID WASTE PROBLEMS

[3]. It is recognized that there are other technical criteria for a sanitary landfill (e.g., no pollution of underground water). This study, however, was concerned with the perceptions and attitudes of the public toward the disposal facility. The important characteristics are those apparent to the layman; his attitude is unaffected by the factors which he cannot sense, unless he is specifically informed about them. So, although this study uses the term "sanitary landfill," it should be remembered that it refers to "landfill which appears to the public to be sanitary," i.e., "is compacted and covered daily," or a perceived sanitary landfill.

Controversy. An important item of information for interpreting data is whether or not a controversy about the solid waste disposal site exists in the community. Controversy was defined as "public discussion with apparent disagreement over solid waste disposal, either about plans or about present operation." News stories indicated by headlines such as "Garbage is Coming! Alarm Arouses Rush," "Groups Plan Fight on Bay Landfill," and "Whitemarsh to Look Anew for Landfills" were one form of evidence of controversy.

Dump. A dump is a disposal site where the solid waste is neither compacted nor covered, and where open burning is permitted. This is a more limiting definition than that a dump is anything that is not a sanitary landfill.

4. THE SURVEY OF HOUSEHOLDERS

Basic data for this study were obtained through an attitude survey of householders. An attitude survey as performed in this study requires the sampling of a population with respect to beliefs or opinions about solid waste disposal. This survey went beyond usual public opinion poll practices in that it probed for awareness, knowledge, kinds of objections, strength of feeling and other information mentioned above. Residents of ten cities (a judgment sample) were sampled. In seven of the cities an area simple random sample [2] of residents living within three miles of the site were interviewed so that data would be obtained from people likely to be concerned about the matter. In the remaining cities - those without a sanitary landfill - a random sample was made of the entire population of the city.*

An attitude was defined as a tendency to accept or reject a solid waste disposal facility. Attitudes may range from a strong acceptance, through mild positive feelings, neutrality, and negative feelings, to strong

---

*Modesto was reported to be a city with a dump, and it was surveyed so that the sample represented the entire city. However, inspection of the operation resulted in its classification as a perceived sanitary landfill.

rejection of the concept [4]. The interviews of the residents were conducted so that respondents could be rated along the entire continuum rather than just at the extremes.

Awareness, in the survey, is indicated by the relative amount of information a householder has about aspects of solid waste management in his environment and about relationships among these aspects. For instance, knowing the difference between a sanitary landfill and a dump is considered an indicator of awareness. Awareness was determined from respondent's answers to such questions presented in the survey interviews as, "Is there a solid waste disposal facility nearby?" and "Three common types of solid waste disposal facilities are called <u>dumps</u>, <u>landfills</u>, and <u>sanitary landfills</u>. Which of these types is being used by this city now?"

4.1. Selection of Cities

The survey gathered data from 1603 residents in ten cities. The rationale for sampling the selected cities was as follows. Since this study was directed to attitudes regarding a specific problem, it was desirable to define populations of interest rather than to conduct a national poll in which many of the respondents would have little or no concern about solid waste problems.

The cities surveyed were selected from a list of 22 candidate cities suggested to the Office of Solid Waste Management Programs. These cities were identified from a review of news clippings to determine where sanitary landfills were being used and where sites have been the subject of controversy. Criteria for the final selection included (a) that the total group of cities should represent different sections of the country, and (b) that cities with perceived sanitary landfills should include both those with controversy and those without controversy. Monterey Park, a suburb of Los Angeles, was selected to represent Los Angeles County, a candidate area. The cities surveyed are shown in Table 1.

Two cities with alternate forms of solid waste disposal were included to provide a comparison, or reference point. The study was restricted to urban areas.

4.2. Selection of Persons to be Interviewed

Because information from this study was to be an aid to understanding problems of influencing the public, it was necessary that data be obtained from people in areas in which there was likely to be awareness of and concern about solid waste disposal. Therefore an area with a radius of three miles around the sanitary landfill was identified in each city with a

TABLE 1

Cities Surveyed

City	Type of disposal		Controversy	
	SLF[a]	Other	Yes	No
Monterey Park, Calif.	X			X
Des Plaines, Ill.	X			X
Sioux City, Iowa	X			X
Schenectady, N. Y.	X			X
Winston-Salem, N. C.	X		X	
Worcester, Mass.	X		X	
Modesto, Calif.	X		X	
San Francisco, Calif.	X		X	
Erie, Pa.		X[b]		X
Little Rock, Ark.		X[c]		X

[a] Perceived sanitary landfill

[b] Incinerator

[c] Open dump

sanitary landfill. Winston-Salem had two sanitary landfills. The newer one was chosen as the focal point of the sample area on the assumption that it was more likely to be perceived as a sanitary landfill than the older one, which had been an open dump. In the cities using alternate methods of disposal (Erie and Little Rock) and in Modesto, which was originally classed as a city with a dump, the sample represented the entire city.

Within the sampling area of sanitary landfills, bands one mile in width around the landfill were defined, nonresidential areas being eliminated. Within the residential areas in each of the three one-mile bands, points were selected at random and in proportion to population density as determined from U. S. Census tract data. Two interviews were conducted on the block identified by each point. The block north of each one designated was the alternate block in case two interviews could not be completed on one block because of refusal, no response, or no more than one occupied house on the designated block. For each of the cities in which the entire city was the sampling unit, 100 blocks for interviewing were selected at random (in proportion to population density) from among all blocks in the city, with alternate blocks designated as in other cities.

The dwelling units at which interviews were conducted were selected in the following manner. The starting corner and direction of travel were systematically rotated. Interviewers called at the third dwelling unit from the corner and, if an interview could not be completed, continued door-to-door until one was. Then, they skipped two dwelling units and repeated the process until a second interview was completed. A sufficiently large number of interviews were conducted in the evening to include persons holding jobs in the daytime and obtain a nearly even male-female division.

Since interviewing was conducted at the end of November, 1969, the number of interviews planned could not be completed for the planned expenditure because of weather and holiday factors. Therefore, interview supervisors in each community with a sanitary landfill were instructed to complete the quotas for the two bands nearest the landfill where more awareness was expected and to spend the remainder of their time in the third band. Thus the actual sample was 1603 persons rather than an originally projected 2000.

4.3. Conducting the Interviews

Forms and procedures were designed for the survey, with each proposed question designed to elicit specified information. The sequence of questions was developed so that there would be no undesired cueing of answers by a previous question, nor other unwanted effects. After procedures for conducting the interviews were developed, a number of interviews were conducted as a pretest. Needed revisions were made before the field interviews.

In each of the cities a local interviewing supervisor designated an interview team from trained persons on his staff. Interviewers were given training on use of the specific forms and procedures for the study. As the interviewing progressed, the supervisors made telephone validity checks of a sample of the interviewees to check the accuracy of the questionnaires returned by the interviewers.

5. THE SURVEY OF CITY OFFICIALS

Officials of each city were interviewed by senior project personnel.

The objectives of these interviews were to:

20. ATTITUDES TOWARDS SOLID WASTE PROBLEMS

1. gain first-hand knowledge of the communities' current solid waste disposal practices;

2. determine the views and reactions of city officials to current solid waste problems in the community;

3. obtain responses of the officials to an appropriate version of the questionnaire used in the survey of householders; and

4. verify that the city solid waste disposal practices put it in the category assigned during the selection process (in each city, personnel visited the disposal site).

Visiting dates generally were arranged through the Directors of Sanitation or the Director of Public Works; appointments with other officials were scheduled into the available block of time. Other officials visited included, to the extent possible, the mayor, a councilman, and officers of other departments that had some responsibility in solid waste management, such as the city manager, planning department director, health department director, and city engineer. Newspaper files were reviewed when they could be made available. In all, fifty-five officials of cities and sanitation districts were interviewed.

The in-depth interviews with these officials were unstructured. The interviewer had a predetermined number of questions or topics based on current concerns in the community. The interview was flexible and the interviewee was encouraged to add information, to introduce new topics, and to explore the subject. This approach was taken because the checklist or questionnaire type of study was not always appropriate. When feasible as part of the interview, the officials were presented a modified version of the questionnaire employed in the survey of householders.

6. OTHER INTERVIEWS

Seven other persons in key positions were interviewed during the course of the study (this does not include those interviewed in the course of other work performed and who, thereby, contributed to general understanding of the solid waste management problem). These seven persons included the presidents of sanitary landfill operations and disposal companies, a newspaper reporter, and a state Secretary of State.

7. STUDY OF ORGANIZATIONS

In order to increase understanding of the process of decision making, organizations which oppose solid waste disposal operations were identified through a study of news clippings in the communities included in the survey. Two permanent organizations were found; opposition organizations existed in other cities, but they were temporary, organized for the express purpose of fighting a facility or practice. Leaders of the two permanent organizations and the leaders of a protest drive that was operating through existing organizations in a community were interviewed as well.

8. DATA PROCESSING

8.1. The Survey of Householders

Data from the questionnaire interviews were coded for keypunching onto IBM cards. Answers to questions requiring a "yes," "no," or scale response were simply assigned numbers. Many of the questions required more extended response, however, and categories for responses were established. For questions in which it was desired to determine how many persons made a specific response (e.g., "Which...is being used...? A dump, landfill, or sanitary landfill?") the answers were coded into predetermined categories. Other questions probed for information and resulted in a range of responses. For these, pertinent categories were established by drawing a sample of questionnaires and grouping the answers under descriptive headings. Naturally, some responses occurred so seldom that it was not deemed advisable to set up a category for each response; thus, one of the coded categories was "other." This procedure sometimes yielded finer distinctions than were needed for purposes of this study, and some of the categories were later combined by the principal investigator.

The coded responses were then keypunched onto IBM cards and subjected to treatment by a computer, using standard programs developed for this kind of analysis. The principal investigator indicated correlations and comparisons desired. The data were processed in several stages. On the basis of initial printouts, more refined analyses were specified. This treatment yielded information regarding the relationships indicated earlier in this paper, e.g., attitude as related to distance from residence to sanitary landfill. Subgroups were identified for separate treatment, e.g., all those who had protested about waste disposal. The data were analyzed by question for the entire sample and, separately, for each city.

One significant and uncommon analysis was the measure of latitudes of acceptance, rejection, and noncommitment [5]. Latitude is one way of measuring the strength with which one holds to his position or attitude. If the strength of one's position is low he will, presumably, be more willing to accept other views than if his position is held with great strength. The knowledge of latitudes is of use in specifying the nature of a communication

20. ATTITUDES TOWARDS SOLID WASTE PROBLEMS

because it indicates the likelihood that a view different from that generally held will be accepted or, at least, considered.

Latitude of Acceptance. This latitude is measured by presenting a series of nine statements regarding the importance of a sanitary landfill to the community. They range from extreme acceptance to extreme rejection. The respondent is asked to indicate all of those statements with which he can agree and to indicate which one best describes his feelings - his basic position. The latitude of acceptance includes all of those statements which a person accepts. When used with groups, as in this study, a statement is considered as accepted by the group and, thus, included in the group's latitude of acceptance, if 50% or more of the respondents find it acceptable.

Latitude of Rejection is measured in a similar fashion. The respondent is asked to indicate the statements which he finds objectionable and the statement which he finds most objectionable. A broad latitude of rejection is believed to indicate a high commitment to the basic position within the latitude of acceptance.

Latitude of Noncommitment includes all statements neither accepted nor rejected by the respondent. In the case of a group, it includes all statements neither accepted nor rejected by at least 50% of the respondents.

Analysis was made of the responses to the semantic differential portion of the questionnaire. In the semantic differential technique [6], respondents are presented with a series of seven position scales. Each scale represents distance between polar adjectives, such as "productive" and "unproductive." Certain sets of adjectives create key scales that are believed to provide a measure of attitudes. The respondent is asked to indicate how he feels about the subject, e.g., sanitary landfill. The patterns of ratings expressed on the key scales indicate positive or negative attitudes. Residents at varying distances from the landfill (living in different bands) were compared in terms of average attitude scores. Statistics used were fairly simple but adequate for the purpose. Percentages were calculated for each response and, where appropriate, means were computed for the variable. The statistical significance of differences in distribution of responses was tested by use of the chi-square technique.

8.2. The Survey of Public Officials

It was not meaningful to restrict analysis of the data obtained from the survey of public officials to quantitative study because the major objective of these interviews was to obtain historical and descriptive information that would be of help in interpreting data from the public survey and because of the wide variety of community situations, the variations in specific interests, and the small samples. Rankings were made of the frequency of responses to questions and, where feasible, these were compared with frequency rankings from the survey of householders. Information from the interviews, from descriptions of procedures (obtained from brochures and other documents from city offices), and from perusal of newspaper files, was subjected to a qualitative review for the purpose of identifying the process by which decisions are made.

9. THE CASE STUDIES

Three cities were selected for study as special cases in contrast to the general summaries made of the other data. The cities chosen and the reasons are as follows:

Worcester, Mass.: critical solid waste disposal problem; history of public controversy;

San Francisco, Calif.: critical solid waste disposal problem; history of public controversy;

Sioux City, Iowa: absence of critical disposal problem; controversy over refuse truck routes.

Information gathered from all sources received qualitative review to develop descriptions of current solid waste disposal practices and the history of the community with respect to current problems and practices. As mentioned earlier, questionnaire information from the survey of householders was summarized for each city; the results became part of the information used in the case studies. All information was considered in making an assessment of the outlook for solid waste disposal in each of the three cities.

10. ANALYSIS OF ORGANIZATIONS

The interviews with the leaders of these organizations were summarized. Responses made by persons who had complained were analyzed to identify the kinds of complaints being made, although, of course, this does not necessarily mean that the complaints were made by members of the organizations. Vignettes and examples of how the organization had attempted to influence city decisions were culled out and incorporated into the analysis of decision making.

11. RESULTS

The survey of residents resulted in 1603 usable responses from the ten cities. In addition, interviews were conducted with fifty-five city officials. Some results will be mentioned here, but much further information and detail are available in the final report [1].

Awareness of residents living within three miles of a solid waste disposal site that a site existed nearby varied considerably from city to city. In cities with sanitary landfills, it ranged from a low of 38.5% to a high of 81.7%. Awareness of what type of facility the city was using for solid waste disposal ranged from 24.3% to 80.9% in cities with sanitary landfills.

Only 7.5% of the residents in the total sample named even one characteristic of a sanitary landfill. But after the concept was explained, differentiating a dump from a sanitary landfill, 86.1% said that it was an acceptable method and 81.8% had a favorable attitude toward the sanitary landfill. There was, however, a difference in attitude toward the sanitary landfill related to the distance one lived from it. Negative attitudes toward the concept of a sanitary landfill were indicated by no more than 10.5% of those in Area 1 (within one mile) but by no more than 5.8% of those living one to three miles away. Only 1.4% of those living from one to three miles away were opposed at a level indicative of ego-involvement, but there were more within one mile.

No significant differences in the characteristics of respondents (indicated by demographic data) existed between those in Area 1 and those in Areas 2 and 3 (from one to three miles away).

The areas differed in their feelings about the value of their property - an indication of attitude. In Area 1, 24.5% said that their property was worth more than similar property in other neighborhoods and 26% said that their property was worth less than similar property elsewhere. In Areas 2 and 3 combined, 31.3% said that their property was worth more but only 12.0% said it was worth less.

During the interviews, respondents were asked to select from a list of seven statements describing degrees of seriousness of the solid waste problem the statement which they felt best described the problem in their city. In the analysis, the cities were ranked from one, the city with the least problem, to ten, the city with the most severe problem, according to the average score assigned by residents. City officials made the same kind of ratings. The ranks of five cities were the same whether based upon scores of the citizens or scores of the officials; these five cities were ranked first, second, seventh, ninth, and tenth.

Both residents and city officials indicated that the number one reason why people would object to sanitary landfills was "improper operations" (which included "smells bad," "unhealthy," and "land settles"). Beyond that they did not give the same ranking to any other reason, although they were quite close, with rankings differing by only one position in a number of cases.

City officials rated "kind of disposal operation" as the most important factor to consider in selection of a disposal site. However, city officials named economic factors very often, and the case studies and newspaper articles indicated that economic factors were considered very important.

Most city officials could make meaningful distinction between a sanitary landfill and a dump, differing from citizens in this respect.

Respondents in the survey of householders were divided into those who had made complaints about a sanitary landfill and those who had not. Those who had objected were more likely to say that their community had liabilities, more likely to live close to the sanitary landfill, and more likely to have lived in the neighborhood before the operation began (42% of the objectors versus 19% of the nonobjectors). The objectors are more likely to say that their property is worth less than similar property elsewhere.

The three case studies included in the final report [1] of the complete project illustrate the variety of problems confronting decision makers, and the range of inputs available for consideration in making the decision. (The case studies themselves were not amenable to tabular presentation of data.) The case studies illustrate the complexity of the problems of management of solid waste disposal. The primary value of the case studies was the contribution to the understanding of the problems of solid waste management; this aided in the development of the communication programs.

REFERENCES

1. J. M. Stormes and B. T. Jensen, A study of public attitudes and problems of solid waste disposal, Report prepared for the Office of Solid Waste Management Programs, U. S. Environmental Protection Agency, under Contract No. CPE 69-107, 1971.

2. L. Kish, Survey Sampling, John Wiley & Sons, Inc., New York, 1965.

3. A. J. Klee, The role of facilities and land disposal sites, The National Solid Wastes Survey: An interim report, U. S. Public Health Service, 1968, p. 28.

4. W. J. McGuire, The nature of attitudes and attitude change, in Handbook of Social Psychology, Vol. 3, Addison-Wesley, Reading, Mass., 1969, p. 136.

5. M. Sherif and C. I. Hovland, Social Judgment, Yale Univ. Press, New Haven, Conn., 1961.

6. J. G. Snider and C. E. Osgood, Semantic Differential Technique, Aldine, Chicago, 1969.

21

STATISTICAL PROBLEMS IN AEROSOL STUDIES

Gerald van Belle

Florida State University
Tallahassee, Florida

0. SUMMARY

A survey was made of the types of statistical distributions describing aerosol systems. The distributions found were: Poisson, negative binomial, exponential, normal, log-normal, Weibull, Maxwell or none at all. Some specific statistical problems are presented for further consideration; most of these involve distributional aspects of the aerosol system. Distribution formulae will not be given here - standard statistical texts can be consulted for these.

1. INTRODUCTION

An aerosol system is a suspension of fine particles. The size of a particle is usually between 10^{-5} cm and 10 cm. The mathematical study of small particles and aerosols has a long history. The statistical study of such systems is of more recent origin and has often confined itself to geometric probability arguments or ad hoc techniques of replacing parameters by their "natural" estimators; e.g., mean and variance of the log-normal distribution are often replaced by estimates derived from sample data, or the theoretical distribution function is replaced by the empirical distribution function.

Statistical properties of estimators are rarely discussed. There is a need to consider carefully most of the estimators used. Tallis [21] and Watson [24] studying related problems (the distribution of sections of particles), showed that many of the common estimators have infinite variance and other undesirable statistical properties. Similar problems crop up in aerosol statistics, as will be shown in Section 3.

One area with rather different statistical problems is that of stochastic coalescence of particles. Whittle [26, 27] and Marcus [15] are concerned with growth and coalescence of particles such as aerosols, while Scott [19] and Warshaw [25] study the related subject of growth and coalescence of raindrops. These problems will not be discussed here.

Articles on aerosols are scattered over many different fields, such as meteorology, industrial hygiene, air pollution, colloid science, public health, and allergology.

2. STATISTICAL DISTRIBUTIONS

In many papers on aerosols no assumption is made about the distribution of the particles; some simple statistics are calculated and perhaps a frequency distribution of mass or of a function of diameter is graphed.

The simplest aerosol system involves only monodisperse (equal-sized) particles, and specification of the number of such particles per unit volume is often adequate. The Poisson distribution fits well, assuming the particles are small and do not interact with each other. Papers by Cornell et al. [3], Gurland [6] and Smith [20] contain examples using the Poisson distribution. Smith [20] considers the complication that particles are not points and derives the corresponding probability of observing empty testing areas or volumes. If the number of particles increases and clumping (coagulation, coalescence) occurs only to a moderate degree, so that individual particles are still identifiable, Gurland [6] has shown that a negative binomial distribution may give a better fit.

For polydisperse systems the size as well as the number of particles varies. No examples were found where both number and size are considered together as random variables. Usually size is partitioned into intervals and the proportion by mass, or the number of particles, in the interval observed.

The most common model for polydisperse systems is the log-normal distribution. It is implicitly assumed that the density or number of particles per unit volume is constant. Representative articles from a variety of areas are Corn [2], Harstad et al. [8], Kerker et al. [14], Takahashi and Kasahara [22], Huang et al. [11]. Corn [2] discusses the problem of testing the goodness of fit of the log-normal distribution; the usual multinomial model for testing is suggested. In many articles

21. STATISTICAL PROBLEMS IN AEROSOL STUDIES

the log-normal distribution is fitted by the use of logarithmic probability paper. The literature on the log-normal distribution is vast; a relevant book is that of Herdan [9], who shows that there are theoretical reasons for using the log-normal distribution. Data are usually obtained by observing the number of particles falling in an interval of specified size (as measured by diameter or radius of a particle). If the geometry of a particle is assumed, conversion can easily be made to mass, weight, volume, or surface area. Various other kinds of transformations are used prior to assuming normality. An example is

$$y(d) = \ln \frac{c_1 d^{c_2}}{d_{max}^{c_2} - d^{c_2}}.$$

Here c_1, c_2 are constants and d_{max} is the maximum diameter of the particles (observed or postulated). One effect of the transformation is to correct the original sample range $[0, d_{max}]$ into the whole real line, $[-\infty, \infty]$.

Another very common distribution is the exponential, or the negative exponential. Examples can be found in Davies [5], Matthews and Rhodes [16], Takahashi and Kasahara [22], and Cadle [1]. A related discussion on raindrop size as having an exponential distribution can be found in Joss and Waldvogel [12]. For most cases the sample space of the random variable (particle size) is broken up into disjoint sets, so that, for example, Takahashi and Kasahara define n(v) as "the number concentration of particles of volume v."

Hilbig and Schwarz [10] use the Maxwell distribution to describe a finely dispersed system ($0.2\ \mu$ to $1.1\ \mu$) while Harris and Dubey [7] use a two-parameter Weibull distribution to describe particle size. The reason Harris and Dubey chose the Weibull distribution is that it integrates in closed form over finite regions.

Sometimes the statistical distribution is generalized on the basis of empirical relationships. Examples of these are the Rosin-Ramler relation and the Nukiyama-Tamasawa equation. References to these are given in Cadle [1].

In coagulation (coalescence) studies the original particle distribution is often unspecified or attention is focused on the equilibrium distribution, either theoretically or experimentally (see, e.g., Quon and Mockros [17], Whittle [26, 27] and Marcus [15]).

3. COMMENTS

Many of these comments may be trivial to the professional statistician, but they may be helpful to the researcher not directly concerned with statistical theory. The comments can be summarized as the necessity for more statistical rigor in sampling problems.

1. A careful distinction should be made between statistic and parameter. Confusing the two leads to faulty interpretations, e.g., the meaning of a confidence interval when sample mean and population parameter are assumed to be the same. A 1969 reference routinely used by chemists and engineers and highly praised in the <u>American Scientist</u> states: "The percentage of measurements which lie within a certain interval is called a confidence interval." Another reference (1965), directly concerned with particle size, specifies a normal distribution with μ (the theoretical mean) replaced by $\bar{d}$ (the sample mean), and theoretical and sample standard deviation similarly interchanged.

2. Statistical properties of estimators of functionals of the aerosol distribution should be calculated. If the distribution is known, or assumed to be known, usually more than one estimator exists, so that statistical properties should be considered. For example, for a given density the proportion of particles with radius less than r_0, $P(r_0)$, can be written

$$P(r_0) = \int_0^{r_0} f(t) \, dt,$$

where $f(t)$ is the density function of the particles. For a set of n radii, $r_1, \ldots, r_n$, $P(r_0)$ is usually estimated by

$\hat{P}_1(r_0)$ = proportion of particles with radius less than r_0.

If the form of $f(t)$ is known to be exponential, for instance, then $P(r_0)$ can be calculated as

$$P(r_0) = 1 - e^{r_0/\theta},$$

where $E(r) = \theta$ and, as shown by Tate [23], a minimum-variance unbiased estimate $\hat{P}_2(r_0)$ of $P(r_0)$ exists, namely,

21. STATISTICAL PROBLEMS IN AEROSOL STUDIES

$$\hat{P}_2(r_0) = 1 - \left(1 - \frac{r_0}{n\bar{r}}\right)^{n-1}$$

where, $\bar{r} = \Sigma r_i/n$. The variance of $\hat{P}_2(r_0)$ is smaller than that for $\hat{P}_1(r_0)$, so if the form of the distribution is known to be exponential a more precise estimate can be obtained.

In some cases "natural statistics" have bad properties. Dallavalla [4] defines S_w, specific surface (surface of the particles per unit weight), by

$$S_w = \frac{6}{\rho d},$$

where ρ is the density and d the mean diameter. The "natural" estimate of S_w based on observed diameters $d_1, d_2, \ldots, d_n$ (assuming ρ known) is

$$\hat{S}_w = \frac{6}{\rho \bar{d}}$$

where $\bar{d}$ is the arithmetic mean of the observed diameters. The expectation of this statistic generally does not exist.

3. Grouping the data tends to obscure the behavior of the distribution, especially at the extremes of the size range. As an example, consider the data in Table 1. Two independent samples of size 100 were generated from the exponential distribution and the log-normal distribution. The mean and variance in both cases was specified to be 1. The data was then grouped into class intervals of unit width. A χ^2 goodness-of-fit test would not detect a difference. If the data had not been grouped, more sensitive tests could have been made (see, e.g., Kendall and Stuart [13]). To have summarized the data by the four statistics, sample mean, standard deviation, skewness, and kurtosis, would not be satisfactory; as Corn [2] writes, "we are not interested in one particle size, but in the behavior of a portion of the distribution."

4. Three brief comments can be summarized as follows. (1) A polydisperse system is best described by a bivariate statistical model, accounting for variation in both size and number. Results from bulk sampling may be useful here (see Scheaffer [18]). This is one example of results from another area being applicable to aerosol studies. (2) There is some question as to what is meant by sample size in particle studies, or: what is the experimental unit? A test volume of polluted air may contain a large number of particles; in fact, so large a number that the only measure obtained is the weight of the particles. But the weight obtained still represents a

TABLE 1

Frequency Distribution of Two Independent Samples[a]

Interval	Exponential	Log-normal
0 - 0.5	42	37
0.5 - 1.0	22	22
1.0 - 1.5	15	15
1.5 - 2.0	8	7
2.0 - 2.5	6	1
2.5 - 3.0	2	2
3.0 - 3.5	3	3
3.5 - 4.0	1	1
4.0 up	1	1
Total	100	100
Sample mean	0.97	0.92
Standard deviation	0.98	0.84
Skewness	2.17	2.23
Kurtosis	9.95	8.97

[a] Each sample of size 100, with mean and variance 1.

sample of one from an infinite population of such volumes. (3) Very little work has been done on the effect of the geometry of the particles on the statistics used. Associated problems involve coagulation and dispersion of aerosols.

ACKNOWLEDGMENT

This paper is based on research supported by Grant AP00880-02, National Air Pollution Control Administration, U. S. Department of Health, Education and Welfare.

REFERENCES

1. R. D. Cadle, Particle Size, Reinhold-Van Nostrand, New York, 1965.

2. M. Corn, Statistical reliability of particle size distributions determined by microscopic techniques, Amer. Ind. Hygiene Assoc. J., 26, 8 (1965).

3. R. G. Cornell, S. F. Welch, and M. S. Hall, A comparison of gravimetric and volumetric pollen samplers, J. Allergy, 32(2), 128 (1962).

4. J. M. Dallavalla, Micromeritics (2nd ed.), Pitman, New York, 1948.

5. C. N. Davies, Distribution of particle size in dust, Nature, 201, 172 (1964).

6. J. Gurland, Some applications of the negative binomial and other contagious distributions, Amer. J. Pub. Health, 49, 1388 (1957).

7. E. K. Harris and S. D. Dubey, Estimating the frequency distribution of dosage from a continuous record of pollutant concentration, Int. J. Air Water Poll., 8, 369 (1964).

8. J. B. Harstad, H. M. Decker, L. M. Buchanan, and M. E. Filler, Air filtration of submicron virus aerosols, Amer. J. Pub. Health, 57(12), 2186 (1967).

9. G. Herdan, Small Particle Statistics, Elsevier, Amsterdam, 1953.

10. G. Hilbig and F. Schwarz, Turbidimetric determination of finely dispersed technical systems, Staub-Reinhalt., Luft, 27(6), 27 (1967).

11. C. Huang, M. Kerker, and E. Matijevic, The effect of Brownian coagulation, gradient coagulation, turbulent coagulation, and wall losses upon the particle size distribution of an aerosol, J. Coll. Interface Sci., 33, 529 (1970).

12. J. Joss and A. Waldvogel, Raindrop size distribution and sampling size errors, J. Atmos. Sci., 26, 566 (1969).

13. M. G. Kendall and A. Stuart, The Advanced Theory of Statistics, Vol. 2 (2nd ed.), Charles Griffin and Co., Ltd., London, 1966.

14. M. Kerker, E. Matijevic, W. F. Espenscheid, W. A. Farone, and S. Kitani, Aerosol studies by light scattering, J. Colloid Sci., 19, 213 (1964).

15. A. H. Marcus, Stochastic coalescence, Technometrics, 10, 133 (1968).

16. B. A. Matthews and C. T. Rhodes, Some observations on the use of Coulter Counter Model B in coagulation studies, J. Coll. Interface Sci., 32, 339 (1970).

17. J. E. Quon and L. F. Mockros, The equilibrium size distribution of an aerosol continually reinforced with particles, Int. J. Air Water Poll., 9, 279 (1965).

18. R. L. Scheaffer, Sampling mixtures of multisized particles: an application of renewal theory, Technometrics, 11, 285 (1969).

19. W. T. Scott, Analytic studies of cloud droplet coalescence I, J. Atmos. Sci., 25, 54 (1968).

20. S. Smith, Theory of "void counting" of polydisperse systems, Ann. Occup. Hygiene, 10, 15 (1967).

21. G. M. Tallis, Estimating the distribution of spherical and elliptical bodies in conglomerates from plane sections, Biometrics, 26, 87 (1970).

22. K. Takahashi and M. Kasahara, A theoretical study of the equilibrium particle size distribution of aerosols, Atmos. Environ., 2, 441 (1968).

23. R. F. Tate, Unbiased estimation: functions of location and scale parameters, Ann. Math. Stat., 30, 341 (1959).

24. G. S. Watson, Estimating functionals of particle size distributions, Biometrika, 58(3), 483 (1971).

25. M. Warshaw, Cloud droplet coalescence: statistical foundations and one-dimensional sedimentation model, J. Atmos. Sci., 24, 278 (1967).

26. P. Whittle, Statistical processes of aggregation and polymerization, Proc. Cam. Phil. Soc., 61, 485 (1965).

27. P. Whittle, The equilibrium statistics of a clustering process in the uncondensed phase, Proc. Roy. Soc., Ser. A, 284, 501 (1965).

DISCUSSION

S. Dawson

Harvard University
Boston, Massachusetts

The statistical problems pointed out in this paper are shown to arise from lack of knowledge of statistical theory on the part of those presenting the studies. Some of the examples given are very dramatic. If a majority of workers in particle counting do not distinguish the concept of a parameter from that of a statistic, there must be very few who realize that an empirically determined best-estimate of an exponential distribution does not have an exponential shape. On the other hand, the use of the dramatically "bad statistics" would appear to be very limited.

This paper then demonstrates the lack of availability of knowledge and the need for it. A next step to making the knowledge available might be to write a monograph that assembles systematically the appropriate theoretical considerations, but one that is written for the nonstatistician. Such a work would permit those doing the studies to improve their knowledge sufficiently to see the need for and methods of making efficient use of consultant statisticians.

An important problem which is not addressed in the paper is the assessment of the nature and limits of inferences that can be drawn about a limited number of imperfect observations of a cumulative distribution function, $P(r_0)$, for an unknown distribution. For example, what can be said of estimates of the underlying probability density function $f(r)$ if the intervals of observation are as in Table 1? This sort of problem would appear to arise whenever an irregular distribution is being measured, such as in most samples of air pollution.

One approach to the problem of drawing inferences about unknown distributions is to rewrite the equation for P as an integral equation.

$$P(r_0) = \int_0^{r_0} f(t)dt = \int_0^{\infty} H(r_0 - t) f(t) dt,$$

where

$$H(x) = 0, \quad x < 0$$
$$ = 1, \quad x \geq 0$$

$r_0 = r_1, r_2 \ldots r_n$, defining the intervals.

Then an estimate of $f(r)$ over the interval $(0, \infty)$ can be obtained from the n observations of $P(r_0)$ by the method of J. R. Franklin (J. Math. Anal. and Appl. 31:682-716, 1970). The statistical properties of this continuum estimate require investigation.

This example of a practical problem of a type which often arises is given, along with an apparently promising direction for a solution, in order to suggest the potential fruitfulness of developing such new statistical tools. This sort of need and opportunity is in addition to the need and opportunity for making the present theoretical tools much more available.

22

EVALUATION OF ACTIVATED SLUDGE TREATMENT
PLANT PERFORMANCE

Jerzy Ganczarczyk

University of Toronto
Toronto, Canada

0. SUMMARY

The performance of activated sludge treatment plants is influenced by the variations in the influent flow and quality as well as by the fluctuation in the technological parameters of the process. Moreover, the current measurement and sampling techniques and procedures applied to control this process are in many applications, at least, questionable and in general not sufficient to produce reliable performance data. On the other hand, lack of proper sensors substantially limits the possibilities for automation of such plants. However, simple statistical analysis of the operational control data of activated sludge treatment plants may improve the extent and quality of information which can be obtained. Specific examples of such analysis are presented and discussed.

1. INTRODUCTION

It is well known that some fluctuation of effluent quality is characteristic of activated sludge treatment plants. Three main factors are involved: the variation of influent flow; the variation of influent quality; and the variation or instability of the technological parameters of the plants. At the same time, activated sludge treatment plants have specific buffering capacities, dependent on their design patterns, operational variables, and the relative volumes of particular units. These buffering capacities have some equalizing influence on the variations mentioned, although often the coefficients of variation for treated effluent quality are higher than those for treatment plant inflows.

The variation of the influent flow and quality of municipal sewage have been extensively studied [5, 12, 13, 15, 17, 23]. Corresponding information on industrial effluents can be determined only for specific cases, as they are more individual in nature. An interesting example of using the Monte Carlo simulation technique for investigating variation of a tannery's wastewater was described recently by Berthouex and Brown [2].

The influence of transient loadings on the efficiency of the activated sludge process has been studied [8, 11, 18]. The paper by Spicka [18], although partially controversial in its theoretical approach, is especially interesting because of a background of full-scale operational experience. The paper by Grady [11] is a purely theoretical study.

In evaluating the significance and utility of activated sludge treatment studies, the flow measurement points and sampling points for the analytical determinations, as well as the frequency and timing of measurement and sampling, are of primary importance and determine the relevance of the data. Therefore, it is of equal importance to establish the optimal scope of the control points of activated sludge treatment plants, with sufficient precision for the reasonable evaluation of their performance and its variation.

The estimation of the quality variations of the activated sludge treated wastewater and the various pollutant loads is not only interesting from the scientific point of view, but also necessary for proper operation of a plant in compliance with regulations for the prevention of pollution in receiving water bodies. Many of these regulations already specify the ranges of acceptable effluent quality, permissible loads, and allowable seasonal variations. Moreover, it is highly probable that in the near future the performance required of treatment plants will be related even more closely to variation of the assimilative capacity of the surface water.

2. ACTIVATED SLUDGE TREATMENT PLANT CONTROL DATA

The control of activated sludge treatment plants is usually based on the flow measurements of the influent wastewater, of the recycling activated sludge, and of the air for the aeration tanks, if such a system of aeration is used, as well as on the determination of:

the influent quality (BOD = biochemical oxygen demand, COD = chemical oxygen demand, TOC = total organic carbon, total nitrogen, total phosphorus, suspended solids, volatile solids, specific organic and mineral substances, coliform bacteria, etc.);

22. TREATMENT PLANT PERFORMANCE EVALUATION

the effluent quality (as above, plus the forms of nitrogen: ammonia, nitrates, nitrites, and organic nitrogen);

the mixed liquor suspended solids, mixed liquor volatile solids, and activated sludge volume index; and

the recycling activated sludge suspended solids.

The frequency of measurements and sampling differs in various plants, and often depends on the local possibilities or local regulations. Carr and Ganczarczyk [6] give an example of such metering and sampling programs in their paper on the Lakeview Sewage Treatment Plant in Port Credit, Ontario.

For this specific case the meter accuracy has been designed with a built-in tolerance of $\pm 2\%$ of actual flow. One of the problems encountered in the day-to-day operation of the plant is that instrumentation soon loses its initial calibration and hence its accuracy. There are reasons for this: grease present in sewage tends to be deposited on the walls of any measuring device placed in the flow, and coats the lines to the pressure bellows, which can easily be clogged. Although most flow metering devices depend on automatic purge water systems to keep lines free, quite often this operation must be performed manually. As this type of maintenance involves much labor and time, plant personnel have a tendency to go by the pump setting alone, regardless of what the meters indicate.

Sampling procedures and techniques at Lakeview are undergoing modification as automatic samplers are introduced. All sampling, whether automatic or manual, is done on a unit basis, usually a 1-liter sample per 8 hours of plant operation (sample size is not proportional to flow). Samples collected are analyzed on a daily basis.

Sampling problems include the operator's ability to collect and decant uniform composite samples and to recognize and take a grab sample of any unusual or exotic wastewater. Raw sewage sampling sometimes causes problems, as large particles will often plug up the sampler lines. Weekend samples are stored in the plant refrigerator for analysis on the following Monday.

3. GENERAL APPRAISAL OF CONTROL DATA

In most cases the frequency of sampling over time in activated sludge treatment plants does not take into consideration the plants' characteristic through-flow patterns, e.g., by taking the samples of the plant influent and effluent at practically the same time, taking the samples at the same hour of the day, etc. Therefore a definite bias occurs in the results obtained. However, this is not a problem when information is based only on daily composite samples. Also, a certain degree of variation must be expected when flow measurement and chemical analysis for activated sludge treatment control are conducted. If in the latter case the average of several determinations is taken as the true determination, it must be remembered that this average is a measurement obtained from one sample. Therefore the subject of sampling error has to be considered. The validity of the results obtained depends on the fairness of the sample and the technique employed in studying that sample.

To secure representative control data it is necessary to obtain characteristic samples selected on the basis of the previously known variations of the activated sludge performance. Of course, they need not be simple random samples, but may properly be stratified samples anticipating the known fluctuation patterns. Therefore, to choose appropriate sampling procedures, extensive study of the plant is needed. It also should be mentioned that the accuracy of most analytical determinations used in the control of the activated sludge process is not known closely enough for application to particular cases. It seems that there is a serious need for an examination of this question.

As was stated earlier, the only way to prevent systematic errors in activated sludge treatment plant control data is to collect the data according to a precise program based on knowledge of the process and the plant. It is also possible to increase, if necessary, the accuracy of most of the analytical procedures used by selecting more advanced analytical methods and using more precise instruments. The only question is how many data are really needed for the solution of the practical task of operating the given plant with the required performance efficiency, and how accurate the data must be. This question by no means is an easy one, but it has to be answered if efficient and economical operation of the plant is the purpose.

If in a given treatment plant there is a shortage of manpower, there is a possibility that the analytical data received will not be sufficient and the plant operation will be heavily based on the operator's intuition. On the other hand, if there is a surplus of manpower or automatic analytical equipment, the flood of control data produces a difficult task of data storage and retrieval.

There is a feeling that in some cases extensive and frequent measurements and analysis for the control of activated sludge treatment plants have only very limited practical significance and are often used only to fill

22. TREATMENT PLANT PERFORMANCE EVALUATION

the monthly and annual reports. At the same time, for immediate plant control, only the sludge volume index and the COD sludge loading factor are determined, on grab samples. Such a method, however, should not be abused and its use is justified only if the particular plant has been previously studied to a reasonable extent and no surprise effects are expected.

4. AUTOMATION OF ACTIVATED SLUDGE TREATMENT PLANTS

An alternative to the conventional control of activated sludge treatment plants is automation. There have been several theoretical and practical attempts to detail the possibilities and requirements for automatic control systems for the activated sludge process [1, 4, 24, 25]. In general, to design a workable automatic control system for activated sludge treatment plants, reliable and accurate methods of automatic measurement and monitoring of parameters would be required for the following at least: flow of wastewater and recycling activated sludge; dissolved oxygen and mixed liquor suspended solids concentrations in aeration tanks; and BOD of influent and effluent.

At present, a satisfactory technique has been developed only for the flow measurements. Continuous measurement of the dissolved oxygen concentration is possible and performed at several plants, although reliability of some equipment used is often questionable. Nevertheless, based mostly on this measurement, the partial automation of these plants has been strongly advocated [1,4].

The automatic measurement of mixed liquor suspended solids concentrations and wastewater BOD is not yet feasible. However, diverse technical equipment is offered which in the wishful thinking of its constructors and producers should be used for this purpose. Its application may be justified only in a few specific situations.

Most existing methods for the automatic determination of mixed liquor suspended solids concentrations in aeration tanks depend on the measurement of the transmission or the scattering of visible light, or some other source of radiation, as in automatic determinations of turbidity. These optical properties of mixed liquor must be related to the actual contents of suspended solids of specific activated sludge. This leads to significant errors, which are due to the substantial concentration and fluctuating flocculation abilities of the material, as well as its other specific features, e.g., distribution of floc sizes. In the conditions of most aeration tanks, this distribution is bimodal [16], making this material completely unsuitable for any optical concentration measurements.

Some correction of these measurements is possible by compensation for mixed liquor soluble color through simultaneous measurements of optical density at two different wavelengths, or simultaneous measuring of scattered and transmitted light. These improvements may have some applications in the monitoring of suspended solids in river water or treated effluents, but in mixed liquor their role is very limited.

Similarly, different attempts to measure BOD automatically have not as yet produced the required information. Also, substituting COD determinations or organic carbon measurements for BOD measurements does not seem satisfactory. In recent years extensive studies have been carried out on the possible application of redox potential measurements and of some enzymatic tests for automation of the activated sludge process. Unfortunately, no practical results were achieved.

For some special studies it is possible to design extensive instrumentation and sampling systems for long-term monitoring of substantial amounts of information about the plant under study [14]. However, for routine control of a plant such instrumentation is not justified. Its operation, moreover, requires much more labor than conventional operation of the plant.

5. ANALYSIS OF PLANT PERFORMANCE DATA

Using different statistical techniques, activated sludge treatment plant performance data can be analyzed and evaluated. The most common approaches are frequency analysis, time-series analysis, and correlation of technological parameters.

A. Frequency Analysis

The basic frequency analysis of an activated sludge plant's performance data is shown in Table 1, covering a period of about three months' operation of the Lakeview Sewage Treatment Plant [6]. This analysis includes calculations of mean values, standard deviations, variances, and coefficients of variation.

The raw sewage BOD shows the largest standard deviation from the mean, while variations become progressively less as the sewage is given first physical and then biological treatment. However, the coefficients of variation of secondary effluent BOD and suspended solids are substantially higher than for primary effluent. A similar pattern was also noticed by Thomann [21] in his study of eight sewage treatment plants, and may be

TABLE 1

Frequency Analysis of Lakeview Treatment Plant
Sewage Flow, BOD and SS Data

Description	Mean	Standard deviation	Variance	Coeff. of variation	No. observations
Plant flow (mgd)	11.85	2.207	4.87	0.186	103
BOD of plant influent (mg/liter)	255.71	82.186	6754.48	0.321	93
BOD of primary effluent (mg/liter)	151.14	46.515	2163.66	0.308	92
BOD of secondary effluent (mg/liter)	26.42	14.075	198.11	0.533	93
Suspended solids in plant influent (mg/liter)	263.23	70.731	5002.85	0.269	102
Suspended solids in primary effluent (mg/liter)	130.19	30.312	918.79	0.233	103
Suspended solids in secondary effluent (mg/liter)	39.75	22.504	507.42	0.566	103

explained by the lower stability of the biological treatment than of the mechanical treatment. It may be due to the complexity and sensitivity to environmental changes of the activated sludge process.

The changes in the coefficient of variation with increasing treatment may also be regarded as a measure of the plant's buffering capacity or ability to accommodate the qualitative and quantitative shock loads imposed on it. Therefore, an increase in coefficient of variation may indicate insufficient buffering capacity in the system studied. However, the data on final effluents may also reflect substantial "background noise" representing inadequacies in the analytical equipment at this range of concentration.

It is also interesting to note that most data on the quality of the performance of activated sludge treatment plants show the log-normal distribution or some other similar skewed distribution. This fact has been confirmed in many studies of wastewater treatment [9, 19, 22], and may be used for a more precise description of studied or projected situations where often normal probability density functions are assumed or considered to exist.

It has proved convenient to present some complicated relationships in the activated sludge process in the form of comparisons of frequency distributions, rather than by using regression analysis or deterministic mathematical models. This is especially true of phenomena whose response time is difficult to determine and/or fluctuating during the observation period. An application of this approach for the chemical neutralization of alkaline industrial wastewater during the activated sludge process was presented by Ganczarczyk [9]. Another example concerns the decrease of the activated sludge volume index during detention in secondary clarifiers [10].

B. Time-Series Analysis

It is possible using time-series analysis to determine quantitatively all significant trends and periodicities in activated sludge treatment plant performance data. Seasonal or any special kind of variability can be removed from the data as given harmonic components in a Fourier analysis. The residual record can be the subject of autocorrelation and spectrum analysis [3]. This method was successfully applied by Thomann to the analysis of water quality data [20] and then to a study of the variability of wastewater treatment plant performance [21].

There are, however, specific difficulties in "data editing" and "trend correcting" when effects of a biological treatment are under study. It seems that sometimes quite substantial differences may be expected as the result of responses of some systems studied to particular external impulses, so rejection of some of the measured data in the process of editing is not necessarily justified if not supported by additional studies.

C. Correlation of Technological Parameters

The study of the relationships among various technological parameters of an activated sludge treatment plant can be used to prove the correctness of the applied metering and sampling systems, and by itself supplies information vital for the most efficient operation of the plant. There are, however, specific difficulties in the realization of this approach.

22. TREATMENT PLANT PERFORMANCE EVALUATION

Most of our knowledge of the activated sludge process is based on experiments carried out under so-called "controlled laboratory conditions." Such studies can be much broader and more diversified than experiments in pilot or full-scale conditions, but are often biased by undue influence of some factors which cannot be properly controlled or modelled. It is usually taken for granted that the behavior of the full-scale plants will be close to that of the models studied, and that relationships among process parameters, observed in smaller scale, will be equally valid in the full-scale plants. However, quite often the fluctuation of wastewater quantity and quality, as well as the instability of some process parameters, together with the inadequacies of metering and sampling, are such that in specific cases expected correlations cannot be realized in practice. The data are characterized by such a scatter of information that their correlation would be without any practical meaning.

The existing situation at the Lakeview Sewage Treatment Plant illustrates correlations between the organic loading of activated sludge and the resulting effluent BOD and sludge volume index as well as the values of the BOD removal rate constant [6]. In this specific case it was proved that, based on routine plant control data, there is no relationship between the BOD remaining and the sludge loading factor (Fig. 1). No satisfactory relationship was observed between the sludge volume index (SVI) and the loading factor during the study period (Fig. 2). Generally, increases in SVI were associated with decreasing effluent quality, but the increasing or decreasing of the loading factor had no noticeable effect on the sludge volume index. High indices were recorded at both high (> 1.0) and low ($0.3 - 0.4$) food-to-organism ratios expressed in g BOD/g MLSS·day. However, for consecutive Mondays, which were reported to have the lowest SVI values, the above relationship seems to follow the general pattern of increased readings above some typical values of the loading factor [10].

Assuming conditions close to complete mixing in the aeration tanks, an attempt was made to determine a value of the BOD removal rate constant K [7] for the Lakeview Sewage Treatment Plant wastewater. The BOD remaining in the secondary effluent was plotted against the BOD removed from the aeration tanks, divided by the time the mixed liquor was under aeration (Fig. 3). From a least squares calculation the measured value of K was found to be 0.00004. This figure is distinctly lower than the range of values quoted by Eckenfelder [7] as being typical for domestic sewage and some industrial effluents. It may therefore indicate that the specific composition of wastewater at Lakeview is more difficult to treat biologically than average sewage. Moreover, the significant scatter of K values presented in Fig. 3 from approximately 0.0001 to approximately 0.001 may indicate not only the imprecision of sampling and analytical procedures but also the fluctuation of the quality of the wastewater studied.

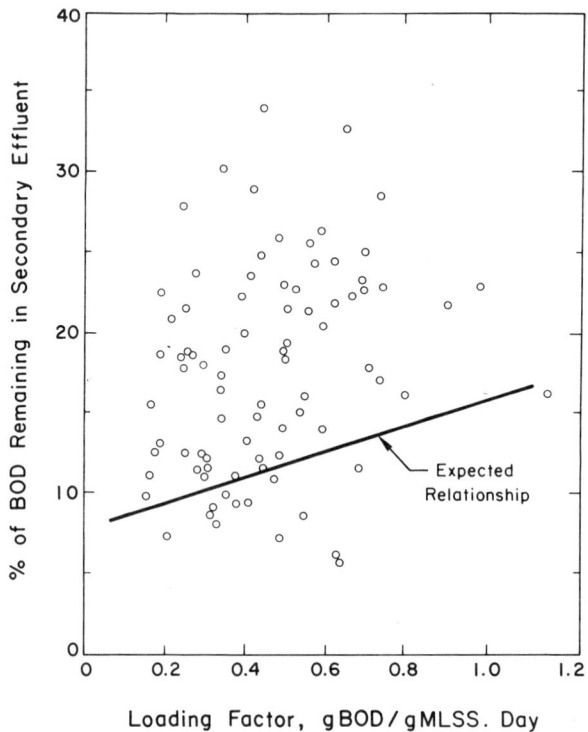

Fig. 1. BOD remaining vs sludge loading factor (86 observations).

6. CONCLUSIONS

1. A realistic program for control of an activated sludge plant has to be preceded by extensive plant studies concerning the variations of influent quantity and quality, as well as the individual through-flow and mixing characteristics of the plant.

2. On the one hand this procedure leads to the minimization of the sampling frequency, and on the other it prevents errors caused by the improper grouping of sampling frequency vs time.

3. The presentation of the raw activated sludge plant control data, or only the maximal, minimal and average values, gives very limited information on plant performance.

22. TREATMENT PLANT PERFORMANCE EVALUATION

4. Further progress is required on the automation of activated sludge treatment plants.

5. Additional studies seem necessary to apply various statistical techniques to improve and extend the conventional evaluation of activated sludge treatment performance data.

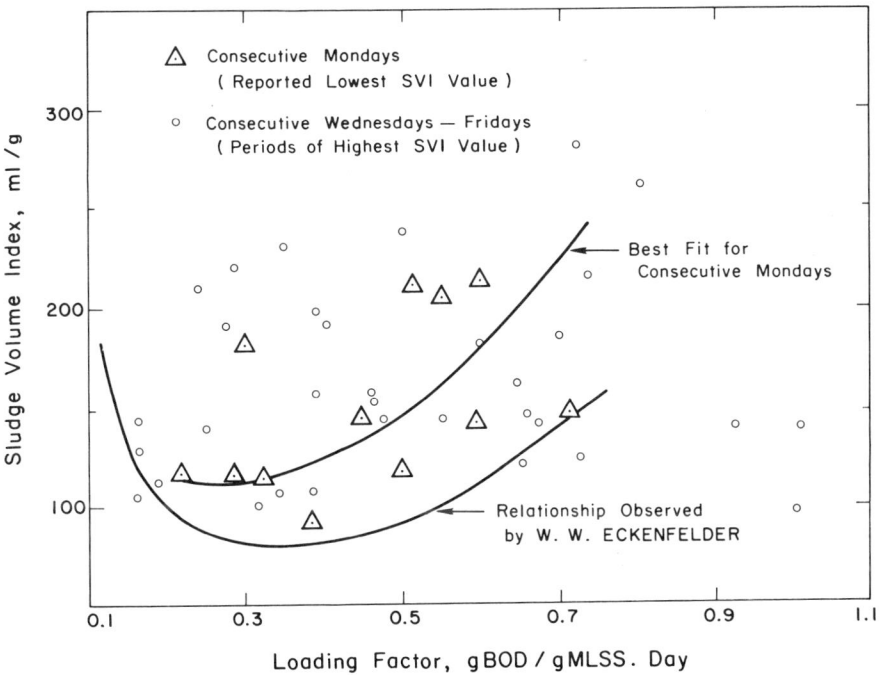

Fig. 2. Sludge volume index vs sludge loading factor.

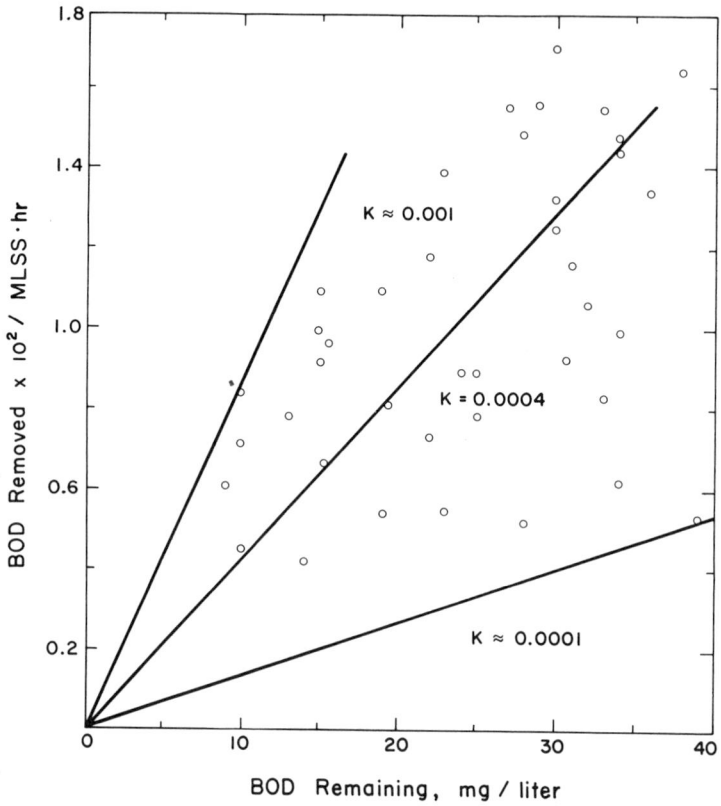

Fig. 3. BOD removal rate constants.

REFERENCES

1. J. Bernard and R. Louboutin, Modulation of the oxygen supply in an activated sludge process, Paper II-6, Fifth Int. Conf. Water Poll. Res., San Francisco, Calif., 1970.

2. P. M. Berthouex and L. C. Brown, Monte Carlo simulation of industrial waste discharges, J. Sanit. Eng. Div., Proc. Amer. Soc. Civil Eng., 95 (SA5), 887 (1969).

3. R. B. Blackman and J. W. Tukey, The Measurement of Power Spectra, Dover Publ., New York, 1958.

4. P. Brouzes, Automated activated sludge plants with respiratory metabolism control, Paper II-5, Fourth Int. Conf. Water Poll. Res., Prague, Czechoslovakia, 1969.

5. C. K. Calvert, Hourly variations of Indianapolis sewage, Sewage Works J., 4, 815 (1932).

6. D. F. Carr and J. Ganczarczyk, A performance analysis of an activated sludge treatment plant, Univ. Toronto, Dept. Civil Eng., Pub. No. 71-600, 1971.

7. W. W. Eckenfelder, Jr., Comparative biological waste treatment design, J. Sanit. Eng. Div., Proc. Amer. Soc. Civil Eng., 93 (SA6), 157 (1967).

8. D. V. Eckhoff and D. Jenkins, Transient loading effects in the activated sludge process, Paper II-14, Section 2, Third Int. Conf. Water Poll. Res., Munich, 1966.

9. J. Ganczarczyk, Performance studies of the unbleached Kraft Mill effluent treatment plant in Ostroleka, Water Res. (Brit.), 3(7), 519 (1969).

10. J. Ganczarczyk, Variation in the activated sludge volume index, Water Res. (Brit.), 4(1), 69 (1970).

11. C. P. L. Grady, Jr., A theoretical study of activated sludge transient response, paper presented at the 26th Purdue Industrial Waste Conf., Purdue University, Lafayette, Indiana, 1971.

12. G. C. Hutchinson and E. R. Baumann, Variation of sewage flow in a college town, Sewage Ind. Wastes, 30, 157 (1958).

13. E. T. Kiliam, Hourly load variations as a factor in sewage treatment plant operation, Water Works Sewage, 81, 7 (1934).

14. W. F. Milbury, V. Stack, N. S. Zaleiko and F. L. Doll, A comprehensive instrumentation system for simultaneous monitoring for multiple chemical parameters in a municipal activated sludge plant, Roy F. Weston, Inc., West Chester, Pa., 1970.

15. M. M. Montgomery and W. R. Lynn, Diurnal variations in sewage flow at Baltimore, J. Sanit. Eng. Div., Proc. Amer. Soc. Civil Eng., 90 (SA1), 73 (1964).

16. D. S. Parker, W. J. Kaufman and D. Jenkins, Characteristics of biological flocs in turbulent regimes, SERL Rept. No. 70-5, Sanitary Engineering Research Laboratory, Univ. California, Berkeley, California, 1970.

17. W. A. Sperry, Hourly variations of sewage concentration and flow, Sewage Works J., 19, 785 (1947).

18. J. Spicka, Fluctuation of effluent quality in activated sludge plants, Paper II-2, Section 2, Fourth Int. Conf. Water Poll. Res., Prague, Czechoslovakia, 1968.

19. A. R. Townshend, Statistical analysis of the effluent quality of biological sewage treatment processes, Proc. 3rd Can. Symp. Water Poll. Res., Toronto, 272 (1968).

20. R. V. Thomann, Time-series analysis of water quality data, J. Sanit. Eng. Div., Proc. Amer. Soc. Civil Eng., 93 (SA1), 5108 (1967).

21. R. V. Thomann, Variability of waste treatment plant performance, J. Sanit. Eng. Div., Proc. Amer. Soc. Civil Eng., 96 (SA3), 819 (1970).

22. C. J. Velz, Graphical approach to statistics, Water Sewage Works, 99(2), 66 (1951).

23. A. T. Wallace and D. M. Zollman, Characterization of time-varying organic loads, J. Sanit. Eng. Div., Proc. Amer. Soc. Civil Engineers, 97 (SA3), 257 (1971).

24. N. Westberg, An introductory study of regulation in the activated sludge process, Water Res. (Brit.), 3, 613 (1969).

25. C. Zaander and B. Johnson, Process control for activated sludge, Proc. 22nd Ind. Waste Conf., Purdue Univ., Ext. Ser., 129, 403 (1968).

23

COMBINING MULTIPLE ATTRIBUTE OUTCOMES INTO AN OVERALL INDEX

Clifford J. Maloney

Food and Drug Administration
Bethesda, Maryland

1. INTRODUCTION

With roots extending back to World War I but with increasing tempo in the period following World War II, a general theory of measurement has developed an extensive body of literature now summarized in a two-volume work, the first volume of which has recently appeared [3]. For environmental scientists the most important lesson of the history of mensuration is, perhaps, that what is or what is not measurable is not a priori obvious; that, where successful, the construction of a scale of measurement is an achievement, but that the mere existence of a scale does not necessarily reproduce any phenomenon of interest.

Mensuration, numbering (at any rate, a distinction between none, one, and many), and explanations of and attempts to subdue natural phenomena are as old as man. It is a striking story, only as yet sporadically and partially comprehended, that progress in any one of these aspects has often influenced and frequently in turn required conjoint developments in both of the other fields. The first two have usually, in the past, been regarded as hardly distinguishable, though the first is the province of natural science and the second of mathematics. It remained for social scientists to establish and for a psychologist [7] to clearly formulate both the existence of the distinction and its nature. In the process it became clear that extending the scope of the application of mathematics to natural phenomena involved both the construction of measurement procedures of a familiar type for new forms of data and the development of additional types of measures, in general by relaxation of mathematical structure. It is believed that this latest process should be of the greatest interest to environmental scientists, since the benefits from their efforts tend to be esthetic and intangible, whereas the costs are measured in the most tangible of terms, money and forbearance. In consequence, the thesis that measurement is inapplicable to the determination of one arm of the exchange subjects the environmental scientists' work to the vagaries of human prejudice; not in his studies, but in attempts to utilize them in the operational decisions stemming from them. These latter ideas are elaborated elsewhere [5].

2. EXAMPLE

This paper describes a statistical technique which it is hoped will prove applicable to some pollution research, although no such application has as yet occurred. The method arose in a medical context, and reference to that publication [4] will be necessary for a full discussion, including a computer program by which all calculations can be made easily and inexpensively.

To suggest a possible type of application which might occur in environmental contamination evaluation, Fig. 1 is a simplification of a plot of rate of soot deposition [6]. Were the plotting attempted for sites at much greater distances from the source it would be found that often the quantities involved would be less than the threshold of the measurement technique. In such cases accidental momentary concentrations might be detectable as they occur, while at other times accidental momentary attenuations would go undetected. The observations, then, would be dichotomous, occurrence sometimes being observable, sometimes not.

The significance of this irregularity of detection would depend on the physical nature of the effect. If the effect is chronic, varying only with long-term accumulation, then the sum of the appearances will be the important measure. If, however, the effect is acute, as in most disease contagion models, then what is summed is not the effect per se, but the probabilities involved. Possibly a reference to the "probability density" of the shell electrons in modern atomic theory will make clearer this second alternative interpretation.

Two characteristics of pollution situations adverse to health sometimes occur. First, many agents are deleterious at very low concentrations not easily measurable, but perhaps detectable as occurring or not occurring. Second, at least sometimes, agents deleterious to health will not occur singly, but several distinguishable agents will be present together in a fixed, though in general an unknown, ratio. The technique to

23. COMBINING MULTIPLE ATTRIBUTE OUTCOMES 355

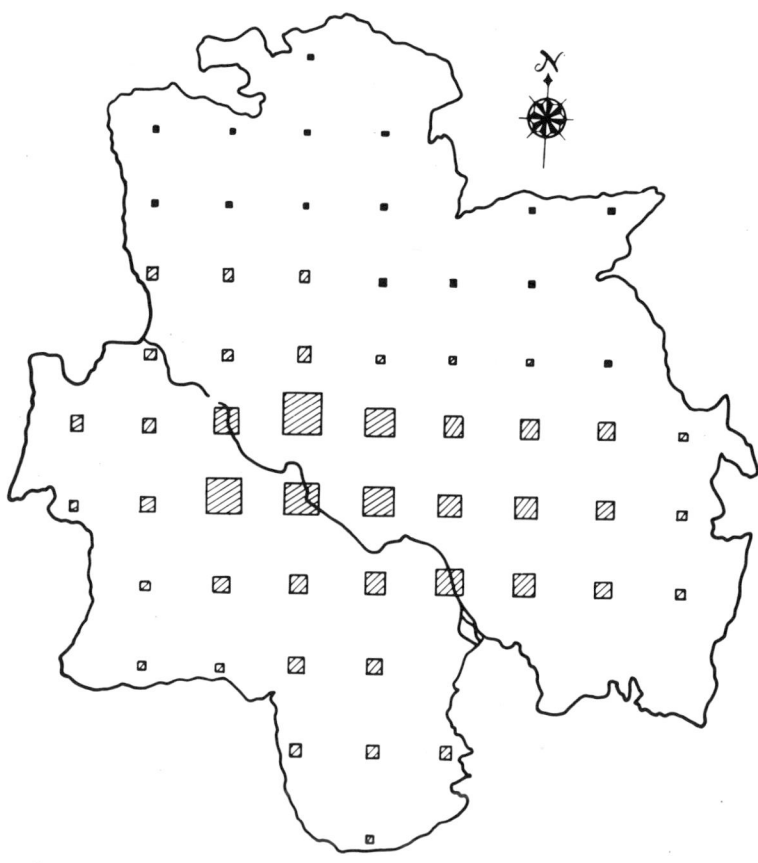

Fig. 1. Map showing rate of soot deposition at Leeds. Side of square proportional to intensity of soot deposition.

be described was developed to exploit all-or-none measures in just such multiple response cases in order to construct a quantitative scale for the causative agents, or, in the case of disease severity, the basic condition.

3. SCALES

A very penetrating examination of types of scales, their characteristics, limitations, and uses was initiated by S. S. Stevens some years ago [7]. Table 1 is a slight adaptation of his Table I. The last line is

TABLE 1

Properties of Types of Scales[a]

Type of scale	Empirical operation	Permissible transformation[b]	Central tendency
Nominal	permuting	equivalence	mode
Ordinal	ordering	monotonic	median
Interval	differencing	affine	arith. mean
Ratio	dividing	similarity	geom. mean harmonic mean
Fixed[c]	none	identity	percentage

[a] Adapted from Stevens [7].
[b] Adapted from Blakers [1].
[c] Taken from Suppes [8].

taken from Suppes [8]. In terms of Fig. 1, the quantities of soot measured at each site possess the characteristics of a ratio scale, since they are necessarily positive and they exhibit a natural origin - no soot. In turn the locations of the several sites follow an interval scale, since there is no natural zero on the earth's surface. If the quantity of soot observed at each site were to be ordered and then the measurements discarded, these ordinal positions would form an ordinal scale. In turn, all sites in the diagram and many beyond would show the presence of soot but, at a sufficiently great distance from the source, failure to detect soot might be as frequent as success. Detection and nondetection would constitute a particular form of a nominal scale - a dichotomy. If there are several, but unordered, outcome classes, e.g., if several mutually exclusive atmospheric compounds are involved rather than a single one such as soot, the scale will still be nominal, but now many-valued.

4. PROBABILITY

The theory of probability supplies a rationale by which, given an assemblage of observations on a nominal scale, a single value of a ratio scale can be derived. For many purposes this constitutes an advantage.

23. COMBINING MULTIPLE ATTRIBUTE OUTCOMES 357

In the example to be discussed below, a cancer patient may or may not exhibit metastasis to a particular organ system. Knowledge of this fact aids the physician in determining how sick his patient is, the sickness level in turn determining in part his choice of treatment. It is customary to allocate cancer patients to one of four "stages" of an ordinal scale. The current technique was applied to a limited set of available data to derive a ratio scale in terms of presence or absence of metastasis to each of seven organ systems: bone, B; nodes, N; skin, S; lungs, L; pleural cavity, P; liver, V; central nervous system, C. Since any patient showing metastasis is allocated to Stage IV of the conventional scale all our distinctions are subdivisions of that stage.

Table 2 classifies the four possible situations in which binomial probabilities apply. In a situation in which the population probabilities are known, or derived from some mathematical model, one is in a position to deduce the number of successes and failures which are to be expected in a given number of trials. This is true whether or not the probabilities are constant from trial to trial or whether or not they vary, so long as they are known. These two situations are called "direct" in Table 2. If the probabilities are known to be constant from trial to trial but the numerical value is unknown, then an actual experiment will reveal the number of successes and failures, and these numbers can be used to obtain an estimate of the fixed but unknown population probability (Table 2, upper right cell). These three situations are described in most elementary texts. The fourth (Table 2, lower right cell) is a specialized topic with an inappropriate title. The technique underlies the method of this paper. It is briefly described in Section 5.

5. QUANTAL REGRESSION

Figure 2 illustrates a statistical estimation problem which arises so frequently in biological contexts that it is often called bioassay even when the application is to another field, such as penetration resistance of armor plate, breaking strength of thread, or flash point of powder - in any situation where there is a variable but known stress or defense against stress and where outcome can be viewed as success or failure. While, again, many-valued outcomes are discussed in the literature, this paper treats directly only dichotomous outcomes. Our sketch of the statistics of quantal assay makes no pretense to either completeness or rigor. For these, reference must be made to monographs and the extensive literature. Our notation conforms to that of Finney [2].

If a group of individuals are subject to no stimulus, then none will respond and the response percentage will be zero. However, as the

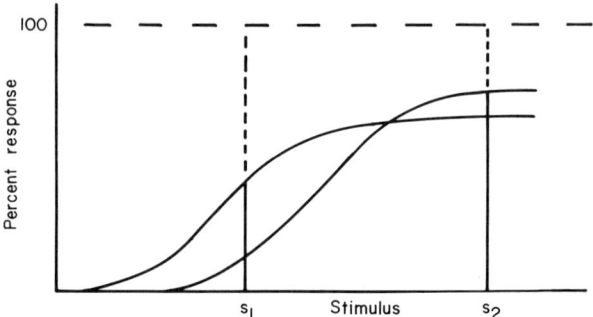

Fig. 2. Hypothetical quantal response curves for two stimuli or treatments.

TABLE 2

Types of Situations in Which Binomial Estimation Applies

Nature of probability	Estimation situation	
	Direct	Indirect
Constant	Np	r
Variable	Σp_i	bioassay

stimulus level is raised, a greater and greater percentage of individuals will be affected until (ideally) all respond. Percentage response is thus a monotone function of level of stimulus. Innumerable complications are possible, e.g., individuals who yield a response even in the total absence of stimulus. Placebo treatments are often used to adjust for this influence. Some stimuli are such that some individuals are immune (so-called nonresponders). All such complications are neglected in this paper, though presumably to handle them would only complicate the analysis and the arithmetic in well known ways [2].

Since the relation between the physically defined stimulus plotted on the abscissa and the expected percentage response on the ordinate is biunique, if a value on either axis is known then the functional relation can be used to ascertain the corresponding value on the other axis. Thus by observing a fraction of animals, insects, cells, or tissue cultures which respond to a certain treatment, the strength of an applied chemical or drug can be estimated. Because of the extreme sensitivity and high specificity

23. COMBINING MULTIPLE ATTRIBUTE OUTCOMES

of biological systems (including man), such measurements can be used at extremely high dilutions and for substances not yet characterized chemically.

Figure 2 shows two such stimulus-response curves. Each curve separately relates a level of intensity of a treatment to a fraction of individuals who will respond. (No change in principle occurs if the number of curves is increased. In the biological example actually treated, there were seven metastatic sites, hence a complete graph would show seven dose-response curves.) When, as in Fig. 2, two (or more) curves are plotted, either they have one or more points in common or they do not. The physical as opposed to the mathematical significance of intersecting stimulus-response curves is that, whereas one of the two or more responses is more frequently elicited by a low stimulus, (one of) the other(s) is more frequent at a higher stimulus. This is often implausible. One could in principle allow for this possibility, were it desirable, though data adequate even for the simpler model (where none of the curves cross) are not easily available. The discussion in this paper will assume the simpler situation, but this is a concession to the paucity of data in real life and not a limitation of the theory.

Any strictly monotone continuous function, bounded or not, can be transformed into an infinite straight line by a pair of transformations, one applying only to the abscissa and the other applying only to the ordinate. In fact, this can be done in many ways.

If two curves do not intersect, and if the same pair of transformations linearize both, then the resulting lines will not intersect; i.e., they will be parallel straight lines. That the same functional form linearizes both (all) dose-response relations being compared in ordinary applications of bioassay is again a physical and not a mathematical assumption. This follows from the fact that such a comparison is only regarded as worth doing if the "modes of action" of all agents are comparable, i.e., if each can be viewed as a dilution or a concentration of any of them. In our application, this is a deduction from the assumption that the order of symptoms with respect to population frequency is preserved throughout the range of interest, not just for the small finite set of symptoms actually observed, but for any two distinct (independent) symptoms, however close in occurrence frequency. In the case of Fig. 1, if n pollutants are under consideration and if we assume that the strengths of each at a given observation station is sought and that the ratio of strengths, while unknown, is fixed, then the condition required for the method will be present. In other words, it is assumed that the absolute quantities of the pollutants are present in an unknown ratio, but that, whatever it is, this ratio is fixed throughout the region. This is the effect of the assumption of parallelism.

While in theory we could employ a different pair of transformations for each of our seven metastatic sites, in practice adequate data to unambiguously discriminate between the many closely similar candidates is never available and of little practical significance. Well known transformations of the response percentage (ordinate) employed by bioassay workers are the probit, the logit, and the like. The dose scale is generally taken to be logarithmic in bioassay, but in the pollution example the abscissa could be untransformed. A detailed discussion of the considerations in the choice and definition of an illness index or sickness score is presented in Ref. 4.

By some such method, our model of Fig. 2 is converted to that of Fig. 3. For simplicity, only four parallel lines, each corresponding to a different metastatic site, are shown in Fig. 3 though seven were actually available in the data. Each of the lines can be represented by an equation of the form

$$Y_i = a_i + bs, \qquad (1)$$

where a_i is the intercept of line i (the a_i may differ), b is the slope (common, since the lines are parallel), and Y_i is the transform adopted for line i of the population percentage response for those patients whose index of sickness severity is s.

The sickness score s characterizes a specific level of disease severity. For a fixed s, the set of equations (1) determines the probability that a patient at that level of disease will manifest each of the n relevant symptoms. Since we are devising a new scale, the slope b is at our disposal and is taken to be 0.1. In the pollution example, the slope might have to be ascertained, since scales for volume and for mass already exist.

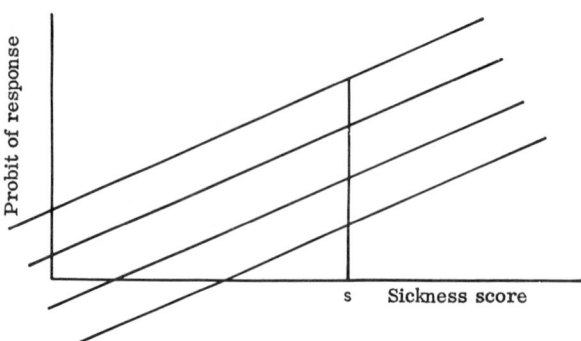

Fig. 3. Four nonintersecting stimulus-response curves transformed to parallel straight lines by a suitable transformation.

6. INDIVIDUAL SICKNESS SCORE

In Section 5 it was explained that a plausible dose-response model applicable to severity of illness (and possibly also in other contexts, including certain problems in pollution research), consists of a relation between degree of illness measured by a sickness score s and the probability of occurrence of one or more all-or-none symptoms, signs, or laboratory findings. How, on the basis of a specified pattern of these symptoms, the illness index is to be assigned to the patient is the subject of this section. Of course, all patients with the same set of symptoms are assigned the same index. From here on, I will speak entirely in terms of the one application so far made, which was to the finding of metastasis in certain of seven organ systems in each of 908 breast cancer patients. This application is made purely to illustrate a statistical technique, and its usefulness in cancer research will require much further study.

To determine how ill a given cancer patient is by this technique it is necessary to have available the equation of each of the lines of Fig. 3. How these equations can be derived on the basis of records of cases of the disease will be the subject of Section 7. In this section knowledge of these equations will be assumed.

It must be admitted at once that the sickness scores are not determined with high precision, since the number of symptoms (in our case the number of organ systems examined for metastasis) is essentially the sample size. In turn, however, since the technique effectively summarizes all the metastasis observations into a single figure, there is no handicap in making much more thorough examinations and in recording observations in much greater detail - and every incentive to do so, so long as precision of diagnosis is sharpened thereby.

If the numerical value x_i is chosen as unity if a particular patient exhibits the symptom i (metastasis to site i in our illustration) and zero if that organ system is free of metastasis, then the likelihood function for that particular combination of symptoms becomes, under independence,

$$L = \prod P_i^{x_i} (1 - P_i)^{1-x_i}. \tag{2}$$

Each P_i is a function of Y_i, which in turn is a function of s by means of Eq. (1). There is therefore just one maximum likelihood equation:

$$\frac{d \log L}{ds} = 0 = \sum \left(\frac{x_i}{P_i} - \frac{1 - x_i}{1 - P_i} \right) \frac{dP_i}{dY_i}. \tag{3}$$

Because of the quantities dP_i/dY_i, Eq. (3) is best solved by trial and error (a detailed example is given in Ref. 4). In practice, however, this solution would be found in advance by computer for all possible symptom combinations and supplied in a table.

In our example of seven metastatic sites only 128 different values of s are possible. Use of the method in medical practice then would consist of a single table search.

Gain to the physician would occur in two ways. First the balancing off of metastasis to certain sites and its absence in others would be objectively performed, relieving the attending physician of attempting to do so intuitively, particularly for the rarer combinations, and it would be based on all available data and not only on his own personal practice. Second, occurrence of rare combinations would be shorn of any enigmatic character and reduced to the value of a familiar parameter - the sickness index s. This latter benefit to the physician could well outweigh the former in his attempt to rationally evaluate the condition of his patient.

7. CONSTRUCTION OF THE SICKNESS SCORE

Section 6 describes how a particular set of symptoms may be used to assign a sickness score to a particular patient. Before this can be done it is necessary to know the value of the a_i in Eq. (1). This can only be done on the basis of accumulated experience. In the current example, records of 908 patients were available. Because the model being fitted to the data assumes that metastases to the several organ systems are ordered in a fixed order irrespective of the illness of the patient (a deduction from the parallelism of Eq. (1)) that order is given by the combined frequency over all records. It is: bone, 468; nodes, 408; skin, 230; lungs, 216; pleura, 179; liver, 97; and central nervous system, 35. Furthermore, the percentage of 908 which each of these numbers is can be taken as a preliminary estimate of the true population percentage occurrence for a patient of sickness index 50. Using these values, a first estimate of each a_i in Eq. (1) is available. From these, the records of each of the 908 patients can be used to calculate a preliminary estimate of illness for each patient. Next, these 908 sickness scores can be used together with the observed percentage responses for each metastatic site - independently but with a common slope of 0.1 - to calculate improved values of the intercepts a_i. All steps are done by computer and converge rapidly to stable values [4].

8. CHECK OF THE MODEL

An intuitive measure of level of illness is afforded by the duration of life remaining. Of course, this index is useless as a guide to treatment,

23. COMBINING MULTIPLE ATTRIBUTE OUTCOMES 363

but if a forecast of months of life remaining is made on the basis of a sickness score derived from extent of metastasis this forecast could indeed be useful. Postexamination survival was available for 829 of the 908 patients used in deriving the index. While of course it would be desirable to apply the derived index to new metastatic cancer patients, a check based on survival of the same patients used to derive the index should be of some value, since no use whatever was made of the survival information in forming the index.

Such a check was carried out, with generally satisfactory results [4]. The chief discrepancies appear to be that metastasis to the lymph modes is less life-threatening and to the liver more so than most sites, which is perhaps in conformity with clinical experience; and unexpected longevity of patients with metastasis to both skin and nodes, which is at least favorable. This illustrates how a disease severity index, like any mathematical model, can organize a body of data and suggest problems requiring further study by its failures quite as much as solve problems by its successes.

ACKNOWLEDGMENTS

It is a pleasure to acknowledge to the late George Kennedy my extreme indebtedness for his energetic and perceptive collaboration since the inception of the project, first at Fort Detrick and later at the National Institutes of Health. Professor Johannes Ipsen and Professor H. Edwin Titus reviewed the argument of the paper, and the former cooperated in preliminary calculations and in reviewing the final draft. Mr. Sidney Spindel developed the program DISSEV and performed numerous calculations. Finally, I am indebted to a number of hospitals for the opportunity to test the procedure on their data.

REFERENCES

1. A. L. Blakers, Mathematical Concepts of Elementary Measurement, (Studies in Mathematics, Vol. XVII), School Mathematics Study Group, Stanford Univ., Stanford, Calif., 1967.

2. D. J. Finney, Statistical Method in Biological Assay, Hafner, New York, 1952; 3rd ed., 1971.

3. D. H. Krantz, R. D. Luce, P. Suppes, and A. Tversky, Foundations of Measurement, Vol. 1, Academic Press, New York, 1971.

4. C. J. Maloney, in "Disease Severity Index," Proc. 17th Conf. Design of Experiments in Army Res. Dev. Testing, October, 1971, ARO-D Rept. 72-2.

5. C. J. Maloney, The limits of measurement (L'Institut National de Recherche Chimique Appliquée), Dixième Colloque sur les Atmospheres Polluées, Paris, 1972.

6. A. Massey, Modern Trends in Public Health, Harper, New York, 1949, p. 159.

7. S. S. Stevens, On the averaging of data, Science, 121, 113 (1955).

8. P. Suppes, Studies in the Methodology and Foundations of Science, Humanities Press, New York, p. 67.

24

A PROPOSAL FOR ECONOMICAL FIRST STAGE SCREENING FOR
TUMORIGENS WITH A POSSIBLE "JOINT ACTION" BONUS

Robert M. Elashoff and Milton Sobel

University of California at San Francisco
San Francisco, California

and

M. A. Schneiderman
National Cancer Institute
Bethesda, Maryland

1. INTRODUCTION

A screening program generally has many parts. The usual first stage has been to delete from the pool of chemicals under consideration those that are very unlikely to be tumorigens. In our setup this means chemicals that do not differ significantly from negative controls. Next, the remaining chemicals are studied in much more detail - perhaps as Shubik [4] suggests. Finally, after several more steps, some definite public policy is articulated with respect to a few chemicals.

Our efforts at formulating and investigating screening designs have been concerned with the first stage of such a screening program. We look upon the first stage in the following way:

1. We want to test as many materials as quickly as possible. Failure to test a material is a source of possible hazard because untested materials are usually (except for a limited number with certain chemical structures) treated as if they carried no risk.

2. We want our test procedures to have good sensitivity. Failure to declare a material that is truly a possible hazard could be a serious error. (An ideal is a "no false negative" system.)

3. We want our procedures to have good specificity. There are potential losses to society as a whole as well as obvious losses to the manufacturer if we overstate the case and declare a safe material to be a potential hazard. (An ideal is a "no false positive" system.)

4. We want to do this on as small a budget as possible.

Thus we ask, "What testing strategy can we use to balance errors of not testing a number of compounds, errors of failing to detect tumorigens under test, and errors stating nontumorigens to be tumorigens?" An introduction to some aspects of the screening literature is contained in Dunnett [1], while work on group screening is given by Sobel and Groll [5], and Watson [6].

There are many chemicals that are candidates for testing as possible tumorigens. (The definition of a "tumorigen," and the resolution of the argument concerning whether all tumorigens are potential carcinogens must be made by the pathologists and biologists. We assume that such a definition can be agreed on, and the arguments resolved.) Earlier data (e.g., Innes [3]) indicate that less than 5% of all the materials will be tumorigens. We further assume that testing will be done much as it is now - maximum tolerated dose, lifetime feeding to small animals (usually rodents), etc. (see, e.g., WHO[7]).

We propose one conceptual change in the present system - the testing of materials two at a time (2AAT) in an animal rather than one at a time (1AAT). As a first approach, we propose segregating chemicals to be tested into groups of four, and testing each member of the group of four with each of the others. (Other designs are possible, e.g., groups of four compounds with each chemical appearing in only two pairs of compounds. The choice of four compounds per group is for laboratory convenience only.) Each chemical is given at its maximum tolerated dose (MTD). This leads to six tests of compounds in each test ($_4C_2 = 6$). If the 1AAT procedure uses n animals per test, we propose using N animals for each pair tested to achieve the same probabilistic requirements as a 1AAT test. In this paper we study how this 2AAT design compares with the 1AAT.

In Section 2 we give a model for the probability that an animal receiving two compounds develops a tumor; Section 3 develops the decision rules to classify chemicals as tumorigens or not; Section 4 develops the

24. FIRST STAGE SCREENING FOR TUMORIGENS 367

probabilistic inequalities to be satisfied; in Sections 5 and 6 a 2AAT and the 1AAT designs are compared; finally, Section 7 gives a brief summary.

2. A PROBABILITY MODEL FOR JOINT ACTION OF COMPOUNDS

Let C_j, $j = 1, 2, 3, 4$, denote compound j; let p_j be the probability than an animal fed C_j at its MTD develops a tumor, and p_{ij} be the probability that an animal fed C_i, C_j in combination at their joint MTD develops a tumor. Then a model for p_{ij} is assumed which is based on knowing p_i, p_j and interaction parameters I_{ij}, S_{ij}. We assume

$$p_{ij} = \begin{cases} (1 - I_{ij}) \max(p_i, p_j) & \text{if } S_{ij} = 0 \\ S_{ij} + (1 - S_{ij}) \max(p_i, p_j) & \text{if } I_{ij} = 0, \end{cases} \quad (1)$$

with $0 \leq I_{ij} \leq 1$, $0 \leq S_{ij} \leq 1$, not both I_{ij}, $S_{ij} > 0$.

I_{ij} and S_{ij} express the strength of an inhibitory interaction and a synergistic interaction, respectively. That is, with no interaction of either sort present, the stronger tumorigen among C_i, C_j (i.e., the one with greatest p) dominates, and p_{ij} is just the p of this stronger tumorigen. But with $I_{ij} > 0$, p_{ij} is somewhat less than $\max(p_i, p_j)$. For example, if $I_{ij} = 0.3$, then p_{ij} is 70% of the probability specified by the dominance model without interaction; if $S_{ij} = 0.3$, then p_{ij} exceeds the dominating probability by 30% of the maximum allowable increase over this dominating probability. We assume that p_0 (the probability of a tumor in animals fed a standard diet under standard conditions) is known.

A special case of Eq. (1) is given by Finney [2] and fits some recent data on tumorigens [8].

3. DECISION RULES FOR SELECTING TUMORIGENS

First, let us consider decision rules for the 2AAT design. For each of the four chemicals in an experiment, a decision as to its tumorigenic nature is made. Consider decisions among three alternatives - tumorigenic, not tumorigenic, needs to be further tested in some other experiment. For chemical C_j, let Y_j be 1, 0, or 1/2 according as the first, second, or third alternative is taken. Let X_{ij} be the number of animals receiving C_i, C_j who develop at least one tumor. Also, let us define

$$X_{i,jk} = X_{ij} + X_{ik} \quad \text{for } i,j,k \text{ all different,}$$

$$X_i = X_{ij} + X_{ik} + X_{il} \quad \text{for } i,j,k,l \text{ all different.} \tag{2}$$

In X_i and $X_{i,jk}$, i is called the major subscript. In X_{ij}, both i and j are called major subscripts.

Each of the 22 statistics X_{ij}, $X_{i,jk}$, X_i can be classified as "significant" or "not significant." For example, X_{ij} is significant if

$$\frac{X_{ij}}{N} - p_0 > C\sqrt{p_0(1-p_0)}. \tag{3}$$

There are 2^{22} possibilities when one observes the significance or nonsignificance of the 22 statistics all together. A decision rule, i.e., a function from the 22-tuple of significant and nonsignificant possibilities to the Y's, will now be presented. (Note that the decision then is based on 22 separate pieces of binary information arising from a natural but by no means exhaustive set of statistics.) The rule is motivated by the desire to satisfy the inequalities of the next section with a reasonable number of animals forming the treatment groups, as well as by the following considerations:

1. If the four chemicals are chosen at random from a large population of chemicals, the probability of two or more of them being tumorigenic, i.e., their p's being greater than p_0, is small.

2. The probability of two or more of the chemicals having an inhibiting or synergistic effect when combined with some other of the four chemicals is small.

The rule is first to locate the chemicals C_i whose $Y_i = 1$, then to locate those whose $Y_i = 0$, and last to assign the remaining chemicals a $Y_i = 1/2$.

A. Assign $Y_i = 1$ if either exactly one statistic is significant and of the form X_i or $X_{i,jk}$, or two or more statistics are significant and all significant statistics have i as a major subscript.

B. Assign $Y_i = 0$ if either no statistic which is significant has i for a major subscript, or the only significant statistic having i for a major subscript is X_{ij} and $Y_j = 1$.

C. Assign $Y_i = 1/2$ otherwise.

24. FIRST STAGE SCREENING FOR TUMORIGENS

When 1AAT design is used, the decision rules are much less complicated. Let X_i be the number of animals that develop a tumor when given a diet with C_i at its MTD. Then, C_i is declared to be a tumorigen if

$$\frac{X_i}{n} - p_0 > C'\sqrt{p_0(1-p_0)} . \tag{4}$$

4. PROBABILITY INEQUALITIES TO BE SATISFIED

We wish to control certain probabilities of correct decision by the appropriate choice of sample size for both the 2AAT and the 1AAT designs. For the 2AAT design it is desired to choose N to satisfy inequalities of the following form:

1. Prob $(Y_i = 0,$ all $i \,|\,$ all $p_i \leq p_0$ and all the $Y_i \neq 1/2;$ all $S_{ij} \leq \bar{S}) \geq P_1$

2. Prob $(Y_1 = 1,$ all $Y_i = 0$ for $i \neq 1 \,|\, p_1 > p^*; p_2, p_3, p_4 \leq p_0;$ all $Y_i \neq 1/2;$ all $S_{ij} = 0;$ for all but possibly one $j \neq 1,$ $I_{1j} \leq \underline{I}\,) \geq P_2$

3. Prob $(Y_1 = 1,$ all $Y_i = 0$ for $i \neq 1 \,|\, p_1 > p^*; p_2, p_3, p_4 \leq p_0;$ all $Y_i = 1/2;$ all $I_{ij} = 0;$ all $S_{ij} = 0$ except for one pair $S_{1j} \leq \bar{S}) \geq P_3$.

For the 1AAT design we chose n to satisfy the inequalities

1. Prob $(Y_i = 0,$ all $i \,|\,$ all $p_i \leq p_0) \geq P_1$
2. Prob $(Y_1 = 1, Y_i = 0$ all $i \neq 1 \,|\, p_1 > p^* ; p_2, p_3, p_4 \leq p_0) \geq P_2.$

5. COMPARISONS BETWEEN THE 2AAT AND 1AAT DESIGNS

Some numerical results have been obtained to compare the 2AAT design with the 1AAT design. As an example of these results, we find that if 72 animals are given each compound in the 1AAT procedure and 36 animals receive each of the 6 pairs of compounds 2AAT (requiring 54 animals per compound tested), then the corresponding P_1, P_2 values are approximately the same when the following illustrative conditions obtain:

a. Negative control materials lead to 15% (or fewer) animals with tumors, as in Innes [3].

b. Tumorigens lead to 30% or more animals with tumors, as in the Innes studies [3] (if tested 1AAT).

c. The mix of four compounds contains either no tumorigens; or 1 tumorigen, 1 nontumorigen inhibitor, and 2 nontumorigens; or 1 tumorigen and 3 nontumorigens.

d. A tumorigen + a nontumorigen noninhibitor gives 30% or more animals with tumors.

e. The response to a tumorigen + inhibitor is the same as for a nontumorigen (i.e., 15% or fewer of the animals have tumors).

These results are based on work not reported in this study. Given these conditions, the proposed 2AAT testing scheme can classify tumorigens and nontumorigens with the same probabilities of making errors as for the 1AAT procedure at a possible 25% less animals and hence less cost. Clearly, the procedures do produce somewhat different results.

If such a 25% savings could be made then 33% more chemicals could be tested on a 2AAT basis than on a 1AAT basis at the same cost, and therefore presumably in the same time. It is therefore worth looking into modifications of the five conditions listed above to see how likely they are to obtain and to see how failure to meet them would reduce the savings.

Here are some possible complications or changes in the conditions that might cause problems:

a, b. Any changes with respect to 2AAT experimentation would also be changes with respect to 1AAT experimentation, leaving the relative positions unchanged when comparing the two systems.

c. The mix could contain two or more inhibitors and/or two or more tumorigens. Inhibitors pose a real problem. If a group of four compounds contained two tumorigens, our decision rules would usually say that further experimentation is needed ($Y = 1/2$).

d. The response to a tumorigen + a nontumorigen is less than 30% (but more than 15%). This, of course, makes the nontumorigen a partial inhibitor, and this problem is subsumed in Statement c.

d'. Another possibility: because of joint toxicity of two materials it is necessary to give less than a MTD of each, so that tumorigen response will be less than 30% (dose-response effect). Preliminary testing of the compounds should enable us to reduce this complication.

d". Two nontumorigens given together produce a tumorigen. This does constitute a problem, but it is also a bonus from the 2AAT scheme that the 1AAT does not allow.

6. INHIBITORS AND MULTIPLE TUMORIGENS

Our discussion of the possible complications which might arise indicates that inhibitions of tumorigenesis induction and multiple tumorigens in a group of four chemicals do produce difficulties for the 2AAT design under study here. Let us first consider the problem of multiple tumorigens.

If there are P % tumorigens in a population, we want to see how often we might have 0, 1, 2, 3, or 4 tumorigens in a group of 4. This will help us see how often we will be able to make a definite decision regarding individual chemicals in the proposed pairwise experiments. Table 1 gives these data for P = 5%, 10% (the estimates from the Innes experiments, as noted earlier).

Consider the 10% case (the worse of the two). In 66% of the experiments (no tumorigens present), the 2AAT and 1AAT experiments will give identical results. If there is one tumorigen present (29% of the time) the 2AAT and 1AAT experiments will give the same results if there are no inhibitors present, or if there is one inhibitor present. We need to make some assumption about the presence of inhibitors, since certain configurations of tumorigens-inhibitors will lead to problems and/or the requirement to carry out further testing.

Before we do so, let us review the decisions that are made with the configurations of Table 2.

We need to compute the probabilities of Configurations C and D to obtain a more complete comparison of the 2AAT and 1AAT approaches. Table 1 shows that the probability of two or more tumorigens in a group of four is 5.2% if P = 10%. This takes care of Configuration D.

Let us suppose that the relative frequency of inhibitors in a population of chemicals is the same as the relative frequency of tumorigens. To our knowledge there are no data on this, but the assumption seems conservative to some biologists at the National Cancer Institute. Given this assumption, the proportion of the times that with one tumorigen among the four, there will be two or more inhibitors present in the remaining three chemicals is 2.8% (from a frequency distribution like that of Table 1 but for groups of 3).

TABLE 1

Percentage Frequency of Tumorigen Multiplicities When P% of the Chemicals in the Pool to be Tested are Tumorigens

No. tumorigens in group of 4	Frequency (%)	
	P = 5%	P = 10%
0	81	65.6
1	17	29.2
2	2	4.9
3	0	0.3
4	0	0.0

TABLE 2

Review of Decision-Making

Configuration	Tumorigens	Inhibitors	Decision
A	0	any number	Most of the time all chemicals are declared nontumorigens. (Probability controlled by P_1.)
B	1	0, 1	Most of the time chemicals are correctly classified. (Probability depends on P_2.)
C	1	2 or more	Most of the time either all chemicals are declared to be nontumorigens or the chemicals undergo additional testing.
D	2 or more	any number	Same as C.

24. FIRST STAGE SCREENING FOR TUMORIGENS

Thus in 2.8% of the times that one tumorigen is present there will be two or more inhibitors present, giving (usually) either an incorrect classification of the tumorigen or requiring further testing for certain chemicals in this group. But the probability of obtaining one tumorigen is 29%. Therefore, the probability of Configuration C is 2.8% of 29.2, or 0.8%. Thus a total of 6% (5.2 + 0.8) of all results may be in error in addition to the errors already controlled for, or else further tests will be required.

The additional errors are mainly false negatives. Most of the 6%, though, relates to the need for additional testing of the chemicals in the group. Some minor sampling error should be included. Thus the 2AAT design under conditions defined in this section does entail an increase in false negatives, while providing a reduction in sample size, permitting more chemicals to be tested, and an opportunity to discover some interactions. The reduction in sample size lowers the error incurred in being unable to test chemicals because of economic considerations.

Two further losses are under consideration with 2AAT testing. The first is the lengthened time it takes to arrive at a decision when the decision rules indicate that 1AAT testing is necessary. The second possible loss has been mentioned earlier: it is the dose-response effect. How much will we have to reduce a MTD if we give two materials at a time rather than one? How much less carcinogenic will a material be at a lower dose? Some mathematical modelling is under way to give some estimates of this loss. It is a little too early to make formal estimates. The data now being developed through the pilot studies at Stanford Research Institute will give a better basis for the assumptions necessary in the mathematical modelling. It is distinctly possible that some borderline tumorigens will be lost. A strategy for retesting borderline materials may be needed. This, of course, is a standard problem in screening. The clearly negative things can be uncovered easily. So can clearly positive materials. The intermediate materials take the most testing.

7. SUMMARY

A 2AAT approach to testing has been described which gives indication of reducing the cost of first-stage screening for tumorigens. In addition, the 2AAT scheme will give leads about joint action of chemicals. The effects of dosage dilution are still to be explored. Current experimentation will help in setting up appropriate mathematical models to evaluate this effect.

REFERENCES

1. C. Dunnett, Screening and selection, in International Encyclopedia of the Social Sciences, Vol. 14, MacMillan and Free Press, New York, 1968, p. 123.

2. D. Finney, Probit Analysis (3rd ed.), Cambridge Univ. Press, London, 1971.

3. Innes et al. Bioassay of pesticides and industrial chemicals for tumorigenicity in mice: A preliminary report, J. Nat. Cancer Inst., 42, 2, 1101 (1969).

4. P. Shubik, Symposium on the evaluation of the safety of food additives and chemical residues: III, Toxicol. Appl. Pharmacol., 16, 507 (1970).

5. M. Sobel and P. Groll, Group testing to eliminate efficiently all defectives in a binomial sample, Bell Syst. Technol. J., 38, 1179 (1959).

6. G. S. Watson, A study of the group-screening method, Technometrics, 3, 371 (1961).

7. WHO Scientific Group, Principles for the testing and evaluation of drugs for carcinogenicity, WHO Tech. Rep. No. 426, Geneva, 1969.

8. J. K. Reddy and D. Svoboda, Effect of Lasiocarpine on Aflatoxin B-1 carcinogenicity in rat liver, Arch. Pathol., 93, 55 (1972).

SOME STATISTICAL ASPECTS OF EXPERIMENTS FOR DETERMINING THE TERATOGENIC EFFECTS OF CHEMICALS

David G. Hoel

National Institute of Environmental Health Sciences
National Institutes of Health
Research Triangle Park, North Carolina

1. INTRODUCTION

Various new chemicals are constantly introduced into the environment both intentionally and unintentionally, and it is therefore of critical importance to determine their potential effects on human health. Ideally, one hopes to accomplish this by studying the relationship of the chemical environment to human diseases. This, however, is usually impossible because of the lack of good experimental controls. In addition, decisions often have to be made concerning the safety of proposed or recently introduced chemicals, for which there is a paucity of human data.

This leads us to the use of the laboratory animal as a biological model for man. From laboratory experiments information is obtained about the biological effects of chemicals on animals; this information provides clues to possible adverse effects on humans. The greatest weakness of using species-to-species extrapolation is the danger of missing human effects which do not occur in the particular animal being studied in the laboratory. A well known example of this was the lack of a teratogenic (birth defects) effect of Thalidomide in "random bred" laboratory mice. In an attempt to overcome this problem, as many different species of animals as possible are tested with each chemical being investigated; but using animals other than mice and rats is expensive and time-consuming, a situation which limits their use.

The usual laboratory tests of a chemical involve the determination of its toxicity levels, and in recent times whether or not it is carcinogenic, mutagenic, or teratogenic. An additional relevant test, survival with complete histopathological examination, has been employed in radiation research, but unfortunately is quite often overlooked in studying other potential health hazards. In most investigations, the decision whether or not a chemical is a possible environmental hazard is based on the tests of carcinogenicity and teratogenicity. In this paper we shall explore some of the statistical aspects of teratological experimentation.

2. THE TERATOLOGICAL EXPERIMENT

In this type of experiment, a pregnant mouse is subjected to the chemical in question at a predetermined period of gestation. The routes of administration are usually subcutaneous or oral. After a given time, usually before delivery, the animal is sacrificed and the fetuses examined. For each mother and litter the teratologist records a considerable amount of information, the main part of which contains the basic data on the number of implants, and the number of dead, resorbed (early death with autolysis) and abnormal fetuses. The specific types of anomalies (congenital malformations) observed are also reported, as are quantities such as average fetal weight, maternal weight gain, and liver-to-body weight ratio.

For the purpose of statistical analysis, several of these data are compared with values obtained from a group of matched control animals. The quantities most often subjected to statistical testing are the incidence of anomalies and fetal mortality. In particular, an anomaly is defined as either a specific type of anomaly of interest or a collection of types of anomalies. Then for a single litter let

a = number of live fetuses with an anomaly

d = number of dead (including resorbed) fetuses

n = number of implants.

Now if

$\frac{a}{n-d}$ = proportion of abnormal to live fetuses

is significantly higher in the treatment group than it is in the control group, the chemical is declared to be teratogenic. (Temporarily ignore $n-d=0$.)

25. TERATOGENIC EFFECTS OF CHEMICALS

Further, if

$$\frac{d}{n} = \text{proportion of dead fetuses}$$

is significantly higher in the treatment group, the chemical is then declared feticidal (causing fetal mortality.) If either situation occurs, the chemical is considered to be a possible environmental hazard.

3. SOME PROBABILISTIC CONSIDERATIONS

In order to examine the definitions of "teratogenic" and "feticidal" we define the following events for a given litter. Let A denote the event that a fetus has at least one anomaly in the set of anomalies of interest, and D the event that the fetus is dead. Next, we assume that the fetuses within the litter behave in a Bernoulli fashion with respect to the events A and D. With this structure it follows that the measure of fetal mortality, d/n, is an estimator of the probability of death, $P(D)$. Also, the measure of fetal anomalies, $a/(n-d)$, is an estimator of the probability of an anomaly given that the fetus is alive, $P(A|\bar{D})$.

When a compound is declared teratogenic this usually indicates a significant increase (treatment vs control) in $P(A|\bar{D})$. Now, $P(A|\bar{D})$ is closely related to and often confused with $P(A)$. However, since the only reported quantities are a, d, and n, it is impossible to estimate $P(A)$ without further assumptions on the probabilistic structure. Besides $P(D)$ and $P(A|\bar{D})$, the probability of normal birth, $P(\bar{A}\bar{D})$, should also be considered. Ideally, then, we want $P(D)$ and $P(A|\bar{D})$ small and $P(\bar{A}\bar{D})$ large. Now it is easy to see that if both $P(D)$ and $P(A|\bar{D})$ are increased then $P(\bar{A}\bar{D})$ is decreased and the chemical is clearly having a deleterious effect. The problem is not so straightforward, however, when $P(D)$ is increased and $P(A|\bar{D})$ is decreased or vice versa. The following hypothetical situation illustrates the difficulty.

Assume a treatment group and a control group have the following probabilities

	Control	Treatment
$P(A)$	p	p
$P(D)$	p'	εp'
$P(D|A)$	p_1	εp_1
$P(D|\bar{A})$	p_2	εp_2

where $0 < \epsilon < 1$. We have given the two groups the same probability of an anomaly, $P(A)$, and decreased the probability of death, $P(D)$, in the treatment group. This decrease in $P(D)$ has also been proportioned between $P(D \mid A)$ and $P(D \mid \bar{A})$. Using these values we find

	Control	Treatment
$P(\bar{A}\bar{D})$	$(1-p)(1-p_2)$	$(1-p)(1-\epsilon p_2)$
$P(A \mid \bar{D})$	$\dfrac{(1-p_1)p}{(1-p')}$	$\dfrac{(1-\epsilon p_1)p}{(1-\epsilon p')}$

and it follows that

$$P(\bar{A}\bar{D}, \text{treatment}) > P(\bar{A}\bar{D}, \text{control}).$$

Now, it is reasonable to assume that there is a positive association between A and D. Specifically, if we assume $P(D \mid A) > P(D \mid \bar{A})$, then it follows that

$$P(A \mid \bar{D}, \text{treatment}) > P(A \mid \bar{D}, \text{control}).$$

Employing this information, the experimenter will probably report that the treatment group yielded a higher incidence of anomalies when, in fact, the two groups have the same value of $P(A)$. Furthermore, the treatment group has a higher probability of normal birth $P(\bar{A}\bar{D})$. What has been done is simply that deaths have been reduced in the treatment group and this may lead the experimenter to declare the treatment teratogenic; hence, a possible hazard. Now, reversing the situation, one would also declare the control a possible hazard because of its increase in fetal mortality, $P(D)$. In order to avoid this circular situation, one should report all of the quantities $P(D)$, $P(A \mid \bar{D})$, and $P(\bar{A}\bar{D})$, with no statement as to whether the chemical is a possible hazard or not. If this procedure is not followed, the experimenter is putting himself in the position of having to decide whether or not death is preferable to life, albeit abnormal.

We conclude this section with some data which illustrate the above discussion. The experiment was performed at two different time periods and the values in Table 1 are proportions averaged over litters. Actually, the values are then estimates of average probabilities, since several litters are involved. Because of a significant increase in $P(A \mid \bar{D})$ for the treatment groups, the chemical was considered to be teratogenic. In light of the decrease of $P(D)$ this increase in $P(A \mid \bar{D})$ is not entirely unexpected.

Finally, it should be noted that in the in the screening of potentially hazardous compounds, one infrequently encounters the situation of a

TABLE 1

Results of a Representative Experiment

Quantity estimated	Time 1		Time 2	
	Control	Treatment	Control	Treatment
$P(A \mid \bar{D})$	0.06	0.18	0.19	0.39
$P(D)$	0.46	0.25	0.40	0.08
$P(\bar{A}\bar{D})$	0.50	0.59	0.48	0.56

decrease in $P(D)$. Typically, $P(D)$ either remains the same or increases, in which case an accompanying increase in $P(A \mid \bar{D})$ clearly indicates a possible safety problem. Nevertheless, one should examine the effect on the probability of normal birth, $P(\bar{A}\bar{D})$, since it may be significantly decreased while neither $P(D)$ or $P(A \mid \bar{D})$ individually appears to be significantly increased.

4. SOME STATISTICAL CONSIDERATIONS

Since it is well known that litters behave differently with respect to the events A and D, the single-litter estimates of Section 3 cannot be used for an experimental group of animals. That is to say, one cannot let d and n represent the total group deaths and implants and then estimate $P(D)$ by d/n. One approach to the problem is the following.

For each litter consider the triplet of random variables (X, N, P), where N represents the number of implants, X the number of dead fetuses, and P an unobservable variable. Further assume that conditionally given $N = n$ and $P = p$, X is binomially distributed with parameters n and p. We also assume that N has a distribution on the nonnegative integers, P has the distribution $G(p)$ on the interval [0,1], and N, P are not assumed to be independent.

We can use the mean probability of death

$$\mu = \int p \, dG(p)$$

as a reasonable measure of fetal mortality.

For a sample from m litters we estimate μ by

$$\hat{\mu} = \frac{1}{m} \sum_{i=1}^{m} \frac{X_i}{N_i}$$

which is well defined since $P(N \geq 1) = 1$.

Van Ryzin [3] has established the following properties of the estimator $\hat{\mu}$:

i. $\hat{\mu}$ is minimum-variance linear unbiased,

ii. $\hat{\mu}$ is strongly consistent, and

iii. $(\hat{\mu} - \mu)/s$ converges in distribution to $\mathcal{N}(0,1)$,

where

$$s^2 = \frac{1}{n-1} \sum_{i=1}^{n} \left(\frac{X_i}{N_i} - \hat{\mu} \right)^2 .$$

Turning now to anomalies, let N represent the number of live fetuses and X the number of anomalies. The estimate $\hat{\mu}$ then corresponds to the usual estimate of $P(A | \bar{D})$. However, we no longer have the condition $P(N \geq 1) = 1$. So assuming that

$$0 < P[N = 0] < 1$$

we then estimate μ as before, except that the sum is taken over litters with at least one live fetus and m is the number of such litters.

Van Ryzin has also established that the previously given properties of $\hat{\mu}$ hold if it is assumed that N and P are independent. Also, if the joint distribution of N and P is unrestricted, it can be shown that we do not have an identifiable estimation problem. Thus we must make some assumption about the distribution of N and P, such as independence, in order to estimate $P(A | \bar{D})$.

The problem of statistically testing (treatment versus control) a measure such as $P(D)$ could be treated by using the asymptotic properties of the estimator $\hat{\mu}$. For small samples, however, a nonparametric test such as the two-sample Wilcoxon test may be preferable. In this case, the test would be performed on the individual litter proportions d/n. These data are clearly not continuous, but will be approximately so except for the large proportion of zero values. This difficulty can be overcome by assuming the data is censored or truncated from below zero and by applying one of the extensions of Wilcoxon's test, such as those given by Halperin [2] and Gehan [1].

REFERENCES

1. E. A. Gehan, A generalized Wilcoxon test for comparing arbitrarily singly-censored samples, Biometrika, 52, 203 (1965).

2. M. Halperin, Extension of the Wilcoxon-Mann-Whitney test to samples censored at the same fixed point, J. Amer. Stat. Assoc., 55, 125 (1960).

3. J. Van Ryzin, Estimating the mean of a random binomial parameter with sample sizes random, Tech. Rep. No. 288, Dept. Statistics, Univ. Wisconsin; Abstr. in Bull. Inst. Math. Stat., 1, 99 (1972).

26

CLOSING COMMENTS*

Gerald van Belle

Florida State University
Tallahassee, Florida

This conference has brought together "producers" as well as "consumers" of statistical analyses, and from the discussion it is clear that there will be an increased demand on the "producers" for quantitative measures of the effects and magnitude of pollution. There is the danger that the wrong aspects will be quantified or inappropriate statistical methods will be advocated, and this is especially disconcerting if these are used in determining administrative or political action. The comments to be made are applicable to several papers presented at this conference, but no particular papers will be mentioned.

A preliminary question is to ask to what audience such papers are addressed. They are clearly "consumer" oriented, not appearing in one of the statistical journals but in Science or one of the pollution journals. As such they will not be read primarily by statisticians but by "consumers" who will use the conclusions as basis for initiating action. Hence, statistical presuppositions will have to be clearly spelled out.

Nowhere is the demand for "scientific" documentation and quantification of the effects of pollution more clearly illustrated than in the concept of "health" or "ill health". It is not enough that I cannot swim in Lake Erie, that air pollution makes my eyes water, that I cannot see the

---

*This paper was originally a discussion of a paper given at the symposium on which this book is based. The editor is grateful to the author for revising it and allowing it to be used to close this book. (Ed.)

Atlantic Ocean from the top of the Chrysler Building; but "Does air pollution shorten lives?" or, "Does air pollution cause ill health?" The first point to note is the narrow, and incongruous definition of "ill health," namely, mortality ratio. Using this definition of ill health creates more problems than it solves. Some of these have been mentioned by other discussants. Major problems are the time delay and social mobility factors. Secondly, the question is asked in a negative context: ill health. There is a marked contrast here between American and European conference speakers in the definitions of health. The Americans have tended to use health = low mortality (or low morbidity), an essentially negative concept. The Europeans have used the WHO definition of health: "a condition of physical, mental, and social well-being" (my emphasis) - a positive concept. It will not do to argue that this definition cannot be quantified. In view of the problems associated with equating ill health with mortality rates, the efforts should be directed to the proper quantification of health and ill health.

A greater danger is that the "producers" of statistical analyses will base their product on arguments of dubious validity, or will assume that the "consumer" is aware of the more common pitfalls. Without quoting from particular papers the following may be noted.

1. The use of a linear regression model to approximate a cause-effect link is questionable. A useful distinction is often made between "control" and "predictor" variables; a regression model usually contains both. The history of spurious correlation should caution the researcher against the indiscriminate assignment of causal status to "independent" variables. More proper would be to assume ab initio that "independent" variables may not be causal variables, and to presume that the burden is on the producer to confer causal status on the basis of physiological or physical arguments. Of course, this is well known, but mere lip service is often paid to this procedure.

2. The use of elasticity coefficients is misleading when the variables are measured in arbitrary units. Since elasticity coefficients involve ratios of coefficients of variation, some remarks by Kendall and Stuart on Gini's coefficients of concentration and variation are relevant: "These coefficients both suffer from the disadvantage of being affected very much by μ_1', the value of the mean measured from some arbitrary origin, and are not usually employed unless there is a natural origin of measurement or comparisons are being made between distributions with similar origin."

3. To recommend to "consumers" the creation of large data banks may induce the impression that the statistician can "analyze" this data at any time. Nothing is further from the truth. The researcher consulting such a data bank finds that the quantities he needs are not there or it is too

26. CLOSING COMMENTS

expensive to extract them. A measurement is an abstraction of the real world and there is an infinity of measurements that can be made. Hence, to measure means first of all to select, and producers of large general purpose data banks are in the least favorable position to make such selection.

4. If in the past indices were confined to the field of economics, it is clear that they are about to be introduced to environmental scientists. There will be indices to determine "air pollution alerts," air quality, water quality, biological activity, and many other matters. Statisticians will have to caution "consumers" that the meaning of index is "pointer," and that the consumer will have to define the target. An ever present danger is that an index becomes an end itself. If this is the case an index value is merely a statistical analysis untouched by the human mind.

The statistical community has an obligation to carefully inform consumers of statistical methods about the limitations of the scales, indices, and models that are used. The quantification of common terms such as air pollution, noise pollution, ill health, and health will require close cooperation between the statistician and the researcher from a substantive area. (For example, a very useful conference could be organized on the topic of indices.) One of the most enjoyable aspects of this conference has been that it has initiated this close cooperation. W. H. Auden's "Under Which Lyre," with the subtitle "A Reactionary Tract for the Times," contains the following advice for those who do not want to work together:

> Thou shalt not answer questionnaires
> Or quizzes upon World Affairs,
>   Nor with compliance
> Take any test. Thou shalt not sit
> With statisticians nor commit
>   A social science.

We did sit down together; we worked together for a week with the result that the disciplines represented by the participants have been enriched.

INDEX

A

Acidification, 173
Activated sludge treatment plant, 339-350
   automation of, 343-344
   control data for, 340-343
   data analysis for, 344-350
   data sampling problems, 340-343
   performance evaluation of, 339-350
Aerosol pollution, 177-191
Aerosol pollution, models of, 329-334
   advice for construction of, 332-334, 337-338
   distributions (statistical), used for, 330-332
   problems with, 329-334, 337-338
Aerosol systems, 329
Agent, environmental, 126-127
Air Quality Display Model (AQDM), 51-58
Ambient air quality standards, 95-96
A posteriori forecasts, 265
AQDM; see Air Quality Display Model
Artificial markets, 65-59, 75-77
Automation of sewage treatment plants, 343-344
Automobile emissions, 95

B

Baseline concept, 11
Belgium, 25-37
   measuring pollution in, 25-37
   smoke pollution in, 25-37
   sulfur dioxide pollution in, 25-37
Bioassay, 357
Biome studies, 10, 11
Biome types, 11
Biosphere, 15
Bronchial tree, geometry of, 189-190

C

Categorical data analysis, 277
CEQ; see Council on Environmental Quality
Chemical contamination, 7
Chemicals
   screening of, 365-373
   testing for effects of, 365-373, 375-381
CHESS; see Community Health and Environmental Surveillance System
Clean Air Amendments, 95-96
Climatic changes, 6, 8
   human activity caused, 6
   indicators of, 8
   monitoring of, 6
Community Health and Environmental Surveillance System (CHESS), 275
Compensation, definition of, 129
Complex regression models, 214-215
Component, environmental, 123
Concentrations, measurement of, 22-23
Council on Environmental Quality (CEQ), 85-86, 90, 94
Currents, modelling of, 213-220; see also Tidal waters, modelling of

D

Data analysis, problems of, 275-281
Data banks, 39-59
　editing of, 40-43
　validation of, 39-59
Death rates; see Mortality rates
Delaware river, water quality studies of, 297-310
Deterioration of the environment, 123-132
　cost-benefit analysis in, 128-132
　definition of, 124
　estimation of, 123-132
　indices of, 127-128
Diffusion of particles in respiratory system, 179
Diffusion models, air quality, 51-58
Diseases and air pollution; see Health and air pollution
Dispersion of pollutants, 22
Dispersion situation, 22
Dose-response models, 357-363
Drogue, 213
Drogue tracks, 219
Dump, definition of, 317

E

E. coli; see Escherichia coli
Effluent charges, 69-70, 83
Emission-concentration relationship, 22
Enforcement, water quality standards, 85-91
　authorities, present, 85-91
　economy, effects on, 90
　legislative proposals, current administration, 87
　statistics, importance of, 88-89
Environment, definition of, 15
Environmental control
　international, 4
　national, 4
Environmental problems, 5-7
　critical, criteria for, 6
　definition, 6
Environmental Protection Agency (EPA), 85-91
　budget, 89
　enforcement of water quality standards, 85-91
　origins of, 94-95
EPA; see Environmental Protection Agency
Escherichia coli, 206-209
　Coliform, 209
　survival rate of, 206
Estuaries, control of water quality in, 193-211
Estuaries, models of, 193-211
　data acquisition and analysis, 200-209
　one-dimensional mixing of, 196-200
　physical problems in creating, 193-211
　tidal range, effects of, 288-289
　variables, relative importance of, 291-293
Exceedence flow, 288
External diseconomies, treating, 65-84
　consumers' role, 80-83
　resistant problems, 70-83
　taxation methods, 69-70, 78-80
　tractable problems, 75-80
Externalities, 65-83; see also External diseconomies

F

Fecal pollution, 206
Federal Water Pollution Control Act, 85-87, 88-89, 93, 96
　amendments to, 85, 87
Feticidal chemical, definition of, 377
Financial damage, definition of, 129

INDEX

389

Forecast a posteriori, 250
Forecasting urban air pollution; see Urban air pollution, model of
Functions, environmental, 123-132
  competition of, 124-127

G

Ground-level concentrations; estimation of, 52-58
Growth level function, 169, 174-175
  parametric description of, 174-175
Growth reduction, in forests, 167-175

H

Hazardous substance pollution, indices of, 106, 108
Health, 9, 16, 110-112
  definition of, 16
  indices of, 9, 110, 112
Health and air pollution, studies of, 275-281
  categorical dependence problems in, 277
  data processing problems, 276
  repeated measurements problems in, 276-277
Health, mental, stench pollution affecting, 135-160
High-exposure stations, 12
Hockey stick function, 278

I

Impaction of particles in respiratory system, 178-179
Index construction, 353-363
Index, definition of, 104
Indicator, definition of, 104
Indicators, environmental, monitoring uses, 7-10
Indicators, environmental; see Indices, environmental

Indices, environmental, 103-121
  classes of, 103
  computation of, 115-120
  economics analogies, 105
  grouping of, 106-115
  ranking of, 113-117
  tabulation of, 115-120
Industrial terrain, diagram of, 139
Inhalation of particles; see Particle inhalation
International cooperation, monitoring, 12-14
International coordination, functions of, 14

J

Jackknife method, 246
Joint action of compounds, model of, 367

L

Landfills, attitudes towards, 315-326
Landfills, sampling opinions of, 315-326
  attitudes, factors affecting, 315-316
  case studies, 324
  data processing, 322-324
  methodology, 317-322
Land management, indices of, 107, 109
Land use, improper, 7
Lapse rate, 22
Latitude measurements, 322-323
Life span; see Mortality rates
Loss from pollution, indices of, 111, 113
Low exposure monitoring stations, 11

M

Measurement theories, 353-363

Measuring stations, air pollution, placement of, 21-24
Mesosphere, 15
Methodology, advice on, 383-385
Microsphere, 15
Migration and mortality rates, 234
Model construction, 353-363
Monitoring, environmental, 5-14
  baseline concept, 11
  basic considerations, 5
  Belgium, 27-31
  data evaluation for, 16
  design of system for, 10-12
  law, function of, 18
  operational and institutional arrangements, 12-14
  organization, international, 12-14
  problems to be monitored, 5-7
  special stations, 12
  station concept, 11
  station placement, 21-24
  variables appropriate for, 7-10
Monodisperse system, 330
Mortality rates, air pollution effects on, 223-242
Mortality and pollution, model of, 223-242, 245-247
  cost implications of, 241-242
  criticism of, 245-247
  economic implications of, 241
  elasticities, analysis of, 236-241
  limitations of, 233-234
  policy implications of, 235-236, 245-247
  racial factors, 237

N

NADB; see National Aerometric Data Bank
Nares, 180
Nasal hairs, particle impaction against, 179
National Aerometric Data Bank, 39-59
  analyses using, 43-59
  applications of, 43-59
  editing of, 40-43
  operation of, 40
  quarterly reports of, 40-49
  validation of, 40-43
National Environmental Policy Act, 94
Natural phenomena, indices of, 109, 112
Network of monitoring areas, 11
No false negative system, 366
No false positive system, 366
Nonconvexities, 72-74

O

Odor pollution; see Stench pollution
1AAT method, 366-373

P

Parallelism assumptions, 359
Parameters, environmental, 104
Particle inhalation, model of, 177-188, 189-191
  criticism of, 189-190
  deposition, calculation of, 179-183
  deposition, factors affecting, 178-183
  redistribution of particles, 183-188
  respiratory system, effects on, 177-188
  retention of particles, 180-183
Penalization due to stench, 136-137
Penalization function, 140
Pesticide control, legislation needed, 97
Physical environment, components of, 15
Plume model, stench pollution applications, 138
Polydisperse system, 330
Population growth, uncontrolled, 7
Population statistics, 110, 112

INDEX

Principal components analysis, application of, 54-58
Property rights, 65-69
Public-opinion surveys, 315-326
Purification, environmental, 129-132
   cost-benefit analysis of, 129-132
   marginal cost of, 129-130

Q

Quality of environment, improvement of, 93-101
   costs of, 98-99
   federal actions, 95-96
   information requirements for, 99-101
   institutions for, 94-95
   needed legislation, 96-97
   progress in, 93-101
   trends in, 97-98
Quantal regression, 357-363

R

Racial factors in mortality rates, 237
Radial growth, 169
Recycling, indices of, 107, 109
Reference growth, 171-172, 173
Refuse Act, 86, 96
Refuse Act Permit Program, 86-87, 88-89
Regional planning, stench pollution considerations, 156-160
Regional stations, 12
Resources, supply of, indices of, 107-108, 109
Respiratory illnesses and air pollution, 223, 275-281
Respiratory system, model of; see Particle inhalation, model of
River pollution, models of, 297-310

S

Salinity, estuarine, 193-211
   river flow related to, 206
Sampling apparatus, air pollution, 25-26
Sampling of opinions, 315-326
Sanitary landfill
   attitudes toward, 315-326
   definition of, 317
Scientific Committee on Problems of the Environment (SCOPE), 3
SCOPE Commission, 3
SCOPE Report Number 1, criticism of, 15-19
Screening, first stage, 365-373
Sensors, air pollution, siting of, 43-51
Settling of particles in respiratory system, 179
Sewage Treatment Plants (STP), performance evaluation of, 339-350
Shadow prices, 128-132
Sickness and air pollution; see Health and air pollution
Sickness score, 360-363
Smoke pollution, 25-37, 167-175
   Belgium, 25-37
   sampling of, 25-37
   trees, effects on, 167-175
Social-aesthetic conditions, indices of, 110, 112
Solid waste disposal, attitudes towards, 315-326
Soufre-fumées, 251
Spatial cultural environment, 15-16
Special monitoring stations, 12
Station concept of monitoring, 11
Stench pollution, model of, 135-160
   assumptions, 137-138
   case studies, 144-155
   chemical plants causing, 135-160

policy implications of, 156-160
testing of, 141-156
STP; see Sewage treatment plant

T

Teratogenic chemical, definition of, 376
Teratogenic effects, 375
Teratogens, testing for, 375-381
Thinness of markets, 70-72
Threshold estimation method, 278-281
Tidal waters, modelling of, 213-220
  applying the model, 219-220
  calibration of the model, 219-220
  extensions of data, spatial, 218-219
  sinusoidal model of current, 216-218
  techniques, 214
  variations of basic current, 215-216
  wind-induced currents, correlation with, 214-215
Toxic Substances Control Act, 97
Tree growth, 167-175
  acidification affecting, 167-175
  factors affecting, 169-171
Tree growth, model of, 167-175
  dead tree problem, 173
  local problem, 168-173
  regional problem, 173-175
  thinning problem, 173
Trees, forest, growth reduction, pollution caused, 167-175
Trend surface analysis, studies using, 52-58
Tumorigens, screening for, 365-373
  comparison of methods for, 369-371
  selection rules, 367-369

2AAT method, 366-373
  benefits of, 370
  complications in use of, 371-373

U

Urban air pollution, 25-37
  assessing, 249-272
  Belgium, 25-37
  data collection, 27-31
  data publication, 27-31
  forecasting, 249-272
  sampling of, 25-37
  statistical studies of, 25-37
Urban air pollution and meteorological conditions, 249-272
Urban air pollution, model of, 249-272
  applications of, 264-272
  detection uses of, 265-270
  equations for, 256-257
  forecasting uses of, 270-272
  policy implications of, 250, 264-272
  significance of variables, 257-261
  wind characteristics in, 253-255
Urbanization, environmental problems of, 7

V

Variables, environmental, monitoring uses of, 7-10

W

Water pollution, indices of, 106, 108
Water quality, determinants of, 340-341